HIGH-FREQUENCY OSCILLATOR DESIGN FOR INTEGRATED TRANSCEIVERS

HIGH-FREQUENCY OSCILLATOR DESIGN FOR INTEGRATED TRANSCEIVERS

by

Johan van der Tang

Eindhoven University of Technology

Dieter Kasperkovitz

Semiconductor Ideas to the Market (ITOM), Breda

and

Arthur van Roermund

Eindhoven University of Technology

KLUWER ACADEMIC PUBLISHERS

BOSTON / DORDRECHT / LONDON

A C.I.P. Catalogue record for this book is available from the Library of Congress.

ISBN 1-4020-7564-2

Published by Kluwer Academic Publishers,
P.O. Box 17, 3300 AA Dordrecht, The Netherlands.

Sold and distributed in North, Central and South America
by Kluwer Academic Publishers,
101 Philip Drive, Norwell, MA 02061, U.S.A.

In all other countries, sold and distributed
by Kluwer Academic Publishers,
P.O. Box 322, 3300 AH Dordrecht, The Netherlands.

Printed on acid-free paper

Printed in the Netherlands.

Id quod ratio debuerat usus docet

Practice teaches what theory
should have taught

Cicero, Tusculanae
Disputationes III

Dedicated to
Nienke, Monika & Marianne

Contents

Preface

OSCILLATORS are key building blocks in integrated transceivers. In wired and wireless communication terminals, the receiver front-end selects, amplifies and converts the desired high-frequency signal to baseband. At baseband the signal can then be converted into the digital domain for further data processing and demodulation. The transmitter front-end converts an analog baseband signal to a suitable high-frequency signal that can be transmitted over the wired or wireless channel. Given the wide range of applications of wired and wireless transceivers, oscillator specifications differ greatly for each transceiver. The challenge for the oscillator designer is to find the right oscillator topology and to define its dimensions in a limited time, so that the oscillator meets the requirements imposed by the transceiver in which it is embedded. This book discusses the analysis and the application of existing and new design insights, design methods, and design tools for a wide range of integrated high-frequency oscillators.

One of the books primary intentions is to serve scientific designers of oscillators as well as industrial ones. Designers of oscillators working in universities or in research are especially interested in an original, innovative and unique design. This can be a new architecture, a new technology or a breakthrough in one of the specifications points like phase noise, tuning range, supply voltage, etcetera. For industrial designers, the reproducibility and cost price are of main importance. In most cases the main target of their designs is to achieve the required specifications with a robust design at a minimum cost price. Often they are also forced to achieve these two goals in a very short time. Fortunately, there are many oscillator design aspects and oscillator properties that are interesting to the scientific as well as to the industrial designer. Assisted by behavioral modeling, elementary properties of a wide range of LC and ring oscillators are analyzed first in this book, followed by an in-depth analysis and discussion of practical properties including many non-idealities imposed by integrated circuit technology. This unique division between elementary and practical oscillator properties has resulted in a clear and scientific analysis of the main differences between each oscillator type and the dominating design issues for each type, including the following aspects:

- LC oscillators versus ring oscillators
- Single-phase and multi-phase oscillators
- Frequency and tuning
- Phase noise to carrier ratio
- Waveform aspects
- Power dissipation
- Technology, process spread, chip area

During the design process the designer is interested in assessing the "quality" of a specific oscillator design with respect to the specifications. This assessment can be performed by Figures of Merit (FOMs). FOMs are powerful tools for the designer that give qualitative and quantitative information about an oscillator design and help the designer to make the right design decisions. Design and benchmark FOMs are introduced and a number of examples are given. Design FOMs provide performance estimators that predict the design margin for a specific oscillator specification. Not only do they allow the designer to make fast high-level design decisions, they are also a means for documenting design knowledge. Important design questions like: "How does my oscillator design compare against the state-of-the-art or to a theoretical performance boundary?", are answered by the benchmark FOMs.

Apart from presenting design methods and tools, as well as a vast collection of selected oscillator designs in a concise and convenient way, this book provides a framework that supports the design process. Scientific designers will benefit from the qualitative insights and the quantitative trade-offs presented in this book to make the right choices for the creation of a "unique" design. The industrial designers will appreciate tools and methods that help them to find the right architecture, the right components and the best trade-offs in a quick and convenient way. We hope that this book will be helpful during the design of many novel, high-performance, robust and elegant oscillators.

Outline

The organization of this book is illustrated in Figure 0.1. After a compact literature overview and a description of the application of oscillators in integrated transceivers in Chapter 1, Chapter 2 continues with a general introduction to oscillators, discusses classification and basic oscillator theory. Various structured design methods of analog

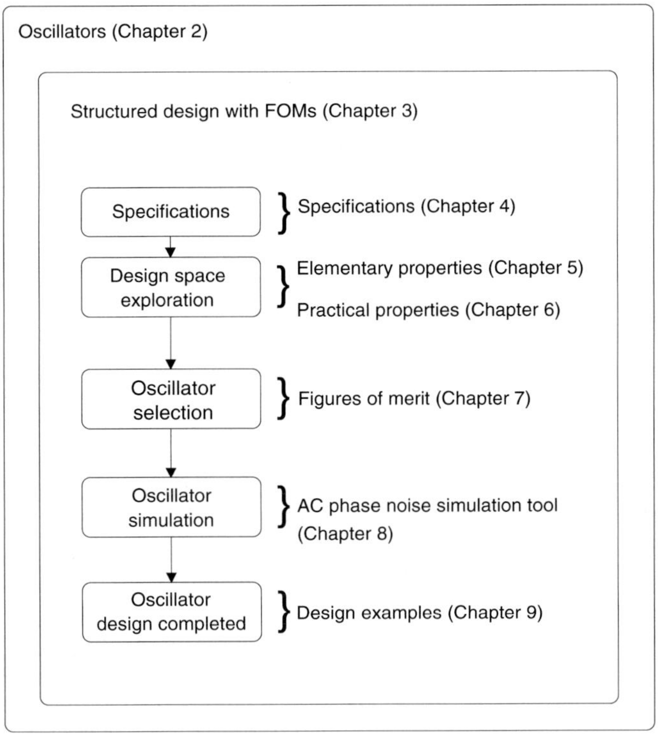

Figure 0.1 Outline of this book.

electronic building blocks have been demonstrated, and a number of interesting approaches will be described in Chapter 3. Most importantly, this chapter highlights the concept of figures of merit, which help the designer with high-level design decisions. In addition, Chapter 3 discusses system, behavioral and circuit modeling aspects that are used throughout the following chapters.

Chapter 4 to Chapter 8 discuss all aspects of an oscillator design flow, as is illustrated in Figure 0.1. It is unlikely that you get what you want, if you don't know what you want. In other words, a clear understanding of oscillator specification is of prime importance. This subject is covered by Chapter 4. In a virtually infinite design space, the oscillator designer has to make a selection of the most promising oscillator configuration. This design space is explored in Chapter 5 and Chapter 6, in which elementary and practical properties of oscillators are investigated, respectively. The discussion of practical properties includes many unwanted and parasitic effects encountered on circuit level, whereas Chapter 5 only discusses properties of oscillators on behavioral level, modeling only elementary properties. The oscillator properties are combined in examples of useful Figures of Merit (FOMs) in Chapter 7, which guide the designer in the design process. Once an oscillator configuration and topology is selected, an important part of the remaining design task is dimensioning and optimization of the design parameters. In this task, transistor level simulation by circuit simulators using accurate transistor models is of high importance. Chapter 8 highlights a fast oscillator phase noise estimation tool, which utilizes standard AC noise analysis of circuit simulators.

There is nothing better than the real thing: working oscillators. In Chapter 9, four integrated oscillator realizations are discussed, complete with measurements. One oscillator implementation is for use in FM radio receivers, two are designed for use in digital satellite receivers and the last oscillator example is designed for optical transceivers.

Acknowledgment

The work described in this book is partly based on the activities of the first author at Philips Research Laboratories (1995-2000). Former management and colleagues at Philips Research are thanked for their support and collaboration during these years. Pepijn van de Ven contributed to the theory on multi-phase LC oscillators. Francesco Centurelli contributed to the design and layout of the 10 GHz ring oscillator. The many technical discussions with Peter Baltus and Cicero Vaucher were very useful as well. Prof. Pietro Andreani provided many useful suggestions that improved the contents of this book. The authors are indebted to many others who contributed in some way to the realization of this book.

JOHAN VAN DER TANG,
DIETER KASPERKOVITZ
AND ARTHUR VAN ROERMUND

Glossary

Symbol	Description	Unit
A	amplitude of a signal	
A	fraction of the maximum amplitude in an N-stage ring oscillator	
A_e	amplitude error	
$B_a(\omega)$	imaginary part of Y_a (negative resistance model)	A/V
$B_p(\omega)$	imaginary part of Y_p (negative resistance model)	A/V
C	capacitance	F
C_{active}	active variable capacitance	F
C_{db}	drain-bulk capacitance	F
C_{fixed}	fixed capacitance in parallel LC resonator	F
C_{gd}	gate-drain capacitance	F
C_{gs}	gate-source capacitance	F
C_{jc}	zero bias collector-base capacitance	F
C_{je}	emitter junction capacitance	F
C_{js}	collector-substrate capacitance	F
C_{j0}	zero bias junction capacitance	F
C_L	input capacitance of ring oscillator stage	F
C_{max}	maximum capacitance of a varactor	F
C_{min}	minimum capacitance of a varactor	F
$C_{maxband}$	maximum capacitance of a switched capacitor	F
$C_{minband}$	minimum capacitance of a switched capacitor	F
C_{MOS}	MOS varactor capacitance	F
C_{node}	node capacitance in a circuit	F
$CNR(f_m)$	carrier to phase noise ratio at an offset frequency f_m	dBc/Hz
C_{ox}	oxide capacitance per unit area	F/m^2
C_p	total capacitance of LC parallel resonator	F
C_{par}	parasitic capacitance in parallel LC resonator	F
$C_{p_{max}}$	maximum capacitance value of C_p	F

$C_{p_{min}}$	minimum capacitance value of C_p	F
C_{pt}	parasitic resonator capacitance	F
C_Q	parasitic capacitance of switched capacitor (off-state)	F
C_s	capacitor modeled with series resistance	F
C_{series}	capacitor in series with varactor	F
C_{switch}	capacitor value that is switched	F
C_{tune}	tunable capacitance (varactor)	F
C_v	varactor capacitance	F
C_μ	collector-base capacitance	F
C_π	$\tau_F g_{m_{bipo}} + C_{je}$, input capacitance	F
$erfc$	complementary error function	
E_{crit}	critical field strength of short-channel MOS	V/m
$f_{1/f}$	1/f-noise corner of oscillator spectrum	Hz
$f_{1/f-device}$	device 1/f-noise corner in a technology	Hz
$f_{carrier}$	carrier frequency	Hz
f_{center}	center frequency	Hz
f_H	upper frequency integration bound	Hz
f_L	lower frequency integration bound	Hz
f_{LC}	$(2\pi\sqrt{C_p L_p})^{-1}$, oscillation frequency of ideal LC oscillator	Hz
f_m	offset/modulation frequency	Hz
f_{max}	specified maximum oscillation frequency	Hz
f_{MAX}	maximum oscillation frequency (technology FOM)	Hz
f_{min}	specified minimum oscillation frequency	Hz
f_{osc}	oscillation frequency	Hz
f_T	transition frequency (technology FOM)	Hz
F	noise factor	
$\delta f(f_m)$	frequency deviation power spectral density	$Hz/\sqrt{Hz}$
Δf	peak frequency deviation	Hz
g_m	transconductance	A/V
$g_{m_{bipo}}$	I_c/V_T, bipolar small-signal transconductance	A/V
$g_{m_{MOS}}$	MOS small-signal transconductance	A/V
$G_a(\omega)$	real part of Y_a (negative resistance model)	A/V
$G_p(\omega)$	real part of Y_p (negative resistance model)	A/V
$h_i(n)$	harmonic current content of nth harmonic	
$h_v(n)$	harmonic voltage content of nth harmonic	
$H(j\omega)$	forward transfer function in a system	
$H_{cl}(j\omega)$	closed-loop transfer function	
$H_{ol}(j\omega)$	open-loop transfer function	
i_c	coupling current in N-stage LC oscillator	A
$i_{carrier}$	rms carrier current	A
i_l	level current in N-stage LC oscillator	A
$\overline{i_n^2}$	mean square noise current in 1 Hz	A^2/Hz
i_{out}	output current	A
i_t	total current into resonator of N-stage LC oscillator	A
I_{bias}	bias current	A
I_c	collector current	A
I_d	drain current	A

I_{lim}	maximum output current value of limiting transconductance	A
I_{peak}	peak carrier current	A
I_{tail}	tail current of a differential pair	A
k	$1.38\,10^{-23}$, Boltzmann's constant	J/K
K_{cco}	CCO tuning constant	Hz/A
$K_{I_{tail}}$	sensitivity of f_{osc} for tail current variations	Hz/V
K_{supply}	sensitivity of f_{osc} for supply voltage variations	Hz/V
K_{vco}	VCO tuning constant	Hz/V
L	inductance	H
L	MOS effective channel length	m
L_{active}	active variable inductance	H
L_p	total inductance of parallel resonator	H
L_s	inductor modeled with series resistance	H
$\mathcal{L}(f_m)$	SSB phase noise to carrier ratio at offset frequency f_m	dBc/Hz
$\mathcal{L}_{bipo}(f_m)$	$\mathcal{L}(f_m)$ of linear bipolar LC oscillator model	dBc/Hz
$\mathcal{L}_{LC}(f_m)$	$\mathcal{L}(f_m)$ of linear LC oscillator model	dBc/Hz
$\mathcal{L}_{NLC}(f_m)$	$\mathcal{L}(f_m)$ of linear N-stage LC oscillator model	dBc/Hz
$\mathcal{L}_{Nring}(f_m)$	$\mathcal{L}(f_m)$ of linear N-stage ring oscillator model	dBc/Hz
$\mathcal{L}_{two-int}(f_m)$	$\mathcal{L}(f_m)$ of linear two-integrator oscillator model	dBc/Hz
m	an integer	
MJ	junction grading coefficient	
MJC	collector-base junction grading coefficient	
n	an integer	
n_{cycle}	number of cycles in a resonator for a specified damping ratio	
N	number of oscillator stages	
N_b	number of switched capacitors	
NBW	Noise bandwidth	Hz
N_{div}	division ratio of main divider	
p_m	oscillator design parameter m	
P	power	W
P_{DC}	DC power	W
P_{RF}	RF carrier power	W
P_{signal}	signal power	W
P_{SSB}	SSB noise power	W/Hz
q	$1.602\,10^{-19}$, charge of the electron	C
q_k	oscillator property k	
q_{max}	maximum charge variation (swing) in a capacitance	C
Δq	total injected charge in an oscillator node	C
Q	quality factor	
Q_b	quality factor of a switched capacitor (on-state)	
$Q_{C_{active}}$	effective quality factor of active variable capacitance	
Q_{cs}	$(\omega C_s R_{cs})^{-1}$, quality factor of C_s	
Q_{nc}	coupling transistor in N-stage LC oscillator	
Q_{nl}	level transistor in N-stage LC oscillator	
Q_{NLC}	$\approx NQ_p\cos(\phi_{res})$, Q of linear N-stage LC oscillator model	
Q_{Nring}	$1/2 \cdot N\sin(\pi/N)$, Q of linear N-stage ring oscillator model	
$Q_{L_{active}}$	effective quality factor of active variable inductance	

Q_{ls}	$\omega L_s/R_{ls}$, quality factor of L_s	
Q_p	$R_p\sqrt{C_p/L_p}$, quality factor of parallel LC resonator	
Q_{res}	quality factor of a passive resonator	
Q_{switch}	transistor that is used for band-switching	
Q_{tot}	total quality factor of compound varactor	
$Q_{two-int}$	unity, Q of linear two-integrator oscillator model	
$Q_{varactor}$	varactor quality factor	
R	resistance	Ω
R_{base}	base resistance	Ω
R_{cs}	effective series resistance of capacitor C_s	Ω
R_{gate}	gate resistance	Ω
R_{ls}	effective series resistance of inductor L_p	Ω
R_p	effective resistance of parallel resonator	Ω
$R_{thermal}$	thermal resistance of a package	K/W
R_{tune}	tuning resistance for active variable inductance	Ω
R_π	$\beta_O/g_{m_{bipo}}$, input resistance	Ω
s	$\sigma + j\omega$, Laplace transform complex variable	
S-parameter	two-port scatter parameter ($S_{11}, S_{12}, S_{21}, S_{22}$)	
S_{IF}	IF signal	
S_{lb}	lower bound of a specification	
S_{LO}	LO signal	
S_n	switched-capacitor number n	
S_{RF}	RF signal	
$Spec_x$	A certain specification x	
S_{ub}	upper bound of a specification	
$S_{v_{out}}$	power spectral density of v_{out}	V^2/Hz
$S_{\theta_{pn}}$	one-sided power spectral density of $\theta_{pn}(t)$	rad^2/Hz
$S^{ds}_{\theta_{pn}}$	double-sided power spectral density of $\theta_{pn}(t)$	rad^2/Hz
$t_{settling}$	settling time of an LC oscillator	s
T	absolute temperature	K
$T_{ambient}$	ambient temperature in the vicinity of an IC	$^\circ C$
$T_{junction}$	junction temperature of an IC	$^\circ C$
$v_{carrier}$	rms carrier voltage	V
v_{in}	input voltage	V
v_{noise}	rms noise voltage	$V/\sqrt{Hz}$
v_{out}	(oscillator) output signal	V
v_{ripple}	ripple voltage on the supply	V
V_{cb}	collector-base voltage	V
VCC	supply voltage of bipolar circuit	V
$V_{control}$	control voltage of active variable inductance	V
V_{DC}	supply voltage of oscillator core	V
VDD	supply voltage of MOS circuit	V
V_{gs}	gate-source voltage	V
V_j	built-in junction potential	V
V_{jc}	built-in collector-base junction potential	V
V_{lim}	input voltage where a transconductance starts limiting	V
V_o	peak LC resonator voltage	V

V_p	peak amplitude of modulation signal	V
V_{peak}	peak amplitude	V
V_{sg}	source-gate voltage	V
V_{supply}	supply voltage	V
V_{switch}	control voltage of a switched capacitor	V
V_T	kT/q, thermal voltage	V
V_{TH}	MOS threshold voltage	V
V_{TMOS}	$V_{gs} - V_{TH}$, voltage where a MOS differential pair approx. starts limiting	V
V_{tune}	tuning voltage	V
W	MOS channel width	m
$X_{in}(j\omega)$	input variable feedback system	
Y	admittance	A/V
$Y_a(j\omega)$	admittance of negative resistance model (active part)	A/V
$Y_{out}(j\omega)$	output variable feedback system	
$Y_p(j\omega)$	admittance of negative resistance model (passive part)	A/V
X_μ	factor that determines the intrinsic part ($X_\mu C_\mu$) of C_μ	
Z	impedance	Ω
Z-parameter	two-port impedance parameter ($Z_{11}, Z_{12}, Z_{21}, Z_{22}$)	
$\alpha(\omega_{osc}t)$	$\Gamma_{\text{eff}(\omega_{osc}t)} = \Gamma(\omega_{osc}t)\alpha(\omega_{osc}t)$, noise modulation factor	
α_{band}	$C_{maxband}/C_{minband}$, band-switch capacitance ratio	
α_{BER}	conversion factor between peak and rms jitter	
α_{cur}	variable current multiplication factor	
α_{ol}	$g_m R_p$, open-loop gain of behavioral LC oscillator model	
$\alpha_{ol_{ring}}$	open-loop gain of a CML ring oscillator model	
α_p	ratio of PMOS and NMOS W/L in an inverter-type ring oscillator	
$\alpha_{varactor}$	C_{max}/C_{min}, varactor capacitance ratio	
$\beta(j\omega)$	feedback transfer function in a system	
β_O	common-emitter current gain	
γ	noise factor of a MOS transistor (2/3 for a long-channel device)	
$\Gamma(\omega_{osc}t)$	impulse sensitivity function (ISF)	
$\Gamma_{\text{DC}}(\omega_{osc}t)$	DC value of effective ISF Γ_{eff}	
$\Gamma_{\text{eff}}(\omega_{osc}t)$	$\Gamma(\omega_{osc}t)\alpha(\omega_{osc}t)$, effective ISF	
$\Gamma_{\text{rms}}(\omega_{osc}t)$	rms value of Γ_{eff}	
ε	constant used for AC phase noise simulation	
η_{LC}	P_{RF}/P_{DC}, LC oscillator efficiency	
η_{ring}	CML ring oscillator efficiency	
θ	phase shift of phase shifter in N-stage LC oscillator model	rad
$\theta_{pn}(t)$	stochastic phase variable	rad
θ_{rms}	rms angular phase deviation of $\theta_{pn}(t)$	rad/$\sqrt{\text{Hz}}$
μ_n	mobility of electrons (NMOS)	m^2/V s
μ_p	mobility of holes (PMOS)	m^2/V s
ν	$\omega/\omega_{LC} - \omega_{LC}/\omega$	
σ_{c-to-c}	cycle-to-cycle jitter	s
σ_X	standard deviation (spread) of component X	%
ϕ	phase	rad

ϕ_0	initial phase at $t = 0$	rad
ϕ_{cur}	phase between coupling current and total current in an N-stage LC oscillator	rad
ϕ_e	phase error	rad
ϕ_{res}	resonator phase shift	rad
τ	time constant	s
$\tau_{CML-stage}$	dominant time constant in a CML ring oscillator stage	s
$\tau_{CMOS-stage}$	dominant time constant in a CMOS ring oscillator stage	s
τ_{delay}	propagation delay in a ring oscillator	s
τ_F	forward transit time	s
τ_t	time constant of pole in two-integrator oscillator model	s
τ_1	time constant of pole in N-stage ring oscillator model	s
ω	angular frequency	rad/s
ω_Δ	angular peak frequency deviation	rad/s
ω_{Grid}	angular grid frequency	rad/s
ω_{IF}	angular IF frequency	rad/s
ω_{LC}	$(\sqrt{C_p L_p})^{-1}$, angular oscillation frequency of ideal LC oscillator	rad/s
ω_{LO}	angular LO frequency	rad/s
ω_m	angular offset/modulation frequency	rad/s
ω_{NLC}	angular oscillation frequency of N-stage LC oscillator behavioral model	rad/s
ω_{Nring}	angular oscillation frequency of N-stage ring oscillator behavioral model	rad/s
ω_{osc}	angular oscillation frequency	rad/s
ω_{RF}	angular RF frequency	rad/s
$\omega_{two-int}$	angular oscillation frequency of two-integrator oscillator model	rad/s
$\Delta\omega_{LC}$	frequency shift with respect to ω_{LC}	rad/s

Abbreviations

AAC	Automatic Amplitude Control
AC	Alternating Current
ACPN	AC Phase Noise
ADC	Analog-to-Digital Converter
ADS	Advanced Design System
AGC	Automatic Gain Control
AM	Amplitude Modulation
As	Arsenide
BER	Bit Error Rate
BiCMOS	Bipolar-CMOS
CAD	Computer Aided Design
CCO	Current Controlled Oscillator
CML	Current Mode Logic
CMOS	Complementary Metal Oxide Semiconductor
CNR	Carrier to phase Noise Ratio
dB	decibel
dBc	dB relative to the carrier
DC	Direct Current
DCR	Data Clock Recovery
DECT	Digital European Cordless Telephone
DRO	Dielectric Resonator Oscillator
DS	Double-Sided
DVB-T	Digital Video Broadcasting Terrestial
EDA	Electronic Design Automation
Ga	Gallium
Ge	Germanium
GSM	Global System for Mobile communication
FDD	Frequency Division Duplex
FDMA	Frequency Division Multiple Access
FM	Frequency Modulation
FOM	Figure of Merit
HBT	Hetero-junction Bipolar Transistor
IC	Integrated Circuit
iDAC	Current Digital-to-Analog Converter

IF	Intermediate Frequency
InP	Indium Phosphide
I/Q	In-phase/Quadrature
ISF	Impulse Sensitivity Function
IRR	Image Rejection Ratio
LNA	Low Noise Amplifier
LNB	Low-Noise Block-converter
LO	Local Oscillator
LTI	Linear Time Invariant
LTV	Linear Time Variant
MOS	Metal Oxide Semiconductor
NBW	Noise BandWidth
Nor-PN	Normalized Phase Noise
NRZ	Non-Return-to-Zero
Osc-No	Oscillator Number
ODE	Oscillator Design Efficiency
OPAMP	Operational Amplifier
PCB	Printed Circuit Board
PLL	Phase Locked Loop
PM	Phase Modulation
PSRR	Power Supply Rejection Ratio
PSS	Periodic Steady State
Q	Quality factor
QPSK	Quadrature Phase Shift Keying
rms	Root-Mean-Squared
RF	Radio Frequency
RO	Reference Oscillator
RX	Receive (-band)
SDH	Synchronous Digital Hierarchy
Si	Silicium
SMD	Surface Mounted Device
SNR	Signal-to-Noise Ratio
SONET	Synchronous Optical NETwork
SOA	Silicon On Anything
Spec.	Specification
SSB	Single-SideBand
Stabi	Stabilizer
TIA	Trans-Impedance Amplifier
TR	TRansient
TX	Transmit (-band)
UI	Unit Interval
UMTS	Universal Mobile Telecommunications System
VCO	Voltage Controlled Oscillator
VHDL	VHSIC Hardware Description Language
XO	Crystal Oscillator

1

Introduction

ELECTRONIC communication nowadays is unthinkable without the use of oscillators. An electronic oscillator[1] is present in almost every electronic communication system, and provides a steady, often tunable, periodic signal, necessary for signal processing functions within the system.

> This book discusses the properties of a wide range of high-frequency integrated oscillators and highlights design methods and circuit techniques for their realization.

In this chapter, some instructive history that led to the wide-spread use of oscillators will be described first, followed by typical examples of electronic systems in which oscillators play a prominent role (Section 1.2). Next, in Section 1.3, a comprehensive overview is given on the literature. The motivation for this book will be illustrated by Section 1.4. We end this chapter with Section 1.5, which clarifies the scope of this monograph.

1.1 History

The verb "oscillate" was first recorded in 1726 and the noun oscillation dates back to 1658 when Christian Huygens worked on the pendulum clock [1]. Electronic os-

[1]The word oscillator is derived from the Latin verb "oscillâre", which goes back to "oscillum", meaning swing. An English dictionary describes the word oscillate: "To swing back and forth with a steady, uninterrupted rhythm" (source: www.dictionary.com).

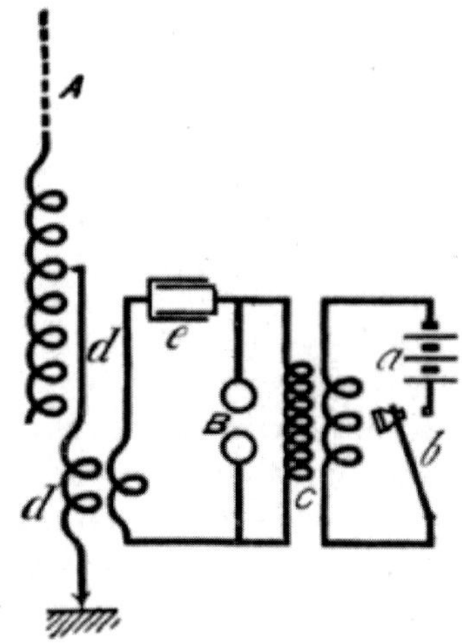

Figure 1.1 Guglielmo Marconi's syntonized transmitter.

cillators gained importance when wireless radio transmission was invented. A breakthrough in the history of wireless transmission was Marconi's invention of "Syntonic Wireless Telegraphy" for which he was granted a British patent in 1898 (no. 12039, filed in 1896). One key paragraph of this patent reads:

> "It is desirable that the induction coil should be in tune or syntony with the electrical oscillations transmitted, the most appropriate number of turns and most appropriate thickness of wire varying with the length of wave transmitted".

By means of the wireless telegraph, cross-Atlantic communication was established in 1901. The "syntonized transmitter", as Marconi called it[2], is shown in Figure 1.1. The capacitor marked "*e*" is tunable, and can be changed to form a resonance circuit with the antenna, marked "*A*". Syntonized transmission did not utilize an oscillator yet, but it did use a tuned circuit, which improved the maximum transmission range. Moreover, it allowed for a simple construction of several transmitters that could receive independent signals. A patent for multiple simultaneous transmission by utilizing tuned coupled circuits was filed in 1900 by Marconi and it is known as the famous "four sevens" patent. In 1902 Marconi summarizes his achievements in a lecture for the Society of Arts and states [2]:

> "I have come to the conclusion that the days of non-tuned systems are numbered"

[2]Source: home.luna.nl/~arjan-muil/radio/museum.html.

Application	Oscillator function
DECT Low-IF receiver	tunable frequency, down-conversion
Bluetooth transmitter	tunable frequency, up-conversion
FM radio front-end	tunable frequency, down-conversion
Satellite tuner	tunable frequency, down-conversion
Satellite LNB	fixed frequency, down-conversion
Baseband digital processing	clock generation
Watch	timing reference
FM radio demodulator	demodulation
Optical transmitter	clock conversion and generation
Optical receiver	carrier regeneration

Table 1.1 Examples of oscillator functions in various applications.

He could not have been more right. In 1914 Marconi transmitted speech over 50 miles by using an RF oscillator, modulated by speech. Since then, oscillators have never left the field of electronic communication.

> The world-wide production of RF products exceeded 900 million units in 2002 and is still rising, all having one or more oscillators incorporated [3].

1.2 Application examples

Oscillators are used in many applications and have various functions. A number of application examples is given in Table 1.1. The first four examples are applications that use Frequency Division Multiple Access (FDMA) to distinguish between multiple users or multiple broadcast channels. A tunable oscillator is used to select one of many channels for the reception and extraction of the information content of the channel. Consider the example of a Digital European Cordless Telephone (DECT) receiver front-end, where for simplicity the Radio Frequency (RF) signal[3] S_{RF} is only the carrier $\cos(\omega_{RF}t)$. In reality the carrier is modulated and occupies a certain bandwidth as shown in Figure 1.2. Multiplication, or mixing, of this signal with an Local Oscillator (LO) signal $S_{LO} = 2\cos(\omega_{LO}t)$ results in an Intermediate Frequency (IF) signal S_{IF} equal to

$$S_{IF} = \cos((\omega_{RF} - \omega_{LO})t) + \cos((\omega_{RF} + \omega_{LO})t). \tag{1.1}$$

Other channels, often equidistantly spaced on a certain frequency grid ($\pm n \cdot \omega_{Grid} + \omega_{RF}$ with n an integer), are also down-converted in frequency. However, a channel

[3]For most wireless telecommunications systems, such as DECT, the RF frequency is in the range of 0.8 to 3 GHz. Many "older" broadcast standards (terrestrial TV, FM radio, AM radio) work below 1 GHz. FM radio for example operates around 100 MHz.

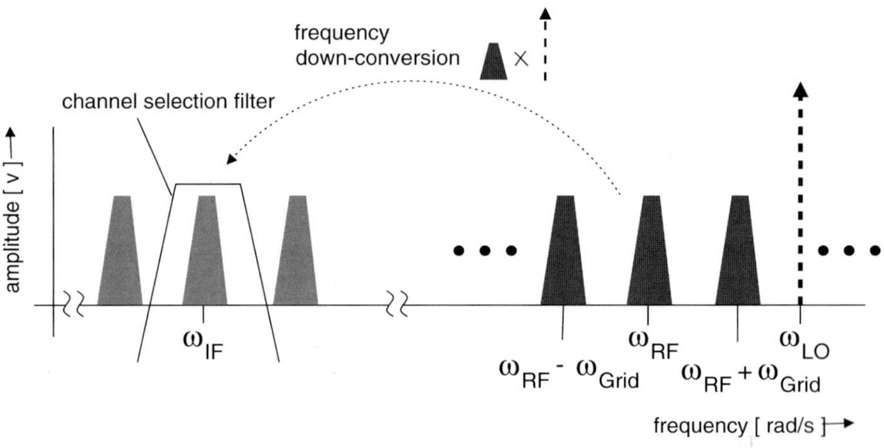

Figure 1.2 An oscillator with frequency ω_{LO} is used to convert a desired channel down in frequency to a fixed IF, where the channel is selected by means of filtering.

selection filter attenuates these channels and the up-conversion product in (1.1) so that only the desired channel at $\omega_{IF} = \omega_{RF} - \omega_{LO}$ remains (see Figure 1.2). If the oscillator frequency is tuned to $\omega_{LO} \pm n\omega_{Grid}$, the channel n at $\omega_{RF} \pm n\omega_{Grid}$ is selected. Hence, many concurrent conversations can be carried out with a DECT handset by many users. This example illustrates the important role of an oscillator in the frequency conversion and channel selection function of a receiver. At the IF, the desired channel can be processed in the analog domain or converted to the digital domain by an Analog-to-Digital converter (ADC).

> Although technology trends allow higher and higher IF sampling frequencies of integrated ADCs, RF sampling generally will not be feasible or not cost-effective for many years to come. For RF sampling the "LO-function" of the oscillator shifts to a clock function requiring extreme oscillator stability.

In a transmitter, such as the Bluetooth example in Table 1.1, the baseband signal is up-converted to a certain RF frequency, filtered by a transmission filter and transmitted[4]. In principle, it is simply the reverse process of down-conversion. There are several reasons why it is necessary to convert an information signal at baseband to a much higher RF frequency, prior to transmission. The most important reasons are adaptation to the transmission medium (for example, to increase radiation efficiency

[4] An alternative method is direct modulation of an oscillator, operating at the RF frequency, the way Marconi did.

or to comply with international frequency standards), reduction of noise and interference, channel assignment, multiplexing of messages over a single channel and to overcome equipment limitations [4].

An example of an application where a fixed frequency oscillator is used for down-conversion, is the Low-Noise Block-converter (LNB) in a satellite dish. Multiple satellite channels in the frequency range of about 10 to 12 GHz are received and down-converted in the LNB to an IF between 1 and 2 GHz. The cable connecting dish and satellite tuner (set-top box) has much lower losses for these frequencies compared to 10 GHz, and can therefore be cheaper. At the IF frequency, channel selection can take place in the satellite set-top box with a tunable oscillator, in a way similar to that we discussed for the DECT receiver.

Most communication systems, like the ones discussed above, can be divided into a front-end and a back-end. The main functions in a receiver front-end are frequency conversion and channel selection. In the back-end, analog signals are converted to the digital domain by one or more ADCs and a lot of signal processing takes place in this domain. The ADCs, and the digital circuitry that performs the signal processing need clock signals, which are generated by an oscillator (see Table 1.1).

Long-term stability is what the consumer wants of the oscillator in a watch. In this application the function of the oscillator is to provide a stable beat to accurately keep track of time. When an oscillator is used for demodulation of FM radio signals, the short-term stability of the oscillator is of major importance. The information of the FM signal is corrupted by fast stochastic variations in the frequency of an oscillator in a demodulator.

The last two examples of oscillators in Table 1.1 are in the field of optical transceivers. In optical transmission systems, a serial non-return-to-zero (NRZ) data stream is transported via the glass fiber. In an optical transmitter, the bits of a parallel baseband data stream are time-multiplexed into a serial data stream prior to amplification and transmission. This operation requires an oscillator that provides a clock signal at transmission speed. In an optical receiver front-end, the serial bit clock has to be regenerated since it is not separately transmitted, but is incorporated in the NRZ data stream. This function is called Data Clock Recovery (DCR); most advanced DCR architectures require an oscillator.

> Oscillators in transceivers are used for many purposes, such as tunable or fixed frequency up/down conversion, modulation, demodulation, clock conversion, clock generation, carrier recovery and timing references.

The examples in this section demonstrate that oscillators are used in many different systems for different reasons. These oscillators have to be designed and optimized to meet the requirements of the system in which they are used. Starting from the early days of oscillator usage in communication systems, the design of oscillators and related theory has intrigued many researchers. This resulted in a rich source of literature on oscillator theory, design and implementations. A comprehensive overview of this literature is given in the next section.

1.3 Literature on oscillators

Inventing the wheel would have been very rewarding, since wherever you are there is something on wheels near you nowadays. Unfortunately somebody had this bright idea somewhat earlier. To prevent the reinvention of existing oscillator theory and to extract important design insights from existing papers and articles, a literature study was conducted.

> Many literature references to numerous key papers on oscillator theory and excellent design examples will be given throughout this book. All references are listed at the end of this book and papers on LC oscillator and ring oscillator implementations have been given separate bibliographies.

A selection of important literature on oscillators is given in Table 1.2. Rayleigh and Van der Pol were among the first to recognize that a practical oscillator requires an amplitude limiting mechanism for amplitude stabilization.

> Barkhausen [7] formulated the necessary conditions for oscillation and oscillation start-up, which were extended for linear and nonlinear feedback systems, such as an oscillator by Nyquist, Bode, and Blaquière.

Most other contributions to the open literature on oscillators in Table 1.2 focus on a description of noise in oscillators. Similar to other electronic building blocks, for example Low Noise Amplifiers (LNA) or filters, the inevitable presence of noise in oscillators limits the attainable Signal-to-Noise Ratio (SNR) in systems. Leeson proposed a heuristic oscillator noise model for oscillators that was widely adopted, probably due to its simplicity. His linear model had an a noise figure in it, which was a fit-factor rather than a priori predictable parameter. Especially during the 90's, where digital wireless communication standards, like Global System for Mobile communication (GSM), were developed and successfully introduced to the consumer market, the need for a better understanding of all, linear and nonlinear, mechanisms contributing to noise in oscillators became apparent. For portable equipment the power dissipation of low-noise oscillators directly influences the battery size, and consumers are not really fond of walking around with battery backpacks. The last five entries of Table 1.2 are examples of important publications on the subject of nonlinear noise mechanisms.

In addition to the qualitative analysis of oscillator literature in Table 1.2, a quantitative overview of the number of publications on oscillators over the past decade

Ref.	Author	Description	Year
[5]	Rayleigh	Nonlin. diff. Eq. / limiting mechanism	1883
[6]	Van der Pol	Nonlin. diff. Eq. / limiting mechanism	1920
[7]	Barkhausen	Feedback theory / osc. conditions	1921
[8]	Nyquist	Stability lin. feedback systems	1932
[9]	Groszkowski	Influence of harmonics	1933
[10]	Kryloff	Equivalent linearization method	1934
[11]	Bode	Amplifier design & Bode diagrams	1945
[12]	Edson	Noise in harmonic oscillators	1960
[13]	Hafner	Noise in harmonic oscillators	1966
[14]	Leeson	Simple oscillator noise model	1966
[15]	Blaquière	Oscillator noise calculations	1966
[16]	Kurokawa	low frequency noise modulation	1968
[17]	Takaoka	Matrix description of additive noise	1980
[18, 19]	Abidi	Relaxation oscillator modeling	1981
[20]	Bates	Stability of multi-freq. oscillators	1984
[21]	Robins	AM and PM noise modeling	1984
[22]	Riddle	Pole location of oscillators	1986
[23]	Vittoz	Crystal oscillator design	1988
[24, 25]	Verhoeven	Relaxation oscillator modeling	1990
[26]	Razavi	Noise CMOS ring osc. / general Q def.	1995
[27, 28]	McNeill	Jitter in ring oscillators	1997
[29, 30]	Hajimiri	General phase noise theory	1998
[31]	Samori	Noise folding in LC osc.	1998
[32, 33]	Darabi & Rael	Phase noise in nonlin. LC osc.	2000
[34, 35]	Huang	Phase noise in nonlin. Colpitts osc.	2000

Table 1.2 Key references on oscillator noise and stability modeling.

is presented in Figure 1.3. The successful introduction of digital wireless standards caused a significant increase in the number of publications. The average was around 250 publications each year in the early 90's and increased to over 350 by the late 90's.

Some of the publications on oscillators report important new insights, like for example the ones listed in Table 1.2. These breakthroughs certainly speed up first-time-right oscillator design, but typically only one oscillator property is taken into account, such as oscillator noise. Most publications typically focus on one specific realized oscillator design for a specific application and report design issues and measurement results of this design. It should be realized that each publication on an oscillator design normally is the result of many man-months of work.

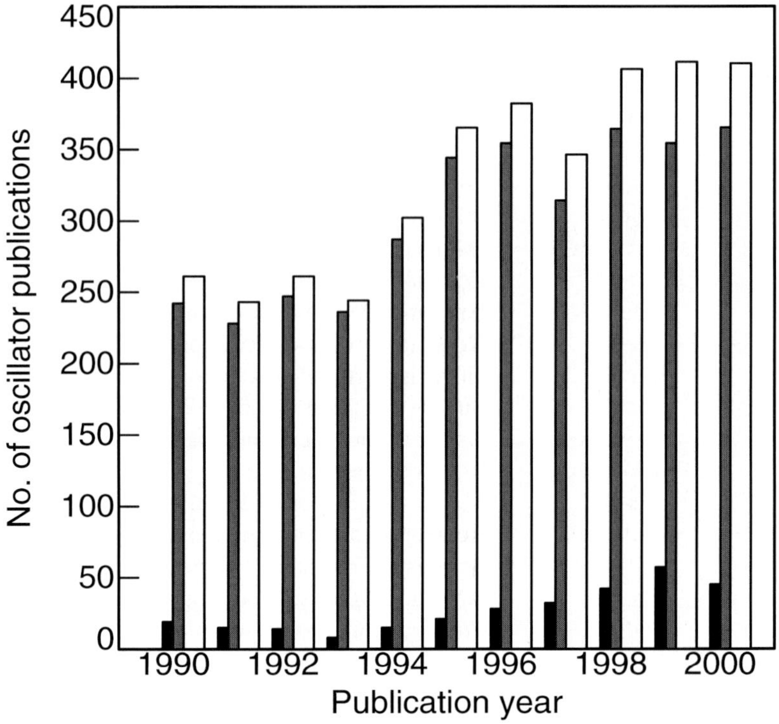

Figure 1.3 Journal and conference publications of the past decade, in the field of electrical engineering with the word VCO (black bar) or oscillator (grey bar) in the title. The white bar represents the addition of the black and grey bar. (source: INSPEC literature database).

The "wireless revolution" in the late 90's prompted the need for a better understanding of oscillators, especially concerning the mechanisms leading to noise in oscillators. The resulting increase in worldwide research led to better oscillator models (see Chapter 6) as well as simulation tools (see the introduction of Chapter 8).

In contrast to the abundance of literature on oscillator designs, the number of publications that discuss oscillator design methods is very small. The number of publications in the past decade with the word "oscillator" and "methods" or "methodology" in the title is less than ten, according to the INSPEC literature database. This book addresses this apparent void in methodical design and describes design methods and tools for a wide range of integrated oscillators, addressing a wide range of oscillator properties.

1.4 The oscillator designer

In practice, design resources are limited and design time reduction is increasingly important, because of the consumer market, resulting in a reduction of time-to-market to stay competitive.

> The oscillator designer has to achieve a certain oscillator performance given limited design resources, where design time often is a dominant limiting resource.

An analogy of the oscillator design process is to view the designer as a traveler who has to travel a certain road. The road is the design process. At the end of the road the reward is waiting: an oscillator design, simulated, manufactured and measured, which meets specification under worst-case conditions. Unfortunately, the oscillator designer has no map of the shortest route to the destination point, and on the way there are many side-roads that may lead to a dead-end street. Obviously, an experienced oscillator designer will have a much shorter traveling time than an inexperienced one, but both their journeys could be shortened by signs along the road. Moreover, there seems no obvious way to dump the brain content of experienced designers into the brain of inexperienced designers.

The work described in this book aims to shorten the traveling time, i.e. the oscillator design process and increase the performance of the oscillator, by providing signs along the road for the oscillator designer and by barricading dead-end streets with road-blocks.

1.5 Scope

The design insight, methods and techniques presented in this book focus on high-frequency oscillators in integrated transceivers.

> Parts of this book discuss architectures and performance aspects of oscillators on a behavioral level, which qualifies these parts for use in a wider application area than integrated transceivers. However, circuit level discussions and design examples are presented specifically for high-frequency oscillators in wired or wireless transceivers.

The following subjects, strongly related to oscillator design in general, fall outside the scope of this study:

- *Layout*
  Once an oscillator design has reached the state of "electrical design ready", the physical layout has to be made. This work focuses on the oscillator design process up to layout. Although important layout issues will be mentioned when

Resonator	Quality factor indication
Ring oscillator	1
Integrated L and C	5 to 20
Integrated C plus bond-wire	up to 40
SOI L with C	up to 60
PCB L and SMD C	up to 80
Crystal	$\gg 100$
...	...

Table 1.3 Examples of resonators with an indication of the quality factor range.

relevant, the development of design methods for making layouts or layout automation falls outside the scope. Design of layouts, and methods and tools for layouts is a huge subject in itself [36].

- *Technology and device design*
  The technology used for implementing an oscillator has a profound impact on the performance of an oscillator design. Usually, an IC technology is optimized, characterized and presented to the designer as a library of devices. Most of these devices are parameterized, like the length and width of a resistor, the emitter area of a bipolar transistor and the W/L of a MOS transistor. It is acknowledged that for LC oscillators the resonator design is of high importance.

> Resonator choice and design is a very important aspect of low noise integrated LC oscillator design. As illustrated by Table 1.3 the designer has various resonator technology options each with a different feasible quality factor range, and a different cost price.

For a fully integrated design with on-chip inductors and capacitors, the resonator should be optimized for maximum quality factor. The optimization of inductors and varactors is considered device design too, and therefore falls outside the scope. In this work, it is assumed that the oscillator designer has access to a technology library, and can concentrate on circuit design rather than device design.

- *Phase Lock Loop (PLL) and tuning system design*
  In most cases, the tuning system of a transceiver consists of a PLL system[5]. In general, a PLL is a feedback system in which the phase difference of two signals (often two oscillators) is minimized [37]. An important part of a PLL is the (main) oscillator, which provides the signal with frequency f_{LO} needed in the signal processing parts of a transceiver. After frequency division by N, the phase of this oscillator is compared to a reference oscillator with frequency f_{ref}. Negative feedback causes phase lock between the main oscillator and the reference oscillator and f_{LO} becomes equal to Nf_{ref}. Generally, the division ratio N is digitally programmable, and f_{LO} can be tuned in frequency steps of f_{ref}, within the tuning range of the oscillator[6]. This book focuses on oscillators and not on complete tuning systems such as a PLL system. Excellent books and publications are available on PLL design [3,37–40].

- *Crystal oscillator design*
  In PLLs, the reference oscillator often is a crystal oscillator, which provides a very good short-term frequency stability as well as excellent long-term frequency stability. For example, based on the absolute frequency stability of the crystal oscillator, the frequency grid of transceivers using FDMA can be defined accurately. Within the framework of this book, crystal oscillators are considered to be low-frequency oscillators and fall outside the scope. A multitude of instructive books and publications are available on crystal oscillator design [23,41–43].

- *Distributed oscillator design*
  The majority of oscillators in integrated transceivers can be designed using Kirchhoff's laws and can be considered quasi-stationary systems[7]. Distributed oscillators will not be analyzed as this group of oscillators forms a minority in integrated transceiver systems (although its application is growing due to the ever increasing performance of IC technologies) and entails an entirely different analysis approach compared to oscillators with components that can be considered lumped. Examples of distributed oscillators can be found in [45,46].

- *Relaxation oscillator design*
  Relaxation oscillators share properties, such as a wide tuning range and ease of integration with ring oscillators. At high frequencies, ring oscillators are preferred to relaxation oscillators in practice, due to their simplicity, equal or better noise properties and their multi-phase character. The design of relaxation oscillators will not be discussed. Many excellent publications are available on relaxation oscillator design [18,19,24,25,47–49].

---

[5] In all application examples in Table 1.1, a PLL system generally is used, except for the watch and the clock oscillator needed in baseband digital processing. In these two applications, a crystal oscillator suffices.

[6] Fractional-N PLL techniques allow to generate frequency steps with a smaller step size than f_{ref} [3].

[7] A rule of thumb is that if the dimensions of a system (or subsystem) are at least ten times smaller than the wavelength of the highest operating frequency, the system can be considered quasi-stationary and Kirchhoff's laws are a good approximation [44].

2

Oscillators

OSCILLATORS are intriguing building blocks. An oscillator is the building block in a transceiver with a built-in timing reference[1]. When the power supply is switched on, DC power is used to somehow translate the information of the timing reference into a periodic signal, which forms the heartbeat of many functions in a system.

> In this chapter a concise overview of basic oscillator theory is given.

The first two sections discuss the ideal oscillator and the non-ideal oscillator, respectively. Based on implementation principle, a practical classification of oscillators is introduced in Section 2.3. Next, two important questions: "what are the conditions for oscillation?", and "how do we obtain stable (steady-state) oscillation?", are answered in Sections 2.4 and 2.5, respectively. Finally, a summary, in Section 2.6, concludes this chapter.

2.1 The ideal oscillator

The basic function of an oscillator is to generate a periodic signal with certain properties. An ideal oscillator generates a signal that only has wanted properties. The signal waveform can have any form and any number of harmonics. The output voltage of an

[1] A timing reference is a system that produces an output signal with a known time behavior [50].

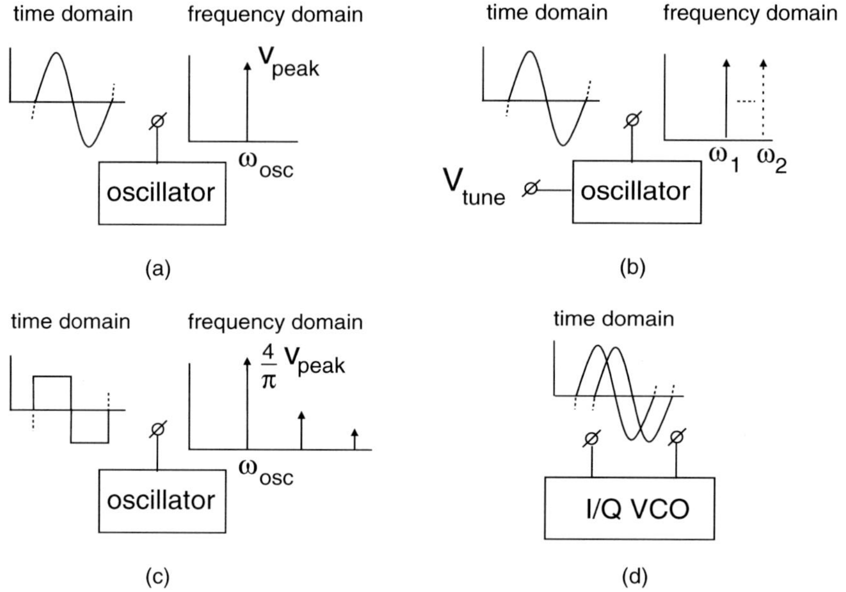

Figure 2.1 A harmonic oscillator (a), a harmonic VCO (b), a square wave oscillator (c), and an I/Q oscillator (d).

ideal harmonic oscillator (see Figure 2.1(a)) with angular frequency ω_{osc} in rad/s and peak amplitude V_{peak} in V, can be written as

$$v_{out}(t) = V_{peak}\cos(\,\omega_{osc}\,t + \phi_0), \tag{2.1}$$

where ϕ_0 is the initial phase at $t = 0$ of $v_{out}(t)$. In the frequency domain, this is equivalent to a discrete spectral line with amplitude[2] V_{peak} at angular frequency ω_{osc}. This means that all carrier power is located in an infinitely small bandwidth around ω_{osc}.

The output signal of tunable harmonic oscillator (see Figure 2.1(b)) can be represented by

$$v_{out}(t) = V_{peak}\cos(\,2\,\pi\,(K_{vco}\,V_{tune} + f_{center})\,t + \phi_0). \tag{2.2}$$

Frequency ω_1 and ω_2 in Figure 2.1(b) represent the minimum and maximum frequency of the oscillator. Tuning voltage V_{tune} controls the frequency, and tuning constant K_{vco} in Hz/V determines the tuning slope. f_{center} is the oscillation frequency with a zero tuning voltage. As the tuning input is a voltage, the oscillator in Figure 2.1(b) is a Voltage Controlled Oscillator (VCO). In the case of a Current Controlled Oscillator (CCO), the tuning constant K_{cco} will have the units Hz/A.

[2] There are several definitions of forward and backward Fourier transforms. In this book we use the transform that yields the amplitude of the time domain signal (i.e. without a factor π, which would yield 4 for the fundamental of a square wave, instead of the factor $4/\pi$ in Figure 2.1(c) [51]).

Equation (2.2) does not explicitly show the "integrating" property of a controlled oscillator and assumes a steady-state behavior. For a non-steady-state description, we have to use the phase property: the phase is the integral of the frequency with respect to time. Use of (2.2) for non-continuous V_{tune} results in phase jumps in the oscillator output signal. The following expression models the instantaneous behavior of a practical oscillator,

$$v_{out}(t) = V_{peak}\cos(2\pi(K_{vco}\int_0^t V_{tune}(t)\,dt + f_{center}t) + \phi_0). \tag{2.3}$$

Now the output phase always is continuous, even if a non-continuous $V_{tune}(t)$ is modulating the oscillator. If we choose $V_{tune}(t)$ equal to $V_p\cos(\omega_m t)$, calculation of (2.3) leads to the familiar time domain equation for a Frequency Modulated (FM) signal,

$$v_{out}(t) = V_{peak}\cos(2\pi f_{center}t + \frac{\Delta\omega}{\omega_m}\sin(\omega_m t) + \phi_0), \tag{2.4}$$

where $\Delta\omega = 2\pi K_{vco}V_p$.

An ideal harmonic oscillator provides a frequency selective signal: in the frequency domain all energy is located at the fundamental frequency. The oscillator in Figure 2.1(c) generates a square wave that is selective in time; timing information is not distributed in time but located at discrete points; the zero crossings. The Fourier series of the output voltage of the oscillator in Figure 2.1(c) is calculated with

$$v_{out}(t) = \frac{4V_{peak}}{\pi}\left(\sin(\omega_{osc}t) + \frac{\sin(3\omega_{osc}t)}{3} + \frac{\sin(5\omega_{osc}t)}{5} + \ldots\right), \tag{2.5}$$

with the initial condition $\phi_0 = 0$.

Instead of generating one output signal, an oscillator can generate several output signals with different phases. A special case of a multi-phase oscillator is an oscillator, which generates a sine and a cosine, see Figure 2.1(d). Many modern receiver and transmitter architectures require these so-called "In-phase and Quadrature" (I/Q) signals in their signal-processing component.

The elementary properties, including many ideal properties, of oscillators in integrated transceivers are discussed in detail in Chapter 5.

2.2 The non-ideal oscillator

Of course nothing is ideal in practice. The oscillator and its properties are no exception. For example, an oscillator will never have exactly the required center frequency (f_{center} in (2.2)), due to the processing spread in an IC technology. Some additional tuning range will always be required therefore on top of the required range.

Figure 2.2 shows the fundamental and two harmonics of a square wave. An important distinction between an ideal and a practical oscillator are phase noise sidebands,

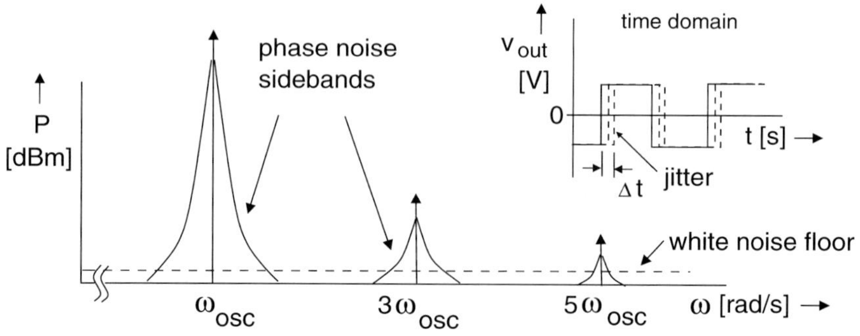

Figure 2.2 The spectrum (fundamental and two harmonics) of a square wave with phase noise sidebands. The inset shows the effect of the sidebands in the time domain: jitter.

as illustrated in Figure 2.2. Noise from the oscillator circuitry and externally generated noise (on the power supply, for example) corrupt the spectral purity of an oscillator signal. This means that the carrier power is now distributed in a finite bandwidth around ω_{osc} and its harmonics. Although most power is present at ω_{osc}, some power is also present at small offsets from ω_{osc} and its harmonics.

> Oscillator phase noise sidebands: the carrier power of the oscillator is distributed in a certain bandwidth around ω_{osc} and its harmonics.

These phase noise sidebands decrease with increasing offset frequency from the carrier frequency. White noise becomes dominant at a certain offset frequency. The white noise floor can originate from the oscillator but often originates from cascaded circuits. In the time domain, if zero-crossings are important, phase noise is referred to as jitter, and is shown in the inset of Figure 2.2. Due to the presence of jitter, the exact moment of a zero-crossing of the square wave (in other words, the phase) is stochastic. The ideal oscillator described by (2.1) can be extended to model this phase uncertainty by stochastic variable $\theta_{pn}(t)$. We can write

$$v_{out}(t) = V_{peak}\cos(\omega_{osc}t + \theta_{pn}(t)). \tag{2.6}$$

In a similar manner, (2.2)...(2.5) can be extended to model uncertainty in the oscillator output phase. In general, an oscillator signal will also have (multiplicative) amplitude noise and V_{peak} will have a stochastic component. However, unlike phase noise, this amplitude noise can be removed by a limiter at the expense of the generation of harmonics.

In a multi-stage oscillator the phase and amplitude relation between the output signals of each stage will be non-ideal. For example, in practice there will be a phase error and amplitude between the two-phases of the I/Q oscillator in Figure 2.1(d). In Chapter 4 we will see that this limits the achievable image rejection in transceivers.

The square wave in Figure 2.2 only has odd harmonics. In general, an oscillator output signal also has even harmonics. In balanced architectures, these are usually significantly lower than the odd harmonics. The application determines whether the harmonics of an oscillator output signal are unwanted or wanted. For example, if an oscillator is used as a clock generator, the harmonics are generally wanted. In that case, zero-crossings in the time domain are less sensitive to noise from the circuits where the clock-signal is used. On the other hand, if signals mix with harmonics and cause unwanted conversion products, these harmonics are unwanted properties of a practical oscillator.

The mentioned properties of the non-ideal oscillator, and others, are extensively analyzed and described in Chapter 6.

2.3 Classification

An oscillator classification can be most useful if it informs the designer about the properties of an oscillator, once he knows its place in the classification. An oscillator classification can be based on many aspects. For example, it may be based on one of the basic properties of an oscillator (frequency, tuning range, or noise performance), or on functionality (single phase output or multi-phase output). Another example is a classification based on the pole pattern of the timing reference in an oscillator, if this reference is a linear system [50].

> The oscillator classification adopted in this book is shown in Figure 2.3, and is based on the principle of implementation.

The oscillator implementation principle gives important information about the properties of an oscillator. Furthermore, oscillator types are easily given a place in such a classification. The components of an oscillator directly point out its place in the classification. Practical oscillators may strictly operate on one implementation principle, for example an oscillator constructed with a lumped resonator. However, a mixture of implementation principles is possible as well. LC oscillators can be connected in a ring structure and thus incorporate properties of a ring oscillators.

The first two branches in Figure 2.3 divide all oscillator types into the two classes "continuous-time" and "discrete-time" operation. A looped-back digital counter is an example of a discrete-time oscillator, as is the relaxation oscillator. Although the latter produces a continuous-time output signal, it entails a state memory with a discrete number of states, and it changes state at discrete time moments [50].

Continuous-time oscillators can either be resonator based or non-resonator-based. A resonator has complex poles and can be distributed or lumped. The lumped type consists of inductors *and* capacitors (2 types of reactive elements) and the same is true for the equivalent circuit of a distributed resonator. Continuous-time operating oscillators without resonators are implemented with capacitors *or* inductors (1 type of

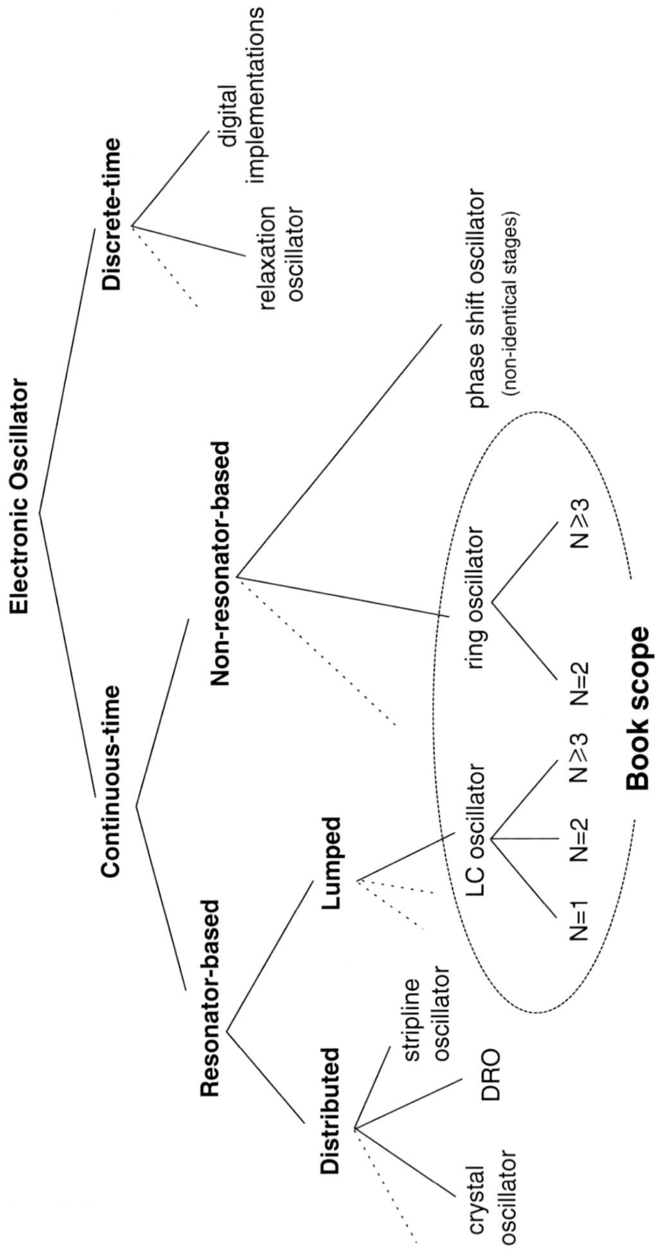

Figure 2.3 Practical oscillator classification based on implementation principle. For the oscillators within the scope of this work, the classification is extended based on the number of identical stages (N).

reactive element). Resistance (of a resistor, e.g. or a transconductor) and feedback are needed to make a non-resonator-based oscillator work. In order words, assuming linear modeling is applicable, explicit feedback is necessary to make complex poles out of the real poles of the reactive elements.

> In resonator-based oscillators, with a quality factor Q>1, energy storage in each oscillation period is possible.

In Chapter 6 we will see that this improves the spectral purity[3] of oscillators. This is indicated by the figure of merit "quality factor" Q, which will be defined in Section 2.4. Q normally is much larger than unity for resonator-based oscillators. Non-resonator based oscillators do not store energy each oscillation period, which generally leads to poorer spectral purity compared to resonator-based oscillators (see Chapter 6).

Oscillators with a lumped LC resonator[4] and ring oscillators fall within the scope of this book as discussed in Chapter 1. For these oscillators, the classification in Figure 2.3 is extended based on the number of identical stages N present in an oscillator. For LC oscillators, three sub-types are distinguished: single-phase LC oscillators (with one stage, $N = 1$), quadrature LC oscillators (with two stages, $N = 2$), and LC oscillators with more than two phases ($N \geq 3$). As mentioned in Section 2.1, quadrature signals are of high interest, since many advanced transceiver architectures require them. The minimum number of stages for a ring oscillators is two. An example of a two-stage ring oscillator is the two-integrator oscillator[5] [52]. Like any even-stage ($N = 2, 4, ...$) ring oscillator, a two-stage ring oscillator provides quadrature signals. All oscillators (single-phase and multi-phase) within the scope of this work are extensively analyzed in Chapter 5 and 6.

> Ring oscillators with $N = 2$, such as the two-integrator oscillator, are fundamentally different from ring oscillators with $N \geq 3$ (see chapters 5 and 6). The latter class of ring oscillators will be referred to as N-stage ring oscillators throughout this book (with $N \geq 3$).

The remainder of this section is devoted to a few implementation examples of oscillator types in the classification. Figure 2.4 shows three resonator-based oscillator configurations from the classification. The LO in RF transceivers is usually implemented using an LC oscillator with lumped resonator (see Figure 2.4(a)). The resonator, constructed with inductors and capacitors, can be on-chip [53], or partially off-chip [54]. Figure 2.4(b) shows an implementation of a distributed resonator-based oscillator. In a Dielectric Resonator Oscillator (DRO), the DR is placed adjacent to a

[3] Lower phase noise sidebands given a certain amount of power dissipation.

[4] The resonator is also referred to as a "tank" circuit.

[5] See Figure 2.17 for its behavioral model.

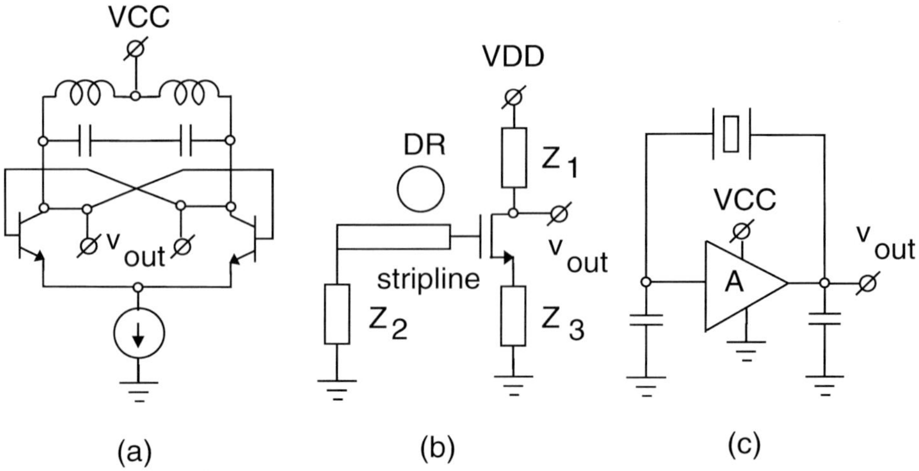

Figure 2.4 Three examples of resonator-based oscillators. A differential bipolar oscillator with lumped resonator (a), A MOSFET DRO (b) and a crystal oscillator (c).

micro strip-line, which implements the coupling between the resonator and the oscillator [45]. The crystal oscillator in Figure 2.4(c) belongs to the same group as the DRO. Similar to the DRO, they have an excellent spectral purity and very good long-term frequency stability too [23,43].

Examples of oscillators without resonators include ring oscillators, relaxation oscillators and the oscillator commonly referred to as a phase-shift oscillator. A possible implementation of these oscillators is shown in Figure 2.5. These oscillator are known for their ease of integration. Figure 2.5(a) shows a ring oscillator implementation. A few cascaded inverters with feedback already can implement such an oscillator. For the single-ended ring oscillator, an odd number of stages is required to avoid latch-up (positive feedback for DC). A bipolar relaxation oscillator implementation is shown in Figure 2.5(b). The principle of this oscillator type is based on the charging and discharging of a capacitor. A current charges a capacitor until a positive voltage threshold is reached and the direction of the current is reversed. Once the negative threshold voltage is reached, the current direction is again reversed, and so on. Figure 2.5(c) shows an entirely different oscillator. In the phase-shift oscillator, three identical RC sections each provide a phase shift of 60°. The transistor inverts its input signals, which makes the total phase shift in the feedback loop 360°. As we will see in Section 2.4.1, this fulfills the phase condition for oscillation.

> The reader should note that there are a great many types of oscillators and the discussed oscillators are only a small sub-set of all electronic oscillators. This is indicated by the dotted lines in Figure 2.3.

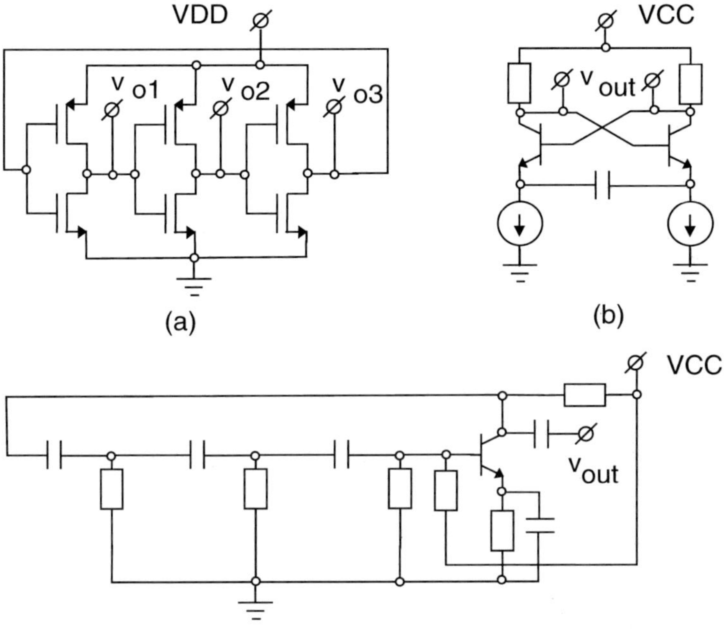

Figure 2.5 A single-ended CMOS ring oscillator implementation (a), a bipolar relaxation oscillator (b), and a phase-shift oscillator implementation (c).

2.4 Oscillation conditions

It is crucial for an oscillator designer to predict whether the oscillator under design is properly dimensioned, so that the oscillator will start and produce a periodic signal. All oscillators in the classification are feedback systems and in most instances feedback modeling is used to assess the oscillation conditions. The feedback model is used throughout this work and is discussed extensively in Section 2.4.1. The negative resistance model can be regarded as a special case of the feedback model and is briefly described in Section 2.4.2.

2.4.1 Feedback modeling

All oscillators in this work can be analyzed by modeling them as feedback systems. Figure 2.6 shows a general block diagram of a linear feedback system with transfer functions $H(j\omega)$ and $\beta(j\omega)$. The conditions needed for oscillation are most easily analyzed using linear models. For many practical oscillators, including the oscillators discussed in this work, linear analysis of the oscillation conditions provides sufficiently qualitative and quantitative insight for oscillator design. However, an oscillator

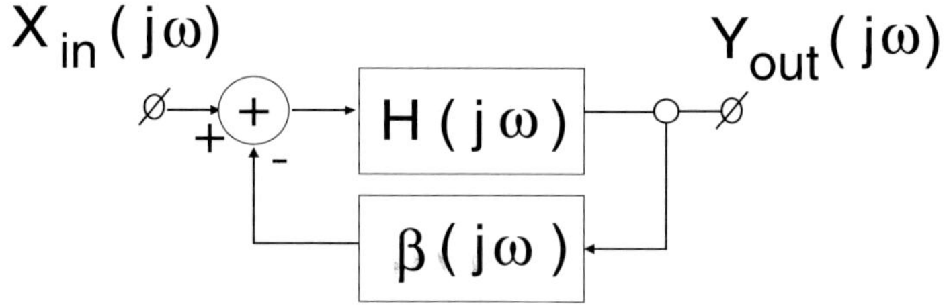

Figure 2.6 Block diagram of a linear feedback model for analyzing oscillation conditions.

may well operate in the weakly or strongly nonlinear region making linear modeling inadequate for capturing the full behavior of these oscillators. The least a designer should do is see how well application of the linear oscillation conditions compares with transient simulations, which reveals the influence of nonlinearities in the oscillator.

Steady-state oscillation conditions

The transfer function Y_{out}/X_{in} of the linear system in Figure 2.6 is the general equation for a feedback system (2.7):

$$\frac{Y_{out}(j\omega)}{X_{in}(j\omega)} = \frac{H(j\omega)}{1 + H(j\omega)\,\beta(j\omega)}. \tag{2.7}$$

The necessary conditions for steady-state oscillation are known as the Barkhausen conditions [7]. The first condition is called the "gain condition" and specifies that the open-loop gain must be unity.

The gain condition for steady state oscillation states

$$|H(j\omega)\,\beta(j\omega)| = 1. \tag{2.8}$$

The second condition for oscillation is referred to as the "phase condition", which is expressed as

$$\angle H(j\omega)\,\beta(j\omega) = (2m+1)\cdot 180^\circ. \tag{2.9}$$

This phase condition states that the total open-loop phase shift must be $(2m+1)$ times 180 degrees[6] where m is an integer value including zero.

[6] This becomes $2m\cdot 180^\circ$, if the minus sign of the adder in Figure 2.6 is changed into a plus sign.

The gain condition is only a good approximation. Due the presence of noise $|H(j\omega)\,\beta(j\omega)| < 1$ holds for all practical oscillators (see Chapter 8). Similar, the phase condition is only a good approximation for most oscillators. This condition is fulfilled only in circuits with linear gain control. For example, due to the presence of nonlinearities in the active part $\angle H(j\omega)\,\beta(j\omega) < (2m+1)\cdot 180^\circ$ holds for LC oscillators (see Chapter 5).

Start-up conditions for oscillation

The Barkhausen conditions state the necessary conditions for stable oscillation, but not for start-up. In any transceiver system it is vital that the oscillators autonomously start oscillating, usually triggered by noise, when the system is switched on. In order to guarantee oscillator start-up, the open-loop gain must initially be larger than unity. Therefore, the "start-up conditions" for oscillation are the following.

The gain condition for start-up is

$$|H(j\omega)\beta(j\omega)| > 1, \tag{2.10}$$

and the phase condition for start-up is

$$\angle H(j\omega)\,\beta(j\omega) = (2m+1)\cdot 180^\circ. \tag{2.11}$$

Once an oscillator has started there must be a mechanism that reduces the loop gain $\alpha_{ol} = |H(j\omega)\beta(j\omega)|$ effectively to unity, since this is a necessary condition for steady-state oscillation. Two mechanisms, automatic gain control and self-limiting, which reduce α_{ol} to one after start-up and stabilize the amplitude, will be discussed in Section 2.5.

Application of start-up and steady-state oscillation conditions

The LC oscillator in Figure 2.7(a) is used to illustrate the start-up conditions for oscillation. Transfer function $H(j\omega)$ is defined as transconductance g_m in the LC oscillator model. The feedback transfer function of the LC resonator[7], $\beta(j\omega)$, formed by C_p, L_p and loss resistance R_p can be described as

$$\beta(j\omega) = \frac{R_p}{1 + j\nu(\omega)Q_p}, \tag{2.12}$$

[7] Most LC oscillators in integrated transceivers use parallel resonators. However, many conclusions derived in this work based on oscillators with parallel resonators can similarly be derived for oscillators with series resonators. This is can be proven by the principle of duality [55].

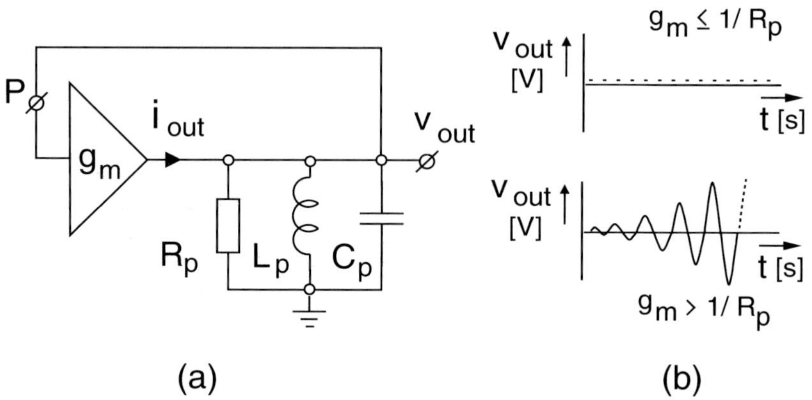

Figure 2.7 Behavioral model of an LC oscillator (a). To start-up, g_m must be larger than $1/R_p$ (b).

where

$$Q_p = R_p \sqrt{\frac{C_p}{L_p}}, \tag{2.13}$$

and the normalized frequency $\nu(\omega, \omega_{LC})$ is

$$\nu(\omega, \omega_{LC}) = \frac{\omega}{\omega_{LC}} - \frac{\omega_{LC}}{\omega}, \tag{2.14}$$

in which

$$\omega_{LC} = 2\pi f_{LC} = \frac{1}{\sqrt{L_p C_p}}. \tag{2.15}$$

At angular frequency ω_{LC}, the open loop gain α_{ol} is equal to $g_m R_p$. As shown in Figure 2.7(b), the oscillator does not start if g_m is smaller than or equal to $1/R_p$ at ω_{LC}. The lower graph in Figure 2.7(b) illustrates oscillator start-up when the start-up condition is met.

The loop gain α_{ol} can be determined by cutting open the feedback loop in the oscillator at a certain point, for example at point P in Figure 2.7(a), and simulating the gain between the transconductance input and output terminal v_{out}. This is easy in a behavioral model. When performing open-loop simulations at transistor level, the same DC and loading conditions must be enforced as in the closed loop situation to obtain the correct results.

The parameter Q_p in (2.12), which is defined in (2.13), is of great physical importance. This parameter is the quality factor of the resonator and is a measure for

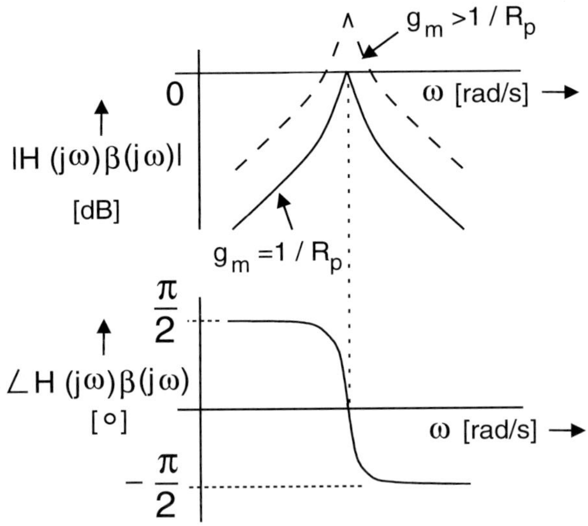

Figure 2.8 Bode diagram of the behavioral model in Figure 2.7(a).

the energy loss per oscillation period. Quality factor Q_p is based on the basic quality factor definition for a resonator[8] given as [56],

$$Q_{res} = 2\pi \frac{\text{Total energy stored in the resonator}}{\text{Energy lost per cycle from the resonator}}. \tag{2.16}$$

Oscillators that are constructed with resonators of a high quality factor (high-Q resonators), may have extremely low noise. In an oscillator, the energy loss has to be compensated by an active element. This active element also generates noise, which is generally proportional to the amount of energy it has to supply to sustain oscillation. In Chapter 6, the relation between the quality factor of an oscillator and its phase noise side bands will be analyzed extensively, but from the previous it already follows:

Q_{res} should be as high as possible for noise minimization.

The capability of (high-Q) resonators to store energy is illustrated in Appendix A.

The oscillation conditions can conveniently be checked using Bode plots[9] [11]. Figure 2.8 shows the Bode plot of the LC oscillator model under discussion. The

[8]The stored energy in the parallel LCR resonator in Figure 2.7(a), is $C_p v_{out}^2$, with v_{out} in rms voltage and the dissipated power v_{out}^2/R_p. Applying (2.16) leads to (2.13).

[9]A Bode diagram of the open-loop gain is a plot of $|H(j\omega)\beta(j\omega)|$ versus frequency on a log-log scale, and $\arg(H(j\omega)\beta(j\omega))$ versus frequency on a log scale.

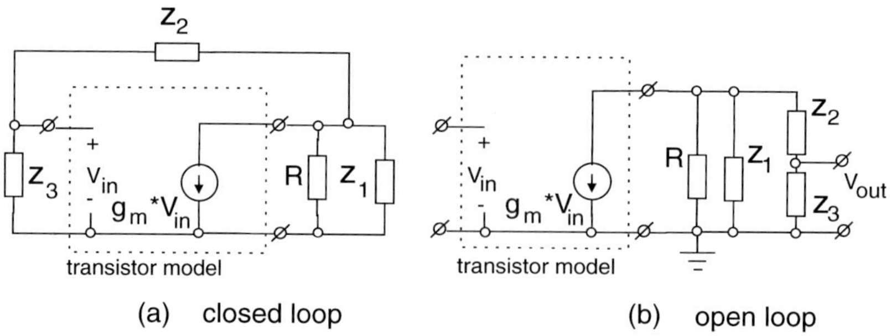

Figure 2.9 General three-terminal oscillator. One of three terminals should be grounded (a) and its open-loop model with one grounded terminal (b).

gain and phase plot shows that the steady-state oscillations conditions are met for $g_m = 1/R_p$. The Bode plot predicts that the oscillator will start oscillating for any setting of $g_m > 1/R_p$. The discussed oscillation conditions are necessary conditions but not sufficient for oscillation: they can also be met by stable circuits as will be shown in the following sections. If in doubt, the root locus[10] method can be used to assess whether the circuit under investigation will oscillate or not [57].

Using the steady-state oscillation conditions, the oscillation frequency of transistor-level oscillator circuit can be derived. As an example, the oscillation frequencies of the classic Colpitts and Hartley oscillator will be derived.

Figure 2.9(a) shows a general three-terminal oscillator. The active part of the oscillator is a transistor (e.g. a bipolar or a MOS transistor) modeled with its transconductance. One of the three terminals should be grounded. All losses are lumped in resistor R. The open-loop model of the three terminal oscillator in Figure 2.9(b) can be used to derive the oscillation frequency. The open-loop transfer function $H(j\omega)\beta(j\omega)$ is equal to

$$\frac{V_{out}(j\omega)}{V_{in}(j\omega)} = -\frac{g_m R Z_1(Z_2+Z_3)}{Z_1(Z_2+Z_3)+R(Z_1+Z_2+Z_3)} \cdot \frac{Z_3}{(Z_2+Z_3)}. \tag{2.17}$$

If Z_1 and Z_3 are capacitors, and Z_2 is an inductor, a Colpitts oscillator is obtained, see Figure 2.10(a), with oscillation frequency

$$f_{osc,Colpitts} = \frac{1}{2\pi}\frac{1}{\sqrt{L_2 C_{tot}}}, \tag{2.18}$$

where C_{tot} is the series combination of C_1 and C_3. Note that if Z_3 is replaced by an inductor, and Z_2 by a capacitor, the gain condition can be met but the phase condition

[10]The function $1+H(s)\,\beta(s)$ is called the characteristic equation and its roots are the closed-loop poles. The root locus is a plot of the locations in the s-plane of all closed-loop poles for the loop gain (in this case) varying from zero to infinity.

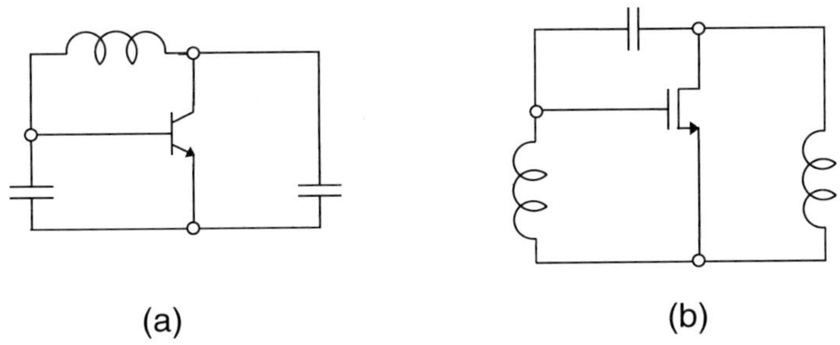

Figure 2.10 The Colpitts (a) and Hartley (b) oscillator.

cannot. Therefore this configuration will not yield an oscillator. A Hartley oscillator is obtained if Z_1 and Z_3 are replaced by inductors and Z_2 is replaced by a capacitor, see Figure 2.10(b). The oscillation frequency of the oscillator is now equal to

$$f_{osc,Hartley} = \frac{1}{2\pi}\frac{1}{\sqrt{C_2\left(L_1+L_3\right)}}. \tag{2.19}$$

Since any of the three terminals in Figure 2.10(a) and Figure 2.10(b) can be grounded, three types of Colpitts and Hartley oscillators exist. Obviously, the Colpitts oscillator can also be constructed using a MOS transistor and the Hartley type using a bipolar device. In practice, the Colpitts oscillator is encountered more often, since it only requires one inductor. Without detailed analysis, it is difficult to say which type is best for which application. In all cases, the choice of the ground terminal determines which parasitics of the active device are shorted and which terminal is best as the output terminal.

Multi-oscillation

When multiple oscillations coexist in steady-state, a multi-oscillation is present [58, 59]. This usually distorts the desired periodic signal, which then becomes useless for application in a transceiver. The oscillation conditions can be used to predict the presence of multi-oscillations. This is illustrated using the circuit diagram of a Pierce oscillator in Figure 2.11. A Pierce oscillator is a Colpitts-type oscillator, with a grounded emitter or source (in this case a grounded source).

Figure 2.12 shows the Bode diagram and root locus diagram of this oscillator. Although the oscillator has two complex conjugated poles for any gain setting, these poles enter the left half plane for very high transistor gain, despite the fact that the oscillation conditions are met (as indicated by the Bode plot). Hence, the root locus method should be used to see whether the oscillator will start. Alternatively, a transient simulation can be performed if calculations become too complex.

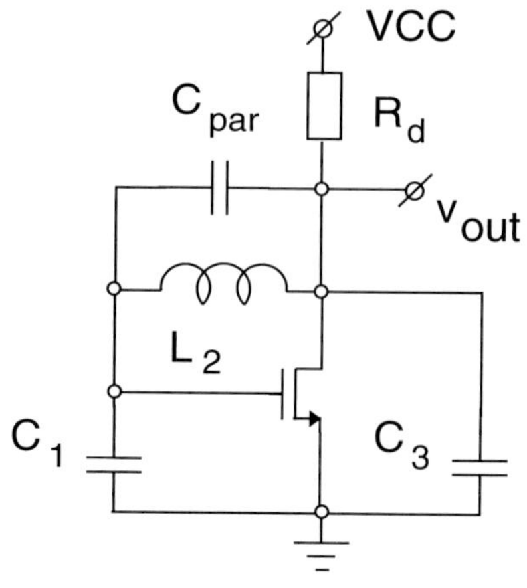

Figure 2.11 Pierce oscillator with parasitic capacitance C_{par} and all losses lumped in the drain resistor R_d.

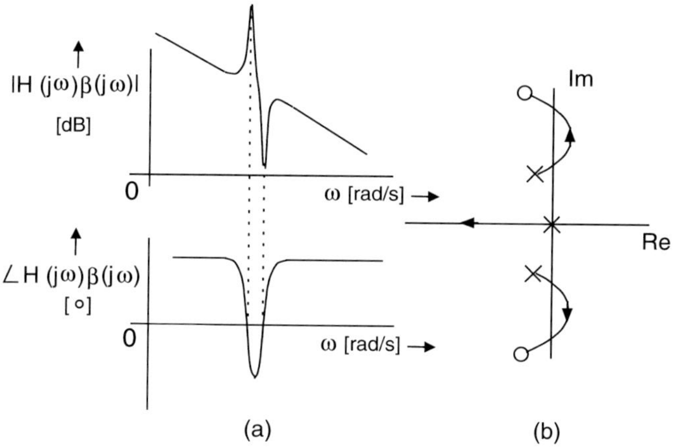

Figure 2.12 Bode diagram (a) and root locus (b) of the Pierce oscillator in Figure 2.11.

In the Bode plots of Figure 2.12(a), the start-up oscillation conditions are met for two frequencies. In a practical oscillator circuit with parasitics this indicates that there is a risk of multi-oscillation [59]. After start-up, the oscillation frequency will be dominant, where the loop gain is the highest but due to non-linearities and a changing bias condition in an oscillator, other potential oscillation frequencies can be excited during an oscillation period. Often parasitics are the cause of multi-oscillations and removal or reduction of these parasitics may remove the presence of multi-oscillations.

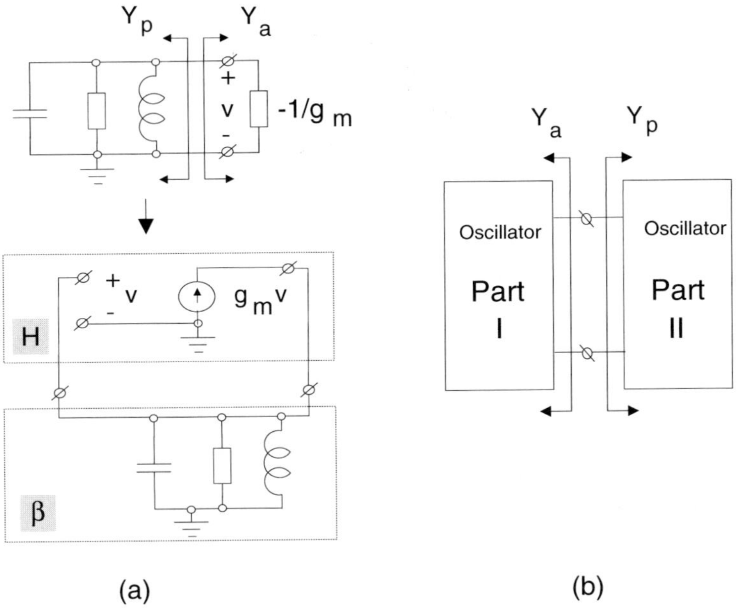

Figure 2.13 The LC oscillator in Figure 2.7 modeled as a negative resistance oscillator (a). The negative resistance model (b).

> In general, the gain at the parasitic oscillation frequencies must be sufficiently reduced below unity to eliminate the unwanted oscillation modes.

Implementation of this measure may include introduction of resistors in the oscillator circuit, which "stop" the unwanted frequency modes ("stop-resistors"). Since resistors add noise and degrade oscillator noise performance as we will see in Chapter 6, this measure should only be used sparingly.

2.4.2 Negative resistance modeling

Negative resistance modeling can be regarded as a special case of feedback modeling. Figure 2.13(a) shows how the LC oscillator in Figure 2.7(a) can be modeled as a negative resistance oscillator. The transconductor in Figure 2.7(a) with positive feedback ($H(j\omega)$ in the feedback model) is modeled as a resistance $-1/g_m$, which compensates the losses in the tank circuit ($\beta(j\omega)$ in the feedback model).

In general, if an oscillator can be divided into two parts with admittances[11] $Y_a(j\omega)$ equals $G_a(\omega) + jB_a(\omega)$ and $Y_p(j\omega)$ equals $G_p(\omega) + jB_p(\omega)$, the negative resistance model can be applied. This is illustrated in Figure 2.13(b). In many practical cases,

[11] Or impedances $Z_a(j\omega) = R_a(\omega) + jX_a(\omega)$ and $Z_p(j\omega) = R_p(\omega) + jX_p(\omega)$ for a series circuit.

like the example in Figure 2.13(a), one part will be the active part (hence the subscript "a" of Y_a) and the other the passive part of an oscillator (hence the subscript "p" of Y_p). The real part of the active admittance Y_a needs to be negative for oscillator start-up. The necessary conditions for steady-state oscillation for the negative resistance model in Figure 2.13(b) can be written as [58] written below.

The steady state oscillation conditions are

$$G_a(\omega_{osc}) + G_p(\omega_{osc}) = 0, \tag{2.20}$$

$$B_a(\omega_{osc}) + B_p(\omega_{osc}) = 0, \tag{2.21}$$

The start-up conditions are almost identical but the equality in (2.20) changes in to an inequality,

$$G_a(\omega_{osc}) + G_p(\omega_{osc}) < 0, \tag{2.22}$$

$$B_a(\omega_{osc}) + B_p(\omega_{osc}) = 0. \tag{2.23}$$

The negative resistance model is a linear model, which is not a major limitation in practice for the application of the oscillation start-up conditions of this model, since the amplitude of the carrier will be very small initially. Similar to feedback modeling, these conditions are necessary conditions and not sufficient and the root locus method should be used when in doubt [57, 58].

Negative resistance modeling is often used when it is applicable to the oscillator type, because it can simplify the design procedure of an oscillator. Application examples are LC oscillators with lumped resonator, microwave oscillators [60, 61] or crystal oscillators [23]. Basically, a resonator has to be chosen and G_a should be sufficiently negative for start-up. If a tunable oscillator is constructed, the resonator must be tunable over the frequency band of interest and G_a must be sufficiently negative for start-up across this frequency band. Finally, the negative resistance oscillator can be optimized for noise, large signal behavior or other properties.

2.5 Amplitude stabilization and settling time

The previous section discussed linear oscillator models and their analysis. None of these models give any information regarding the amplitude of the oscillator output signal. When the start-up condition for oscillation is met, the poles of a harmonic oscillator are in the half right plane. This is illustrated in Figure 2.14(a) for the LC oscillator model presented in Figure 2.7(a).

The time response of a feedback system with these pole locations corresponds to a growing sine wave. Without some sort of amplitude stabilization mechanism in a

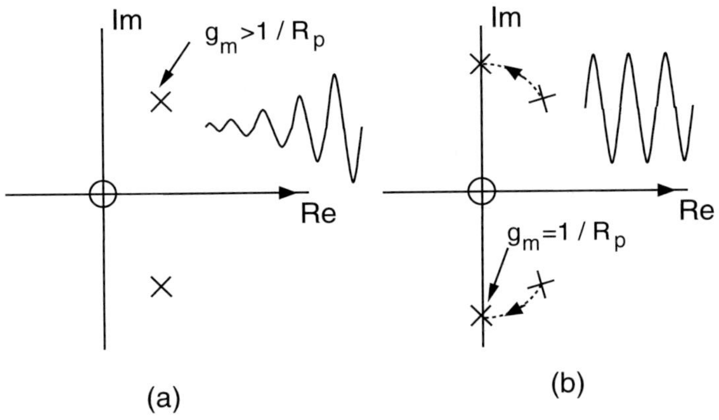

Figure 2.14 Start-up conditions are fulfilled resulting in an exponentially growing sine wave (a). The effective g_m in the LC oscillator model is reduced by an amplitude stabilizing mechanism, fulfilling the steady-state oscillation conditions (b).

practical oscillator, the oscillator signal would keep on growing and growing. This mechanism causes the poles of the oscillator to move to the imaginary axes, which corresponds to a steady oscillation level in the time domain. In the case of the LC oscillator from Figure 2.7(a), an amplitude control mechanism reduces g_m after start-up from a value larger than $1/R_p$ to exactly $1/R_p$ (Figure 2.14(b))[12].

> Any practical electronic oscillator contains an amplitude control mechanism that reduces the loop gain after start-up, such that steady-state oscillation is reached. Two control mechanisms, self-limiting and automatic gain control, are discussed below.

2.5.1 Self-limiting

One option for amplitude stabilization, called self-limiting, is to use the nonlinear characteristic of an active element in the oscillator. Since this method does not need additional circuitry, it is often used in high-frequency oscillators. In the literature the simplest self-limiting oscillator is often referred to as the "Van der Pol" oscillator [6], in which the active element has the transfer function $I_{out} = aV_{in} - bV_{in}^3$.

[12]This is a simplification. Poles are only defined for linear systems. During start-up an oscillator may be modeled using poles and zeros. However, as soon as the carrier is so large in an oscillator that nonlinearities start having an influence, the pole/zero description is no longer valid. Provided an oscillator is nearly linear in its steady-state, it can be described by a equivalent linearized differential equation. For such an oscillator it can be shown that the poles in Figure 2.14(b) have a noisy nature and that their mean value is located in the left half plane, very close to the imaginary axis [22]. Hence, the average g_m in steady-state will be a tiny amount smaller than $1/R_p$.

The simple form of the limiting function of the Van der Pol oscillator allows us to gain insight into the start-up behavior of a self-limiting oscillator. Let us, once more, investigate the oscillator model in 2.7(a). Assume that the transconductance has a transfer function $I_{out} = g_m V_{in} - bV_{in}^3$, and recall from page 24 that the open-loop gain α_{ol} is equal to $g_m R_p$. Assuming that α_{ol} is larger than one, but at the same time very close the one[13], the oscillator model obtained leads to a differential equation that can be solved analytically[14]. The solution of the steady-state maximum peak voltage is

$$V_{peak} = \sqrt{\frac{4}{3}\left(\frac{\alpha_{ol}-1}{bR_p}\right)}. \tag{2.24}$$

The complete solution of the oscillator output voltage then is

$$v_{out} = \frac{V_{peak}}{\sqrt{1+e^{-(t-t_0)(\alpha_{ol}-1)\omega_{LC}/Q_p}}}\cos(\omega_{LC}t+\phi_0), \tag{2.25}$$

in which t_0 is an integration constant and depends on the initial conditions. The oscillator phase ϕ_0 is the phase at the time t_0 and thus also results from the initial conditions. During start-up, when v_{out} still is small, (2.25) can be approximated by

$$v_{out}|_{start-up} = 2 \cdot V_{peak} e^{(t-t_0)(\alpha_{ol}-1)\omega_{LC}/(2Q_p)}. \tag{2.26}$$

Figure 2.15 illustrates the derived formulas and shows the start-up of two Van der Pol oscillators running at 1 GHz, one with a resonator quality factor Q_p of 5 and one with Q_p of 10. A comparison between the two simulation results shows a longer settling time for a higher Q_p. Equation (2.25) allows calculation of the settling time. If the settling time of a Van der Pol oscillator is defined as the time needed for the oscillator output signal to grow from $0.1V_{peak}$ to $0.9V_{peak}$ then $t_{settling}$ by good approximation is equal to

$$t_{settling} \approx \frac{6Q_p}{(\alpha_{ol}-1)\omega_{LC}}. \tag{2.27}$$

The expression for $t_{settling}$ also shows that a high Q_p will result in a longer settling time. On the other hand, a large open-loop gain shortens the start-up transient.

In IC technology, the limiting characteristic of a MOS or bipolar differential pair is readily available to implement self-limiting. In Figure 2.16(a) a cross-coupled differential pair is shown, which is used to realize a simple LC oscillator. Since the

[13]This limits the solution to systems that produce harmonic oscillators. These systems are referred to as quasi-linear or high-Q systems, because the stored energy is large compared to the energy loss per cycle [56].

[14]Two possible methods are "variation of parameters" and "equivalent linearization [9, 56].

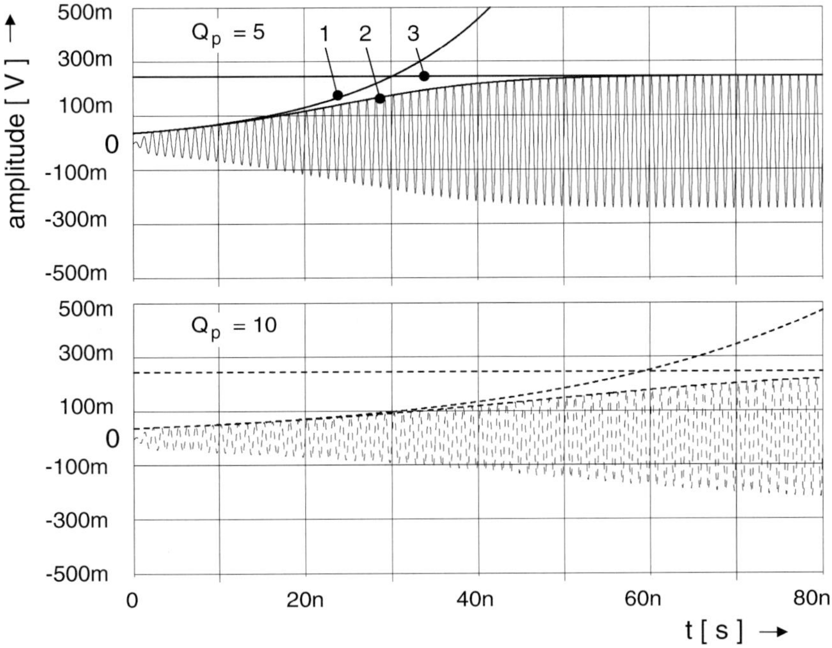

Figure 2.15 The simulated starting behavior of a 1 GHz Van der Pol oscillator with quality factors 5 and 10. The solid lines marked 1, 2 and 3 are the plotted numerical values versus time of (2.26), (2.25) and (2.24), respectively.

transfer functions of a bipolar or MOS differential pair can be approximated with the first two terms of their Taylor expansion ($I_{out} = a_1V + a_3V^3$, with a_3 a negative number), the previous discussion on the Van der Pol oscillator is representative of the oscillator in Figure 2.16(a). The transfer function of this cross-coupled bipolar differential pair is $I_{out} = I_{tail}/2 \cdot \tanh(V_{in}/2V_T)$, where V_T is the thermal voltage kT/q. Its derivative to V_{in} ($g_m(V_{in}) = I_{tail}/(4V_T) \cdot \mathrm{sech}\,(V_{in})$) is shown in Figure 2.16(b). When the oscillator starts up, V_{in} is small and g_m is equal to $I_{tail}/(4V_T)$. As can be seen in Figure 2.16(b), g_m decreases for an increasing V_{in}. Since the oscillator output signal traverses the $g_m(V_{in})$ versus V_{in} characteristic, the average g_m is smaller than the small-signal g_m needed for start-up at $V_{in} = 0$. The oscillation amplitude will stabilize when the average g_m compensates all losses (modeled by R_p) exactly, and the amplitude condition for steady-state oscillation is met. If the oscillator signal becomes large enough, the transistors will act like switches and the output peak amplitude will become $2/\pi \cdot I_{tail}R_p$. The amplitude depends linearly on I_{tail} (current limited region), unless the supply voltage (or another saturation mechanism) becomes the dominant amplitude limitation (voltage limited region).

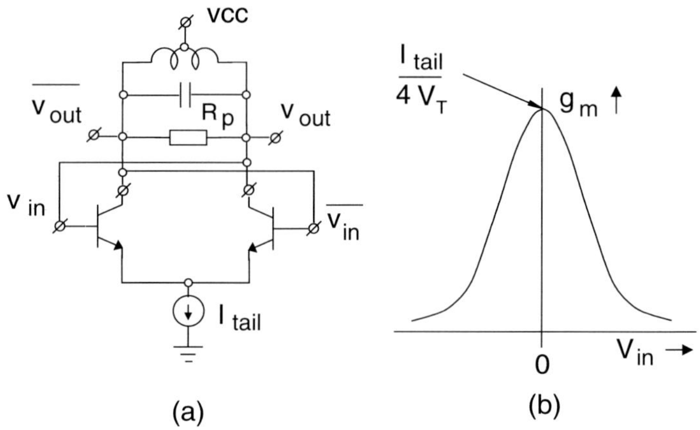

Figure 2.16 LC oscillator with differential pair (a) and the derivative of the transfer function of the differential pair (b).

> The voltage limited region is not a good operating region due to increased noise and harmonics. In addition an increase in tail current does not yield an increase in carrier level anymore.

Increased noise may, for example, arise from junctions that go into the forward region, whereas the strong nonlinearities involved with the voltage clipping result in harmonics.

2.5.2 Automatic gain control

A second method for amplitude stabilization is to use an Automatic Gain Control[15] (AGC).

> AGC: the oscillation amplitude is measured and used in a negative feedback control loop that stabilizes the oscillator amplitude to a set value after start-up.

Main reasons to use AGC are as follows [62]

- Fast and reliable start-up. During oscillator start-up the loop gain in an oscillator can be made very large to reduce the oscillator settling time (see (2.27)). Once settled the steady-state loop gain can be made any value that suits the application. For example the loop gain can be set for very low harmonic content in the oscillator output signal.

[15] Also referred to as Automatic Amplitude Control (AAC).

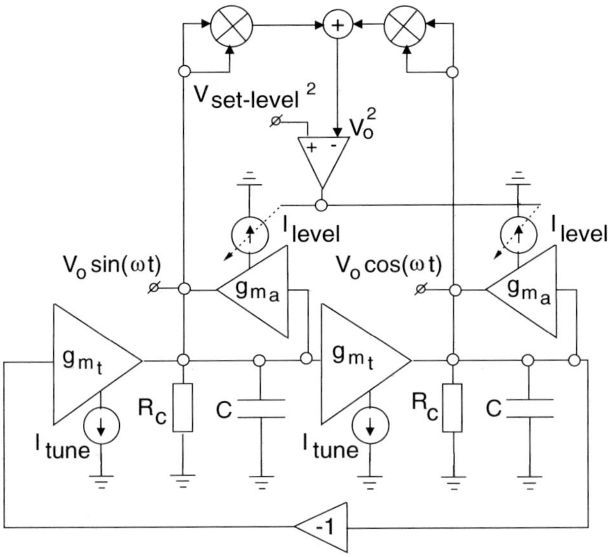

Figure 2.17 A two-integrator oscillator with AGC-control.

- Combining optimum biasing for noise with reliable start-up. The requirements for optimum noise performance of an oscillator and reliable start-up may well be contradictory. An AGC can be used to orthogonalize these two demands.

- Obtaining a well-defined output level. The negative feedback control loop of an AGC guarantees that the output of the oscillator can be equal to an amplitude set-level within a very small error band. Hence, the oscillator output level can be made very stable over a large temperature range and robust against IC process spread.

- Power dissipation reduction. Consider the example where a large start-up current is needed for the quick start-up of a crystal oscillator. After start-up, this current and thus the power dissipation can be reduced significantly during steady-state operation.

Disadvantages of AGC are an increased oscillator complexity, and additional noise sources. The bandwidth of the AGC control is an important design parameter. If the bandwidth is too large, noise and spurious signals may modulate the oscillator via the AGC loop resulting in a poor phase noise performance

Figure 2.17 shows the behavioral model of a two-integrator oscillator with AGC. Note that after start-up (when steady-state is reached) this AGC-loop can be modeled by a linear model. If the set-level is such that the transconductances g_{m_t} and g_{m_a} work in the linear region, this oscillator is practically linear. This is especially interesting for applications where harmonics need to be very low. The frequency g_{m_t}/C can be varied by controlling g_{m_t} with current I_{tune}. Transconductance g_{m_a} is needed to compensate the losses modeled by R_c and can be used for amplitude control. As the oscillator provides I/Q signals, AGC can be implemented using two mixers and a comparator, since $V_o^2 \sin(\omega t)^2 + V_o^2 \cos(\omega t)^2 = Vo^2$. Notice that frequency and amplitude control for this oscillator are orthogonal.

In the case of an LC oscillator, AGC can be implemented using a peak detector that measures the amplitude and is used as an error signal in a negative feedback loop, which in turn controls the tail current of the oscillator [62].

2.6 Summary

An oscillator generates a periodic signal based on an internal timing reference. An ideal oscillator only has desired properties. Among other things, important properties include frequency, amplitude, tuning range, number of output phases, and harmonic content of the output signal. Tunable oscillators in integrated transceivers either are voltage controlled or current controlled. Depending on the application of the oscillator, more than one output phase can be required. Quadrature oscillators generate two phases, an in-phase signal ($A\cos(\omega_{osc}t)$) and a quadrature signal ($A\sin(\omega_{osc}t)$). These elementary properties of oscillators are more extensively analyzed in Chapter 5.

Practical, non-ideal, oscillators have a number of unwanted properties. Examples are phase noise sidebands, limitations on frequency and tuning range and spread on parameters due to temperature and IC process spread. In the time domain, phase noise sidebands result in uncertainty in zero-crossings of the periodic signal, and is called jitter. The practical properties of oscillators are extensively analyzed in Chapter 6.

A practical classification based on the implementation principle of oscillators has been introduced in this chapter. Two important principles used in integrated transceivers are continuous-time resonator based and continuous-time non-resonator-based operation. Oscillators investigated in this book are LC oscillators with lumped LC resonator (resonator-based) and ring oscillators (non-resonator-based). Feedback modeling can be used to assess the oscillation conditions and the start-up conditions for oscillation. When a Bode diagram of an oscillator shows that the start-up conditions for oscillation are met for more than one frequency, this can be an indication of the presence of multi-oscillation. In this case, the oscillator oscillates on multiple frequencies, hampering use of the oscillator in a transceiver. For some oscillator types, negative resistance modeling can be convenient to investigate the oscillation conditions.

Every practical oscillator has a mechanism that stabilizes the amplitude of the oscillator output signal. One mechanism, often used in high-frequency oscillators, is self-limiting. This method makes use of the nonlinear characteristic of the active oscillator part. Alternatively, automatic gain control can be used.

3

Structured design with FOMs

WHAT is structured[1] design? An appropriate meaning of the word "structured" in the context of this chapter is "having a well-defined structure or organization". In this chapter we explore various existing structured design methods for design of analog building blocks (Section 3.1 and 3.2). Most approaches are generic in the sense that they are of use for the design of analog building blocks in general, including oscillators. The exploration of design methods and methodologies allows us to formulate a number of requirements (or better, desired features) of a structured design method for oscillators. None of the reviewed structured design approaches help the designer in an insightful manner during the important design phase of oscillator type and topology[2] selection.

> In Section 3.3 a structured design approach is discussed that is based on Figures of Merit (FOMs), which aims at shortening the oscillator design time and it provides qualitative as well as quantitative insight.

FOMs assist the designer in a useful way during the important design phase of oscillator type and topology selection. This phase has a profound influence on the design time and thus on the cost of design. In terms of the analogy between an oscillator designer and a traveler introduced in Section 1.4, FOMs allow the designer to quickly assess whether a possible design route is a high-way, motor-way, country road, desert

[1] From the Latin word "struere"; "to construct". Source: www.dictionary.com.

[2] We use the word circuit topology, or short, topology, in a broad sense in this work: the topology of a circuit represents the way all devices are interconnected but also the type of the devices that are interconnected. Note that in graph network theory often the way to interconnect only is called "topology" [63].

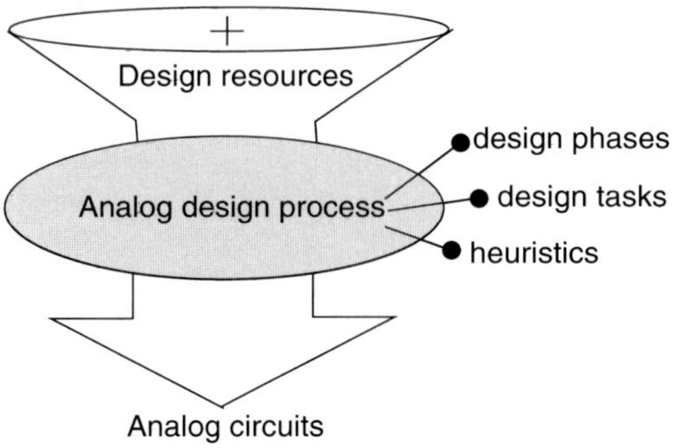

Figure 3.1 Functional specifications and design resources form the input of the analog design process.

path or dead-end street. FOMs are likely to be a very effective element of a FOM-based design methodology. However, a design methodology, a design flow defining each design step for oscillator design is outside the discussed scope of this work; it would require a much more formal approach, and is a huge subject on its own. System level, behavioral level and circuit level modeling play an important role in the formulation of useful FOMs. These hierarchies in modeling and modeling conventions are discussed in Section 3.4. Section 3.5 concludes this chapter with a summary.

3.1 Analog circuit design

Analog circuit design differs substantially from digital circuit design. In the past decades, the design of digital systems has largely been structured and automated. Complex digital systems can be described with a high level language such as the VHSIC Hardware Description Language (VHDL)[3], which can then be synthesized to gate level [64]. This synthesis process is based on well-defined explicit rules. Automatic placement and routing complete the digital design trajectory. The resulting systems can have tens or even hundreds of millions of transistors on one die. The digital designer still has to interact with the Computer Aided Design (CAD) tools in every phase of the design, but in terms of man-hours spent per device, digital design has reached a state of enlightenment compared to analog design.

The design of high-frequency oscillators is an analog design process. This process is illustrated in Figure 3.1. First, the two inputs to this process, functional specifications and design resources are described in Section 3.1.1. In Section 3.1.2, a division of the analog design process in three phases is introduced.

[3] IEEE Standard 1076-1987.

> An important reason why the analog design is far less automated and less efficient in terms of man-hours per device, is that it uses a lot of heuristics and rules in contrast to the well-defined rules and boolean algebra of digital design.

In digital design, the redundancy obtained by using only 0's and 1's, makes it relatively easy to divide a large design problem into many small sub-problems. This complexity reduction is a lot tougher for analog design challenges, and one approach is to use heuristics. A number of heuristics used in analog design are investigated in Section 3.1.3.

3.1.1 Functional specifications and design resources

The functional specifications form the design question for the designer who in turn answers with a design meeting these specifications. The majority of functional specifications, denoted $Spec_x$, can be written in the form $Spec_x \in [S_{lb}, S_{ub}]$; the specification is constrained in an interval. The lower bound S_{lb} or upper bound S_{ub} can be $\pm\infty$ or can be expressed in finite tolerances. Furthermore, the interval can be open or closed or be a single value. In other words it can be an inequality constraint or a indexFunctional specifications!equality constraint equality constraint, respectively [65]. Examples are the power dissipation should be lower than 100 mW, the power supply voltage is 3 V $\pm 10\%$, and the tuning range should start at 1 MHz and extend to at least 1 GHz. In integrated circuits these specifications have to be met over a specified set of environmental conditions like temperature range and with a specified amount of process variation. Most common functional specifications for oscillators are extensively described in Chapter 4.

Design resources are considered to be the items that enable the design process to reach its desired end: an analog circuit that meets the functional requirements. These resources and their range are as follows:

- **The designer**:
  - very experienced: high salary, difficult to find, exception.
  - inexperienced: not unusual.
- **Design time**:
  - large: not available in high-tech companies, short product life-cycle.
  - small: common practice.
- **Technology**:
  - high performance: exotic, not well characterized, costly.
  - low performance: main stream, short turn-around time.

- **Tools**:
  - Design flow:
    - fully automated: nearly bug-free, all verification steps automatic such as back-annotation of parasitics, large initial costs, expensive to maintain.
    - a lot of manual labor: several cheap tools, designer has to perform all interfacing issues between tools in the flow, slow design cycle, small initial cost.
  - Architecture and circuit libraries:
    - many: long design history, many designs on the shelf in up-to-date technology, many layout examples, well documented.
    - hardly available: short design history, few design examples in ancient technology, not documented architectures and circuits.
  - Front-end (circuit simulators, etc) and back-end (layout, etc.) tools:
    - high performance: short simulation times, many analysis types, costly, expensive computing servers needed.
    - low performance: public domain, limited application range, runs on standard personal computers.

All design resources have a profound influence on the design process, and are implicitly or explicitly specified. A technology is often explicitly specified by the customer or the manager of the designer. The properties of the design resources enable the oscillator design but at the same time limit the set of solutions. A clear example is the technology that sets the practical boundaries. At the same time, without the technology, a practical implementation would not be possible.

> After capitalization, all design resources can be translated into one single resource **"Money"**, by their contribution to the initial costs of a product.

The importance of these initial costs compared to the final costs, depends on the product. For example, the final product may be aimed at the consumer market (high-volume, low profit margin) or at the professional market (low-volume, high profit margin). If the total budget of an oscillator design is no issue, a large number of experienced designers can be used and state-of-the-art technology and tools can be made available.

3.1.2 Design phases

A break-up into three phases of the overall analog design process, the process from specification to tested hardware, is shown in Figure 3.2 (derived from a similar partitioning in [66]).

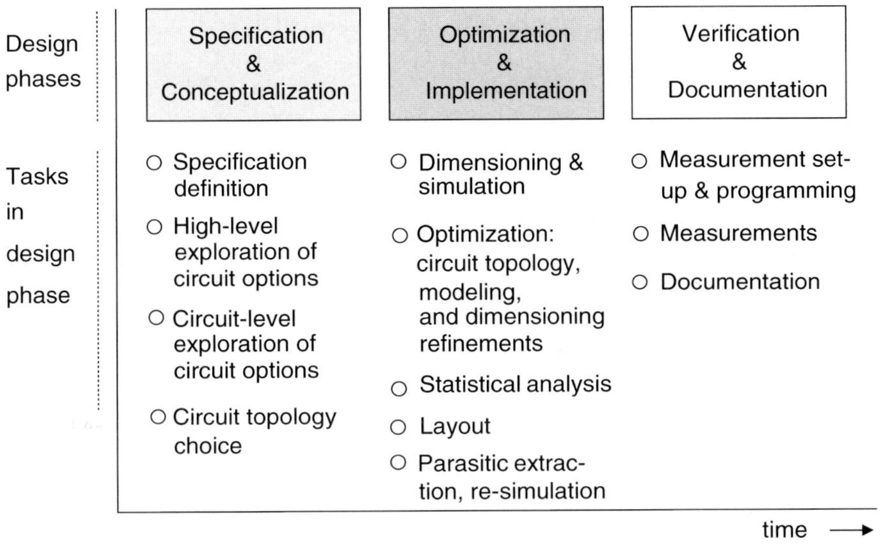

Figure 3.2 Design phases and tasks in a design process of an integrated analog circuit, such as an oscillator.

The first phase is designated *"specification and conceptualization"*. An important step in this phase is to define the specification unambiguously. Next, an inexperienced designer may start with a literature study to explore circuit options, whereas an experienced designer might instantaneously know a good circuit topology for the job, because he used it successfully in an earlier design job. Behavioral modeling (high-level modeling) can be used advantageously to acquire a good impression of the first order behavior of circuits. Before the second design phase is entered, a circuit topology is selected.

The second phase of the analog design process is the *"optimization and implementation"* phase. All devices are sized, and all specifications are simulated and compared to their target values. In this phase optimization takes place until all specifications are met. The optimization process may include topology changes and modeling refinements, and will certainly involve dimensioning refinements. In order to predict the yield, a statistical analysis usually is part of phase two. When the electrical design of the analog building block is ready, the layout of the circuit has to be made. Interconnect parasitics alter the electrical behavior of the implemented circuit. This can be taken into account by extracting layout parasitics, re-simulating and adjusting the circuit characteristics and the layout, to regain the wanted behavior. The same can be done to include effects of a chip-package. The layout is now ready for mask making and processing.

During and after processing of the analog IC, the third and last phase of the analog design trajectory starts. This phase is designated *"verification and documentation"*.

A measurement set-up or set-ups have to be arranged, and the functionality, performance, yield and reliability of the analog building block have to be verified. If all verification steps match expectation, the documentation can be finalized and the analog design is completed; ready for mass production.

The three grey levels in Figure 3.2 indicate the relative importance of the design phases with respect to the time spent per phase. Phase two is the most dominant phase. Both the novice and the experienced designer spend about 50% to 60% of the total design time simulating, optimizing and realizing the layout[4]. The novice generally spends significantly more time, around 30%, in phase one, compared to the experienced designer (15% to 20 %). Novice designers do not have a large set of previous design examples, circuit options and circuit tricks, which makes it more difficult to reach a first circuit topology. The least amount of time is spent in phase three, about 15% to 20%.

> A bad choice in phase one can have a huge negative impact on the total design time. It forces the designer to repeat phase two several times with many iterations and circuit modifications, until a topology has been found that meets the functional specification.

Obviously, the number of iterations between the phases and between the tasks within a phase should be as small as possible, to minimize the design costs. In other words, although most of the total design time is spent in phase two, design phase one has the dominant influence on the total design time. In practice phase one heavily relies on design heuristics. A number of them are explored in the next section.

3.1.3 Design heuristics

An analog designer makes a number of heuristic decisions during design. This is especially true for the first design phase in Figure 3.2. Several "analog design" heuristics are described by the previously referenced paper of Bowman [66]. An extended and commented list is given below from the perspective of an oscillator designer.

- *Hierarchical circuit design.* The design of an oscillator starts with its specification. At that stage the oscillator is like a black box, of which only the desired properties are known. Next a circuit topology has to be chosen. Finally all devices have to be dimensioned, such that the oscillator meets the specifications. There can be a number of levels between system, circuit and device level, but these three levels already show the natural hierarchy in the design process.

[4] In [66], R. Bowman describes in his paper "Analog integrated circuit design conceptualization" the results of a survey of 75 analog designers. The question was "how much time is spent in each design phase?" A distinction was made between the novice (less than 3 years design experience) and the experienced designer.

- *Specification is prioritized starting with the most difficult specification.* The moment an oscillator designer gets a set of specifications, he begins to assess and prioritize it. For example, if the tuning range should be four decades this immediately attracts his attention. This specification may be the most difficult to achieve and will have the biggest influence in design phase one; see also the next design principle.

- *The first focus is on the few specifications that strongly influence the design decisions made in an early stage of the design process.* To follow up on the example above; if the tuning range of four decades is the dominant one, the designer will focus on this oscillator property first. If this is not realized, the oscillator will not cover the frequency band of interest, and other performance aspects become irrelevant.

- *Rules-of-thumb estimations are applied to simplify the design process.* Common sense aided by experience is an important tool in oscillator design. Part of this common sense, technology awareness and over the years design experience condense in rules-of-thumb. Some examples:

  - A ring oscillator generally has a larger tuning range than an LC oscillator.
  - Bond-wire inductance is about 1nH per mm.
  - A dominant source of $1/f$ noise often is the tail current source of an oscillator.
  - The carrier swing should be maximized.
  - Saturation effects should be avoided.

- *Reuse of knowledge*

  - *Accumulation of circuit options and tricks.* An experienced oscillator designer has made many oscillators for various applications. When he sees the specifications for a new design, he will immediately relate the specifications to a realized design, and circuit options pop up in his mind. During circuit level optimization (phase two in Figure 3.2), the experienced designer opens his backpack of circuit tricks to refine the oscillator design. The backpack of an inexperienced designer still has lots of room.
  - *Optimization and application of circuit tricks are performed at sub-circuit level where only a handful of devices play a role.* A high-frequency oscillator does not usually consist of many transistors. But if a more complex oscillator is designed, the designer usually partitions the oscillator into sub-circuits. This decomposition into (in general) more basic circuit topologies makes the complexity manageable and insight is more easily gained. Most circuit tricks are applied at this level.

- *Similar circuit topologies are used in various technologies.* Oscillator implementations often have the same or similar topologies in different technologies. An LC oscillator implemented with a resonator and a cross-coupled pair or a Colpitts oscillator are basically the same in bipolar, CMOS or, for example, GaAs technology. Of course, this not always the case and can be quite sub-optimal. Each technology also has its own merits and these should be exploited. In bipolar technology a PN-junction varactor may be a good choice, whereas in CMOS technology a MOS-varactor can be preferable, for example.
- *Combinations of known sub-circuits.* Cross-coupled pairs, Colpitts, Hartley, ring oscillators are all quite ancient. On a circuit level, implementations may seem new to an oscillator designer. However, they often are a combination of existing circuit techniques, or are not novel in the sense that they do not add a new circuit to the total pool of known circuit topologies.

The preceding list of design principles and heuristics is certainly not exhaustive. In fact, consciously or unconsciously used heuristics vary from designer to designer. Nevertheless, most common ones are given and the list can be used advantageously to derive a structured FOM-based design approach for oscillators, which blends in with common design practice.

3.2 Structured and automated design methods

Circuit analysis using CAD tools like Spice [67], Pstar[5], Spectre [68] or Aplac[6] is indispensable in modern analog circuit design. As mentioned, most of the design time is spent on computer-aided circuit analysis (part of phase two in Figure 3.2).

> A number of structured design methods, heavily based on computer-aided circuit analysis, is discussed in this section.

The list of structured methods illustrates the state-of-the-art in analog design automation. Analog design automation, like the FOM-based structured design of oscillators that will be described in Section 3.3, aims at enhancing the productivity of the designer. Once automated, a design task can be shortened by adding more computing power. The structured methods highlighted in the following sections are applicable to analog circuits in general, including oscillators.

[5] Philips' in-house circuit simulator.

[6] Aplac is a circuit design and simulation tool. See http://www.aplac.com.

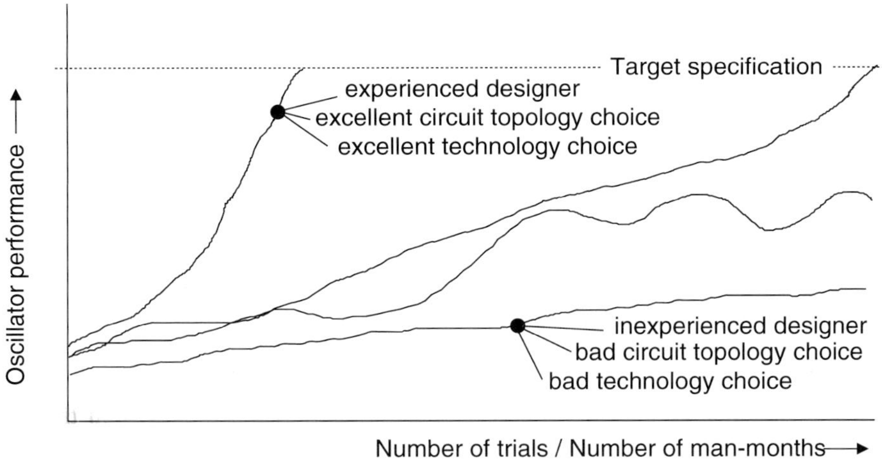

Figure 3.3 Example of possible outcomes of the trial-and-error design process.

3.2.1 Trial-and-error

Every designer is familiar with structured design. The design strategy called "trial-and-error", "cut-and-try" or "turning the knobs" is apparently very important since it is wide-spread. One (or somewhat more unstructured: more than one) design parameter is varied and circuit performance is monitored. If the performance improves, the value of the design parameter is increased until a specific functional specification is met or other design aspects start to move away from the functional specification All design parameters can be varied in this way until the total circuit performance meets the functional specification. The fact that this process can be performed in a mathematically structured way is demonstrated by today's circuit optimizers, which use sophisticated algorithms in their pursuit of the design requirements. As with any method, practical limits of the chosen circuit topology and technology may prevent the circuit performance from ever reaching the functional specification.

Figure 3.3 illustrates the trial-and-error design process. In combination with the knowledge of an experienced designer, and a good circuit topology and technology choice, the "trial-and-error" method can be a fast way to complete an oscillator design. This case is represented by the upper trace in Figure 3.3. On the other hand, an inexperienced designer may never reach the target specification, regardless of the number of trials.

Arguably, whatever design method will be used for the oscillator design trajectory, there will be always trail-and-error elements in it. The ideal case of a design process of a high-frequency oscillator without any iteration, with an optimally sized circuit in only one simulation run, is not realistic[7]. However, as pointed out by Figure 3.3, with-

[7]Unless it is a copy of an earlier design.

out making sensible circuit and technology choices, this design method has distinct disadvantages:

- Little or no insight is gained. For every new design the same dimensioning process starts over again. No reuse is made of previously gained experience. Only the increased experience, gained from the final results of a new design, as added value compared to the previous design process. Thus, the reuse of earlier design experience, and the extraction of generic reusable knowledge during a design is very low for trial-and-error design.
- A very large number of time consuming and costly iterations may be necessary. As illustrated in Figure 3.3, a bad topology choice results in a long design time.
- Possibly, the design process may not converge.
- The design knowledge built up by the designer is difficult to transfer and maintain within the design team. When the experienced designer leaves the design group and is replaced by an inexperienced designer, the design time increases a lot.

3.2.2 Optimization tools

Once an oscillator circuit topology is chosen and all devices are given an initial value, automatic optimization tools can be used.

> Automatic optimization tools can perform the tedious task of adjusting all design parameters until specification is met under all process and temperature conditions.

Most electronic design automation (EDA) tools or circuit simulators have an optimizer built in or have an interface to an optimization tool. Examples are "Cadence Analog Circuit Optimizer[8]", the optimizer in Aplac [70], and Adapt[9].

A simplified flow chart of automated circuit optimization is shown in Figure 3.4. The key to the optimization process is the optimization algorithm, which determines in what direction and how much the design values are adjusted, prior to new simulation runs. Examples of used algorithms are simulated annealing [72], genetic optimization [73], minimax optimization [74], and Nelder-Mead optimization [75]. There are a few interesting optimization approaches specifically for LC oscillator optimization. For example, geometric programming can be used to optimize LC oscillators [76]. Another approach uses a graphical optimization method for the same purpose [183].

Similar to manual trial-and-error design, the optimizer may not find a solution to the design problem. Several causes can prohibit a solution. The specification may be unrealistic or the circuit topology in combination with a chosen technology can not

[8] http://www.cadence.com. The EDA market (in 2001) is dominated by Cadence Design Systems. Cadence has 70% market share, Mentor Graphics 20 %, Avant! 8% and others 2% [69].

[9] Philips' in-house circuit optimizer [71].

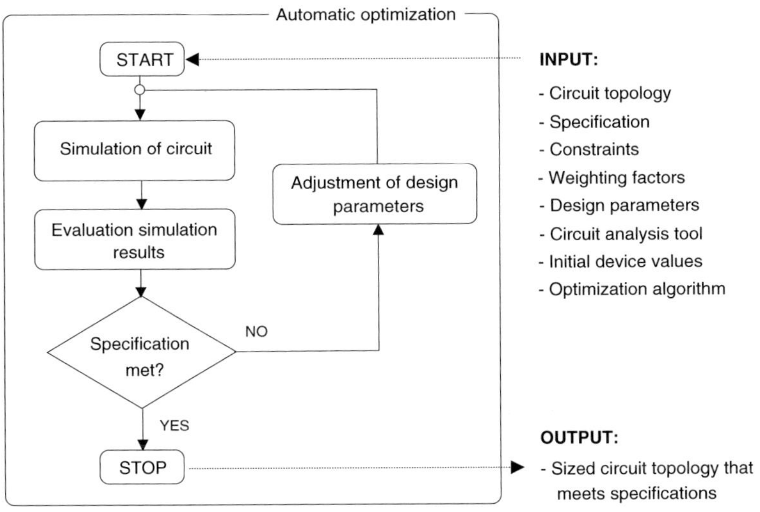

Figure 3.4 Simplified device-sizing flow using automatic circuit optimization.

meet the specification. In other words, the solution to the optimization problem lies outside the practical solution space. Assuming that the specification is realistic and cannot be lowered, the optimization process then has to be reentered with a refined initial topology or with an entirely different circuit topology. Of course, a carefully selected circuit topology choice can prevent or reduce the number of iterations, and will also yield a faster optimization process. Another reason why an optimizer may fail to find a solution, is simply because it can not find it. An optimization method can only search through part of the solution space, and can therefore overlook a solution.

An important remark should be made about the analysis types used for simulating a circuit in the optimization process. The circuit simulator should be able to simulate all specifications[10]. This is not always trivial. For example, simulation of the noise behavior of oscillators is generally not supported by standard circuit analysis types, such as DC, AC, or transient analysis. It is only recently (in the late 90's), that phase noise analysis for oscillators has become available in circuit optimizers. An example is the phase noise analysis implemented in the commercially available circuit simulator SpectreRF [77].

3.2.3 Expert systems and synthesis environments

Expert systems and synthesis environments not only aim to automate and shorten the simulation and optimization task, but also to include topology selection and/or layout

[10]Either that or we need an accurate model to predict the performance of a specific oscillator property.

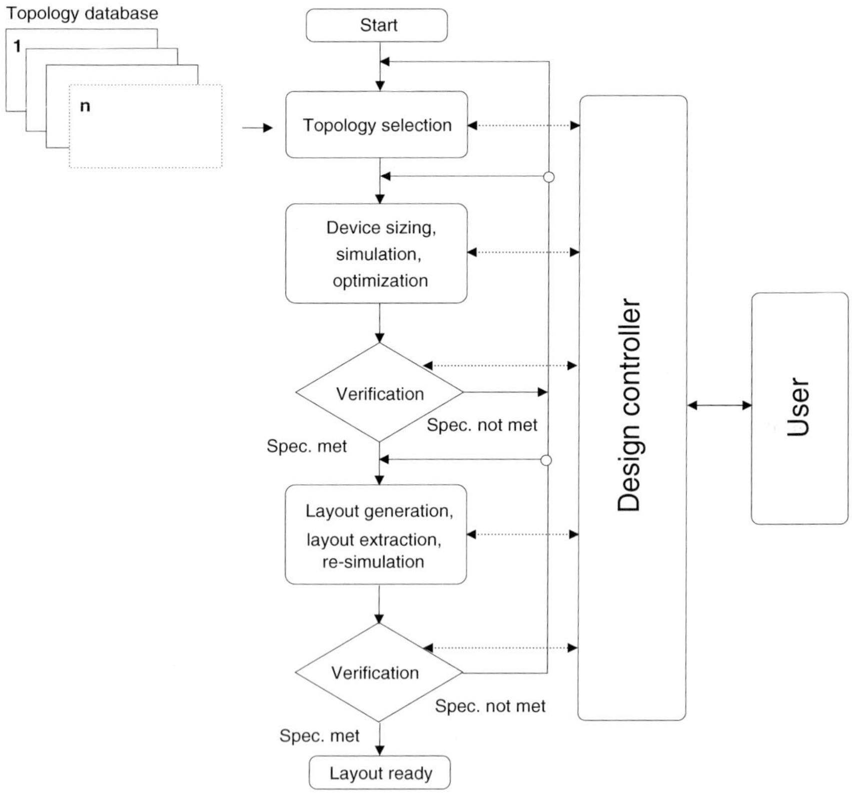

Figure 3.5 Simplified hierarchical design flow of expert systems and analog synthesis environments.

generation. A graphical representation of the most common elements of analog design expert and synthesis systems is shown in Figure 3.5.

> A distinctive difference of an expert system compared to automatic optimization tools is the presence of a topology selection mechanism, which chooses a topology from a topology library.

However, in general this library is small and the selection criterion is as simple as selecting the next topology lined-up if a first topology fails to meet the functional specification. Once the topology is selected the optimization process starts, described in the previous section, resulting in a sized and verified topology. To complete the automatic design flow, layout generation, extraction of parasitics, re-simulation and verification can be part of a synthesis environment. The whole design process and interaction with the user is controlled by the design manager/controller.

There are many publications on analog design optimization environments aiming to implement an automated design flow similar to the one in Figure 3.5. Detailed literature overviews are given in [65,78–80]. All published environments have intriguing names and, interestingly, most often demonstrate their capabilities on operational amplifier (OPAMP) design. Normally, these systems are not available in the public domain or commercially either. A few representative examples follow.

- BLADES [81].
  This automated design approach for analog circuits is expert system based. The use of the system is limited to OPAMPs and it does not include layout generation. A so-called circuit design manager is the design engine of the system. One part of the manager is the circuit topologizer. This is a configuration processor that generates circuit topologies based on sub-circuits present in the knowledge database of BLADES.

- OPASYN [82].
  This synthesis framework is also dedicated to OPAMPs. It includes layout generation. In contrast to BLADES, topologies are not generated from sub-circuits, but selected from a small topology database. The design parameters of the selected circuit topology are first optimized using an analytical model. This is relatively fast but the accuracy is limited. After this, the results are verified and optimized with full accuracy models.

- AMGIE [65].
  This synthesis environment for CMOS analog integrated circuit covers all functionality shown in Figure 3.5. AMGIE can only design circuits that are present in the so-called cell library. This library can contain both standard and custom cells. Designers can add to this library, to extend the possibilities of the synthesis environment. Like OPASYN, device sizing is first performed on analytical expressions of a selected topology, followed by a verification and optimization with accurate device models. After device sizing the layout is automatically generated by the performance-driven place-and-route tool LAYLA [83].

- CYCLONE [84].
  This synthesis tool is an automated layout-aware RF LC-oscillator design tool, capable of delivering an optimized LC-oscillator implementation, starting from specification and including layout. It provides finite element simulations and optimization of RF coils, as well as VCO circuit sizing. Module generation, placement and routing is done within the mentioned tool LAYLA. Two common RF LC oscillator topologies are incorporated. The tool does not support ring oscillator design.

3.3 FOM-based structured design

The overview of structured and automated analog design methods, given in Section 3.2, shows that a number of tools are available, which can shorten the analog design phases pictured in Figure 3.2. However, no structured design method or automated design tool was found in literature that can handle the wide range of LC and ring oscillators within the scope of this work. Furthermore, a practical, insightful and appealing method for topology selection, which has a major influence on the total design time, was not encountered.

In this section FOM-based structured design is described, which aims at reducing the number of iterations between the design phases and design tasks of an oscillator design. FOMs, once formulated, yield fast and insightful oscillator type and topology selection. Moreover, FOMs diminish the disadvantages of the trial-and-error design method, while incorporating design heuristics commonly used by experienced designers.

3.3.1 Structured design requirements

A number of requirements –or maybe better "desired features"– for structured oscillator design can be formulated, keeping in mind the common analog design practice and available tools described in the preceding sections. To be a real guide for the oscillator designer, the structured oscillator design should:

- **Strive for orthogonality**. The structured design method should tell us how the oscillator design parameters (the degrees of design freedom, like quality factor, tail current, number of ring oscillator stages, etcetera) are linked to oscillator properties (frequency, phase noise, etcetera), which determine the performance. If possible, oscillator properties should be orthogonalized so that they can be optimized independently. For high-frequency oscillators complete orthogonalization is less desirable from a performance point of view. In that case, qualitative and quantitative insight into the dependency between parameters and relevant oscillator properties is the next best thing.

- **Fit in a hierarchical approach**. As pointed out, the design process of oscillators already has a natural hierarchy. Within the hierarchy, the number of iterations should be kept to a minimum.

- **Connect with designers**. By focusing on common circuit parameters and definitions to formulate a structured design approach for oscillators, the chances of being useful and appealing to oscillator designers are maximized.

- **Provide qualitative insight**. This helps the designer to prioritize the oscillator specification, to gain insight into circuit options and tricks, and to acquire awareness about the most important oscillator design parameters.

- **Provide quantitative insight**. Since design requirements are in the form $Spec_x \in [S_{lb}, S_{ub}]$, the design method should help the designer to determine oscillator performance relative to quantitative borders. Furthermore, quantitative insight is needed for the oscillator topology selection.

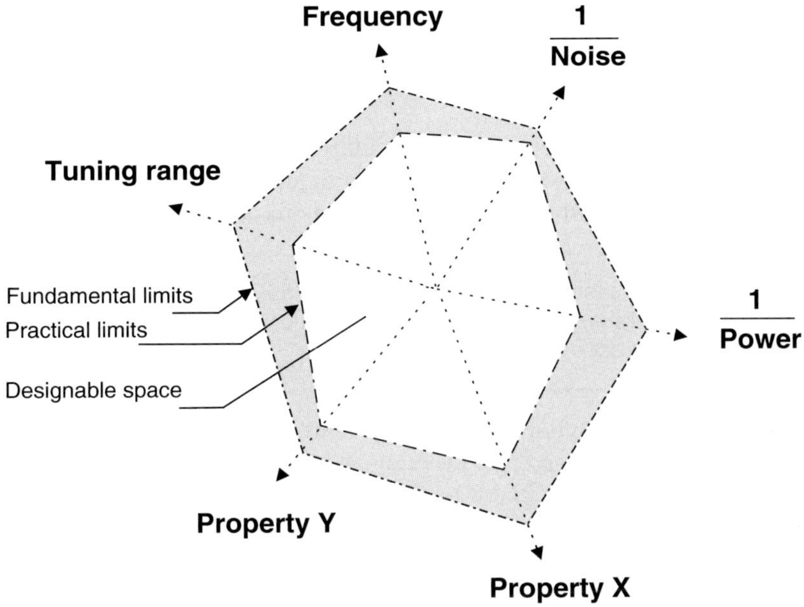

Figure 3.6 Simplified solution space of all oscillator designs with an oscillator property assigned to each axis. The dashed-dotted lines connecting the various properties represent the fundamental en practical limits of property combinations.

○ **Lead to a shorter oscillator design time**. The benefits of using a particular structure design method for oscillator should result in a shorter oscillator design time.

In the following section, structured design with FOMs will be highlighted, which addresses the above requirements.

3.3.2 Figures of merit

> The design space of an oscillator can be visualized by an N-dimensional hyper-cube, in which each oscillator property is assigned to a separate axis [66].

In Figure 3.6, a simplified two-dimensional representation of this hypercube is shown [85]. It is very important to realize that this is a highly suggestive representation in two dimensions of a higher dimensional solution space. The multiple axes connected by lines are drawn to illustrate the bounded character of combinations of properties of the solution space by fundamental limits and practical limits, and by no means represent actual relations between the properties. Certain combinations of oscillator properties will be impossible to realize in an oscillator design.

One oscillator property is assigned to each axis. The arrow at the end of each axis points in the direction of increased performance of an oscillator property. In the direction of the arrow it is increasingly difficult for an oscillator designer to achieve the required performance of one of the properties. Six axes are shown including the most important oscillator properties, but there are more, as there are more properties.

> Design solutions always are inside the fundamental boundaries and practical boundaries. The area inside the practical limits can be regarded as designable space [66].

Fundamental limits refer to limits imposed by nature as we know it, for example the speed of light c or the charge of an electron q. All the limits set by nature impose fundamental limits on the oscillator performance. Practical limits are imposed by the design resources, like technology. The IC technology used for oscillator implementation has non-ideal devices and properties, such as finite transition frequencies, interconnect capacitance, etcetera. These non-ideal elements limit practical oscillator performance.

In Figure 3.7 a solution –that is, an oscillator design– is shown within the designable space, which meets the functional specification. There are two other possibilities. If the functional specification is beyond the fundamental limits, there is no oscillator design that can meet this specification. If the functional specification is beyond the practical limits, there is no oscillator design that can meet the functional specification, given the IC technology used for implementation. As another IC technology may be much more advanced, practical limits can shift towards the fundamental limits. Hence, usage of a more advanced technology can bring the oscillator specification back into the designable space.

Functional specifications quantitatively specify the performance of every oscillator property of interest. If the definition of all specifications (these are discussed in Chapter 4) is unambiguous, this clearly defines what is expected of an oscillator design. For example, if the frequency range specification is from 1 GHz to 2 GHz, the oscillator requirement concerning the tuning range of the oscillator is completely clear to the oscillator designer. This is not the case for the fundamental and practical limits. It is very hard to prove where the practical and fundamental limits are exactly located [11] for an oscillator property (e.g. one of properties on the axis in Figure 3.7). However, to meet the requirements we defined for structured oscillator design, we need to have quantitative information about the oscillator solution space.

Figures of merit provide quantitative information about the performance of one or more oscillator properties. By combining more than one property, a FOM can reduce the dimension of the designable space for oscillators. In other words, FOMs can

[11] For example, we would need to know the global optimum of an oscillator design in a certain technology to have the practical limit of the topology/technology-combination quantitatively available.

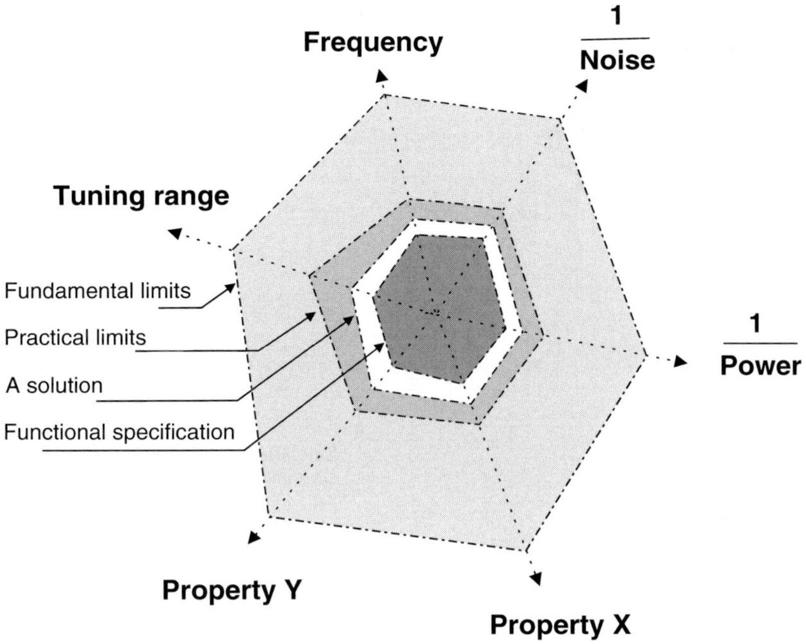

Figure 3.7 Solution space of all oscillator designs. Within the practical limits there is a solution that meets the functional specification.

be regarded as hyper-planes in the N-dimensional hyper-cube, which represents the designable space. Therefore FOMs can limit the solution space making it easier for the designer to find a oscillator implementation that meets the functional specification. It is useful to distinguish between two types of FOMs, design FOMs and benchmark FOMs.

Design FOMs

> Design FOMs provide the oscillator designer with qualitative and quantitative insight into oscillator properties. They predict the design margins for oscillator specifications, and show the trade-offs involved in these margins.

Design FOMs are defined as the quotient of a function f with the design parameters $p_1 \dots p_m$ as arguments, and a functional specification, represented here by a function g with oscillator properties $q_1 \dots q_k$ as arguments. Function f, is an analytical function that estimates, when evaluated, the oscillator performance in terms of the functional specification represented by g, making use of behavioral level modeling for example.

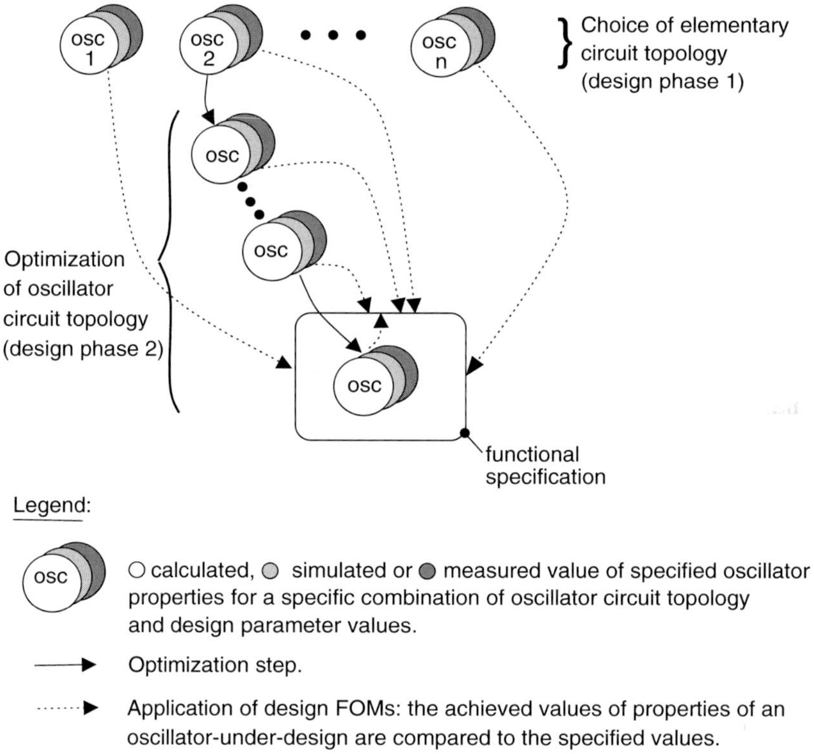

Figure 3.8 Illustration of the use of design FOMs during the design process. Design FOMs are used to select the most promising circuit topology. During subsequent optimization steps, design FOMs indicate the distance with respect to the functional specification: the design margin.

Alternatively, the function f may simply be the simulated or measured oscillator performance, in terms of the functional specification represented by g. It follows that a design FOM always is dimensionless. In summary, a design FOM is defined as

$$FOM_{design} = \frac{f_{ana|sim|meas}(p_1 \cdots p_m)}{g_{spec}(q_1 \cdots q_k)}. \tag{3.1}$$

Figure 3.8 illustrates the application of design FOMs during the design of an oscillator. Design phase one and two from Figure 3.2 on p. 41 are represented in Figure 3.8. In design phase one, a choice has to be made between numerous possible circuit topologies. In design phase two, the design parameters of the chosen circuit topology have to be optimized, which can include circuit topology modifications or choosing an entirely new topology, such that the functional specification is met. Design FOMs provide the designer with quantitative information on the performance of an oscillator property with respect to its specified value, and are thus useful in design phase one and two. In dBs, if the realization[12] of a design FOM is negative, the functional specification is not met, if it is 0 dB the specification is exactly met, and if it is positive the amount of positive design margin is given in dBs. All relevant design FOMs for an oscillator will therefore always be zero or positive (when expressed in dBs) when the electrical design of the oscillator is ready. This situation is illustrated in Figure 3.8 by the final design, which is drawn inside the rectangle that represents the functional specification.

Benchmark FOMs

Important design questions like: "how does my realized oscillator design compare against the state-of-the-art, or to a theoretical performance boundary ?", are answered by benchmark FOMs.

Benchmark FOMs normalize performance aspects of an oscillator, calculated by function g, to allow a fair comparison with other oscillator designs, or to a theoretical performance limit. This type of FOM is defined as the quotient of a function g with the oscillator properties $q_1 \dots q_k$ as arguments, and a normalizing function h with arguments design parameters $p_1 \dots p_m$. Function g is a function of one or more oscillator properties. As an equation we can write

$$FOM_{benchmark} = \frac{g_{performance}(q_1 \cdots q_k)}{h_{norm}(p_1 \cdots p_m)}. \tag{3.2}$$

The application of benchmark FOMs is illustrated in Figure 3.9. As mentioned, benchmark FOMs are used to compare performance aspects of a realized oscillator design with other designs (for example designs found in literature). For example, the normalization function h normalizes the measured phase noise performance for certain design parameters, such as, power and oscillation frequency. This can be considered relative benchmarking. Alternatively, the normalization function h in a benchmark FOM may be a theoretical performance limit. In this case the performance of an oscillator is benchmarked in an absolute sense to a theoretical bound. For example, the normalization function may relate to the best achievable phase noise performance. Similar to relative benchmarking, absolute benchmarking can be used to compare realized oscillator designs. However, unlike relative benchmarking, absolute benchmarking can also be useful during design phase one: to check whether the specified performance is possible to achieve.

[12]The result of the calculation of a FOM with all arguments given a value.

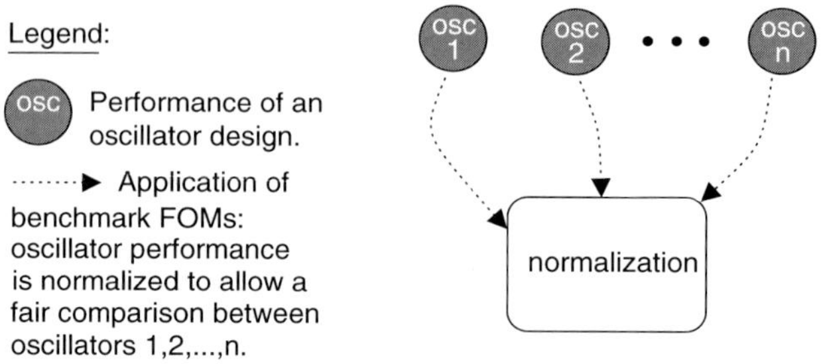

Figure 3.9 Illustration of the use of benchmark FOMs. Performance aspects of a number (n) of oscillator designs are compared after normalization of this performance.

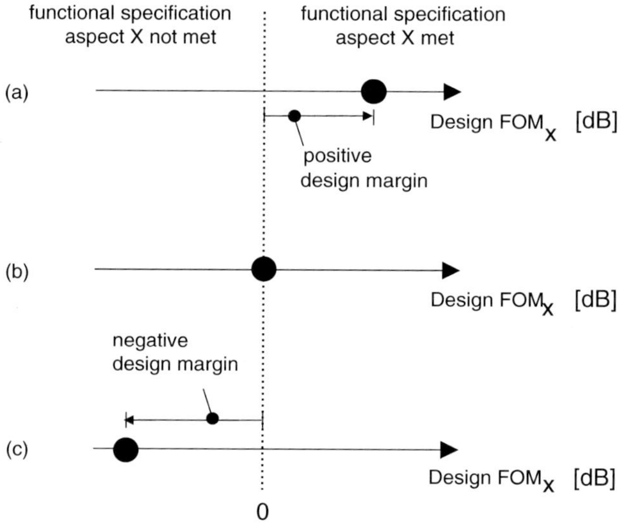

Figure 3.10 Three possible realizations of a design FOM.

Benefits of using FOMs

Figure 3.10 illustrates the use of design FOMs. In this figure, a design FOM, which covers a certain functional specification, is set out on one axis. Three possible locations of the design FOM are shown in Figure 3.10(a), (b) and (c), respectively:

(a) A realization of a design FOM has a positive value. The functional specification X covered by this FOM (for example tuning range) is met and the FOM value indicates the positive design margin.

(b) A realization of a design FOM is 0 dB. Specification aspect X is met and there is no design margin.

(c) A design FOM realization is negative. Specification aspect X is not met: the design margin is negative. If best case values of design parameters $p_1 \ldots p_m$ are used in the function f of the design FOM, the oscillator class or topology from which the function f is derived can be can be discarded by the oscillator designer.

Clearly, the situation pictured in Figure 3.10(c) is the most powerful one, assuming best case values are used to calculate the design FOM realization. If the design FOM realization is negative, no feasible oscillator circuit topology exists that is covered by the model, which was used to derive the function f in the design FOM. The oscillator designer can move on to explore more promising solutions to the design problem. The sooner the designer takes a 180 degrees turn in a dead-end street the better, as that saves time. Once the designer has verified that a solution exists with positive design FOMs for each performance aspect, this design can be further optimized. Apart from the ability of design FOMs to guide the designer in the design process, they also provide a great means for transferring and maintaining design knowledge.

> Once a useful design FOM is defined, it enables inexperienced designers to make design decisions as quickly as the experienced designer.

The experienced designer who has formulated the FOM may have left the design group:

> FOMs serve as documentation to transfer qualitative and quantitative design insight.

In the previous section the implicit assumption has been made, that a negative realization of a design FOM with best-case design parameters always is beyond practical limits. In other words, the design FOM overestimates practically achievable performance. The above assumption represents the ideal case: the designable space is limited by the FOMs, and not a single solution is discarded that could potentially meet the functional specification. For many FOMs this ideal case may be valid: if the function f in the FOM neglects effects or parasitics encountered in reality, this FOM will always overestimate practically achievable performance. However the above assumption does not necessarily hold, and is too restrictive. In general, FOMs can limit the designable space but at the cost of discarding potential solutions: part of the actual designable space is not searched. Provided that the FOMs guide the oscillator designer to solutions within the designable space bounded by the FOMs, this is acceptable.

The benefits of using benchmark FOMs are clear: without normalization of oscillator properties like oscillation frequency, power consumption, phase noise performance, etcetera, it would be impossible to make a fair comparison between various realized oscillator designs. In addition, a benchmark FOM can be very useful to a designer during design if the FOM performs an absolute benchmarking. If a realization of this benchmark FOM is positive (the specified, simulated or measured value is greater than the theoretical bound), the designer knows that the given specification is impossible to achieve, or that the simulation or measured value is inconsistent (i.e. highly unlikely).

> In Chapter 7 a number of examples will be given of design FOMs and benchmark FOMs.

Clearly, the most difficult step in defining FOMs is the derivation of suitable analytical f-functions in design FOMs or analytical h-functions in benchmark FOMs. To derive a useful function f an oscillator model is required. Once a function based on this model is defined, its accuracy can be assessed by comparing the calculated FOM value with simulated or measured values substituted in the FOM, instead of the realization of a calculated f. To develop the design insights and required knowledge for useful FOMs, the oscillator design space is explored in Chapter 5 and 6. Some conventions and several modeling levels used in these chapters are discussed next.

3.4 Modeling framework

In Section 3.4.1, system level equations will be introduced, which will be used in Chapter 6 to investigate phase noise in linear oscillator models. Section 3.4.2 introduces several behavioral building blocks used in Chapters 5 and 6 . Analysis of the models constructed with these building blocks will lead to the definition of several design FOMs and benchmark FOMs for LC and ring oscillators. A brief discussion on circuit level modeling in Section 3.4.3 concludes the modeling aspects used in this book.

3.4.1 System level modeling

In Chapter 6 we will learn that there are several mechanisms that give rise to the oscillator phase noise sidebands discussed in Chapter 2. Many phase noise generation mechanisms arise from nonlinear oscillator aspects. However, noise shaping of noise sources in an oscillator is always present even in the absence of nonlinearities.

> Noise shaping leads to phase noise and can be calculated using a linear feedback system model [26]. This model is reviewed and extended in this section.

The influence of oscillator nonlinearities on the accuracy of the noise shaping equations will be dealt with in Chapter 6.

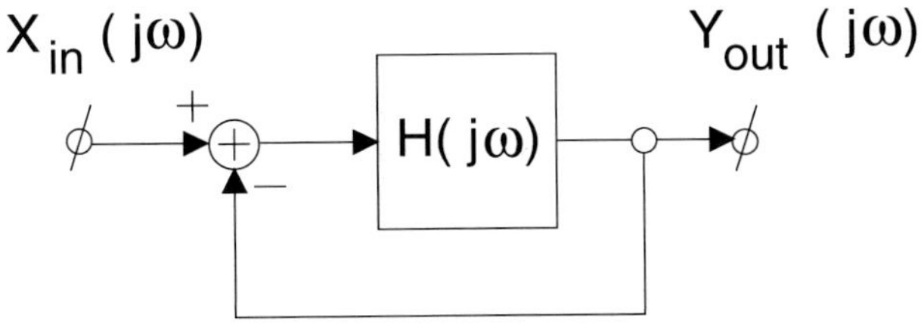

Figure 3.11 An linear oscillatory feedback system with unity feedback.

Single stage feedback system

Consider the linear feedback system in Figure 3.11. Output $Y_{out}(j\omega)$ is the spectral density of the oscillator and input $X_{in}(j\omega)$ is a noise power density, arising from white noise sources in an oscillator for example. The system has unity feedback and an open loop transfer function $H(j\omega)$. The closed loop transfer function is given by

$$\frac{Y_{out}}{X_{in}}(j\omega) = \frac{H(j\omega)}{1+H(j\omega)}. \tag{3.3}$$

When $H(j\omega) = -1$ at $\omega = \omega_{osc}$, the system obeys the oscillation conditions stated in Chapter 2. For frequencies close to the oscillation frequency $\omega = \omega_{osc} + \Delta\omega$, a Taylor expansion can be performed on $H(j\omega)$. Neglecting higher order terms, $H(j\omega)$ can be approximated as

$$H(j\omega) \approx H(j\omega_{osc}) + \Delta\omega \frac{dH(j\omega_{osc})}{d\omega}. \tag{3.4}$$

Substitution of $H(j\omega_{osc}) = -1$ and (3.4) in (3.3), and taking its absolute value leads to [26],

$$\left|\frac{Y_{out}}{X_{in}}(j(\omega_{osc}+\Delta\omega))\right|^2 = \frac{1}{(\Delta\omega)^2\left|\frac{dH(j\omega_{osc})}{d\omega}\right|^2}, \tag{3.5}$$

in which $|\Delta\omega\, dH(j\omega_{osc})/d\omega| \ll 1$ has been assumed. The term $dH(j\omega_{osc})/d\omega$ can be expanded by defining $H(j\omega) = A(\omega)exp[j\phi(\omega)]$. Now (3.5) can be rewritten as [26],

$$\left|\frac{Y_{out}}{X_{in}}(j(\omega_{osc}+\Delta\omega))\right|^2 = \frac{1}{4Q^2}\left(\frac{\omega_{osc}}{\Delta\omega}\right)^2, \tag{3.6}$$

with a quality factor Q defined as

$$Q|_{\omega=\omega_{osc}} = \frac{\omega_{osc}}{2}\sqrt{\left(\frac{dA}{d\omega}\right)^2 + \left(\frac{d\phi}{d\omega}\right)^2}, \tag{3.7}$$

where $A = |H(j\omega)|$ and $\phi = \arg(H(j\omega))$.

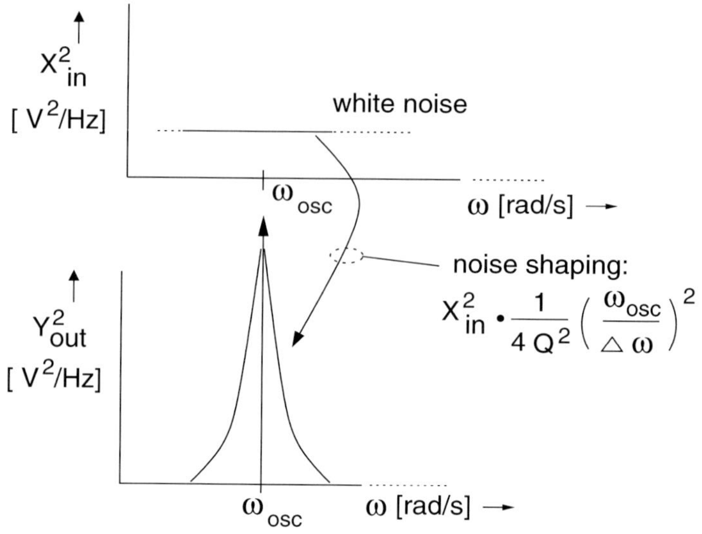

Figure 3.12 White noise is selectively amplified by the closed loop gain of the oscillator, and shaped into phase noise.

Equation (3.6) shows that a noise power density X_{in} will be selectively amplified by the closed loop gain of the oscillator. This process is illustrated in Figure 3.12. The white noise is shaped into phase noise sideband around the carrier at frequency ω_{osc}.

It is instructive to note that the quality factor definition in (3.7) reduces for a parallel LC resonator at resonance to

$$Q_p = \frac{\omega_{osc}}{2}\left|\frac{d\phi}{d\omega}\right|, \tag{3.8}$$

since $dA/d\omega = 0$ at resonance. Above formula is equivalent to the Q_p definition in (2.13) on p. 24. However, $dA/d\omega$ is for many oscillator types and implementations nonzero, and then the more general Q definition in (3.7) should be used. For example, as we will see in Chapter 6, $d\phi/d\omega$ is zero for the two-integrator oscillator and $dA/d\omega$ dominant is therefore.

Feedback system with *N* identical stages

As shown in the oscillator classification in Chapter 2, many oscillators of interest consist of more than one stage. Specifically, I/Q oscillators have two identical stages in general. If more than two output phases are required, more stages can be added. For N-stage oscillators with identical stages, it is practical to express the quality factor definition in (3.7), in terms of the transfer function of one stage. This simplifies calculation of the quality factor of an N-stage oscillator.

Consider the N-stage linear feedback system shown in Figure 3.13. The system has unity feedback and the open loop transfer function $H_{N-stage}(j\omega)$ is a cascade of N

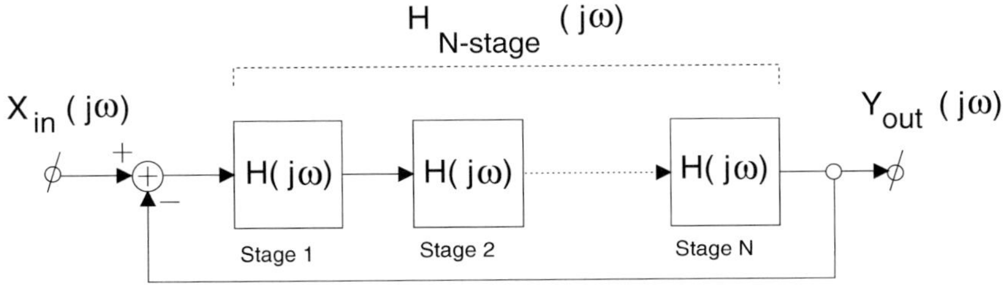

Figure 3.13 An N-stage oscillatory linear feedback system.

identical stages with transfer function $H(j\omega)$. Obviously, the same calculations and assumptions for this system can be made as for the single stage oscillatory feedback system. Hence (3.6) is also valid for this system, provided that (3.7) is redefined. For the system model in Figure 3.13, A in (3.7) should be redefined as

$$A_{N-stage} = |H(j\omega)|^N, \tag{3.9}$$

and ϕ in (3.7) should be redefined in as

$$\phi_{N-stage} = N \cdot \arg(H(j\omega)). \tag{3.10}$$

Substitution of (3.9) and (3.10) in (3.7) yields the quality factor of an N-stage oscillator in terms of sub-system $H(j\omega)$,

$$Q_{N-stage}|_{\omega=\omega_{osc}} = \frac{\omega_{osc}}{2}\sqrt{\left(\frac{dA_{N-stage}}{d\omega}\right)^2 + \left(\frac{d\phi_{N-stage}}{d\omega}\right)^2}, \tag{3.11}$$

which can be rewritten in terms of A and ϕ as,

$$Q_{N-stage}|_{\omega=\omega_{osc}} = N \cdot \frac{\omega_{osc}}{2}\sqrt{\left(A^{N-1} \cdot \frac{dA}{d\omega}\right)^2 + \left(\frac{d\phi}{d\omega}\right)^2}. \tag{3.12}$$

Since (3.12) is derived at system level, analyzing $H(j\omega)$ as a black box, (3.12) can be used to assess the noise shaping in N-stage LC and ring oscillators.

3.4.2 Behavioral level modeling

A behavioral model captures part of the behavior of an oscillator circuit, in a model using network elements[13]. The abstraction of an oscillator provided by behavioral modeling is very powerful for several reasons. First of all, calculations on behavioral

[13]Network elements are ideal models of resistors, capacitors, etcetera [86].

models are relatively simple and give insight into the first order behavior of the oscillator. This, of course, requires insight into what oscillator properties are dominant and what properties can initially be neglected, when the behavioral model is constructed. Second, a behavioral level is partially circuit topology independent. Therefore, design FOMs derived with a behavioral model are relevant for a large pool of circuit topologies. To define oscillator behavioral models unambiguously, behavioral building blocks used in this book are described in the following section. Next, conversion from a differential circuit level model to a single-ended behavioral model is discussed.

Behavioral modeling building blocks

In behavioral models, the same symbols for passive network elements are used as in circuit level modeling. However, in circuit level modeling these symbols model physical components. On a behavioral level, passive network elements used in the behavioral models like resistors, inductors, capacitors and varactors, are ideal and noiseless. If an inductor, capacitor, varactor or LC resonator has losses, these losses are explicitly introduced in the behavioral model by adding one or more resistors. In addition, resistors are by convention in this work noiseless in behavioral models. All noise sources will be represented by separate noise current sources, or by one noise current source that represents all noise contributors.

In addition to passive network elements, a number of symbols that represent ideal transconductors, switches, phase shifters, etcetera are required. The symbols used in this work, and their meaning or equivalent circuit are listed in Appendix B. We have already encountered several examples of behavioral models in Chapter 2, for example Figure 2.7(a) on p. 24 and Figure 2.17 on p. 35.

Differential circuits and single-ended behavioral models

An oscillator in an integrated transceiver is only one of the many building blocks of a system on a chip. For maximum robustness and minimum interference generation, an integrated oscillator generally is a differential structure. Since, a differential circuit is truly symmetrical, assuming perfect matching and a symmetrical layout, calculations on one half of the circuit are sufficient to capture its behavior (half-circuit concept [87]). Hence, it is efficient to use single-ended behavioral models, instead of differential behavioral models, to represent differential oscillator circuit topologies.

Obviously, equations and FOMs derived from a single-ended behavioral model should be identical to the results obtained using its differential counterpart, and vice versa. Figure 3.14(a) shows the circuit diagram of a differential LC oscillator. The transconductance of each transistor is g_{m_Q} and the generated output current noise is $i_{n_{diff}}$. The single-ended behavioral model of this oscillator is shown in Figure 3.14(b). Simple calculation shows that the transconductance of the cross-coupled pair is equal to $-g_{m_Q}/2$.

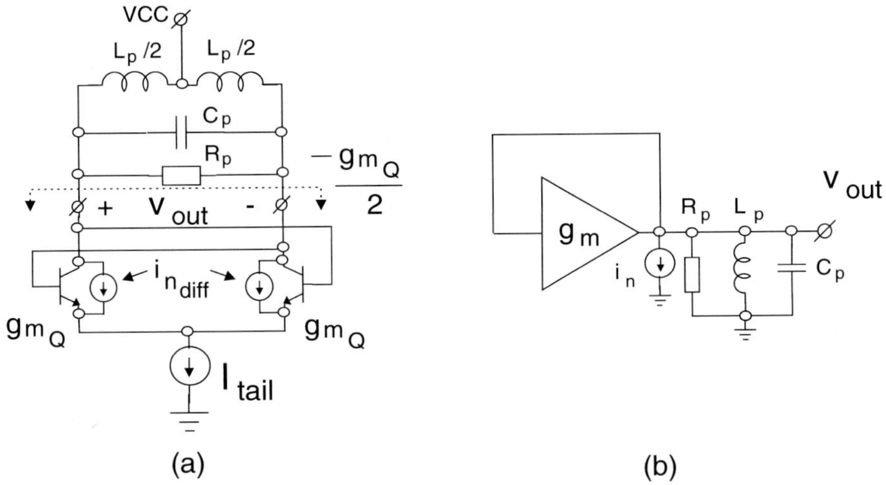

Figure 3.14 Differential LC oscillator circuit diagram (a), and its single-ended behavioral model (b).

The transconductance of the single-ended model , expressed in terms of g_{m_Q}, is equal to

$$g_m = \frac{g_{m_Q}}{2}. \tag{3.13}$$

As explained, noise in behavioral models is explicitly modeled in this work (i.e. the resistor in the behavioral model is noiseless, and represented by i_n). Noise current source i_n in Figure 3.14(b) therefore represents the noise generated by resistor R and current sources $i_{n_{diff}}$. It can be shown that the differential noise power produced by the identical current sources $i_{n_{diff}}$ is $\overline{i^2_{n_{diff}}}/2$ [88].

The single-ended current source i_n models the noise in the differential circuit correctly if

$$\overline{i^2_n} = \frac{4kT}{R_p} + \frac{\overline{i^2_{n_{diff}}}}{2}. \tag{3.14}$$

3.4.3 Circuit level modeling

At circuit level, an oscillator is represented by a circuit diagram, in which each element represents a physical component such as a physical transistor, physical resistor or physical capacitor, etcetera. Although each component is designed to implement one

Bipolar modeling	MOS modeling
Gummel-Poon	BSIM (level 1)
Mextram 503	MOS level 9 / BSIM3
Mextram 504	MOS level 11 / BSIM4

Table 3.1 Three modeling levels of bipolar and MOS transistors.

primary physical function, physical components also exhibit many unwanted properties and second order effects.

At circuit level the accuracy of FOMs derived with behavioral level modeling can be improved. To be able to do this, an equivalent circuit model has to be drawn of the circuit diagram. For example, a resistor in a circuit diagram can be replaced by a resistor model with capacitive losses to ground. The physical resistor is replaced by a model built with network elements, which approximates the physical resistor. Similarly, the transistor symbols in a circuit diagram have to be replaced by a transistors model, before calculations can be performed. In Table 3.1 a number of transistor models are listed. Only the models on the first row qualify for use in hand calculations although even these "simple" models are normally reduced to simplify hand calculations. Bipolar and MOS transistor "back-of-the-envelope"-models can be found in [87,89]. Mextram 504[14], MOS level 11[15] and BSIM4[16] are very advanced and complex transistor models and these should be used for circuit simulations when available. These models are optimized to capture, among other effects, the high-frequency and saturation effects of integrated transistors, whereas Gummel-Poon and BSIM (level 1) simply are not complex enough to predict measured behavior accurately. The values of the parameters in the transistor models are provided by the IC technology of choice.

3.5 Summary

High-frequency oscillator design is an analog design process. Inputs to this process are functional specifications and design resources. The functional specifications specify the performance an oscillator should achieve, whereas the design resources provide the means to meet these specifications. An analog design process, like the design of an oscillator, can be divided into three phases. In the first phase, "specification and conceptualization", the design space is explored and a topology is chosen. In

[14] http://www.semiconductors.philips.com
[15] http://www.semiconductors.philips.com
[16] http://www-device.eecs.berkeley.edu

the second phase, "optimization and implementation", the oscillator is dimensioned and a layout is prepared. The main activity in the final phase is "verification and documentation". Especially in the first phase, design heuristics are used by the analog designer to reduce the complexity of the design problem.

Three structured design methods are discussed. Trial and error, optimization tools and expert systems/synthesis environments. Although all reviewed structured methods aim to speed up design phase one and two, a practical, insightful and appealing method for topology selection, which has a major influence on the total design time, was not encountered in literature. Well-defined figures of merit provide quantitative information of the performance of one or more oscillator properties. They limit the oscillator design space, guiding the oscillator designer in his search for a cost-effective oscillator that meets the functional specification.

The concept of design FOMs and benchmark FOMs is proposed as a structured design method for high-frequency oscillators. A design FOM provides the designer with quantitative information on the performance of a oscillator property with respect to its specified value. If the realization of a design FOM is negative in dBs, the functional specification is not met (negative design margin), if it is 0 dB the specification is met exactly, and if it is positive the amount of positive design margin is indicated. Therefore all relevant design FOMs for an oscillator will always be zero or positive if the electrical design of the oscillator is ready.

Benchmark FOMs are useful to compare performance aspects of an oscillator design with the state-of the-art. This can be viewed as relative benchmarking. Alternatively the normalization function h in a benchmark FOM may be a theoretical performance limit. In this case the performance of an oscillator is benchmarked in an absolute sense to a theoretical bound. If the value of such a benchmark FOM is positive (the oscillator performance is greater than the theoretical limit), the designer knows the given specification is impossible to achieve, or the simulation or measured value is inconsistent.

System, behavioral and circuit level modeling are discussed in the final part of this chapter, as part of the modeling framework that will be used to explore the oscillator design space (in Chapter 5 and 6) and to define FOM examples (Chapter 7).

4

Specifications

AT THE BEGINNING of an oscillator design, the functional specifications and design resources must be completely clear. Deriving a well understood and complete set of oscillator specifications from the application in which the oscillator will be used, can be a challenging task in itself. In any case, time spent on this task in the beginning of the design process is worth the effort, since it is difficult to "hit a moving target" (specification).

There are many different specifications for oscillators. Given a certain application, some will be very important and some will be less important.

> In this chapter, the most important oscillator specifications are defined, discussed and illustrated with examples.

In Section 4.1, a distinction is made between nominal specifications and design specifications. Sections 4.2 and 4.3 discuss frequency and tuning range, and phase noise to carrier ratio, respectively. The time domain equivalent of the phase noise to carrier ratio specification, jitter, is highlighted in Section 4.4. Other oscillator specifications that are also of importance in integrated transceivers follow, concluded by a summary in Section 4.12.

4.1 Nominal specifications versus design specifications

The functional specifications for an oscillator in a transceiver are derived from system level considerations. These "nominal functional specifications" specify the minimum requirements for a functional system. Design functional specifications are far harder because of the limitations imposed by the design resource technology. Spread on the integrated devices should be taken into account. Furthermore, addition of a safety margin is a common design practice, to allow for modeling inaccuracies or simply to increase the yield of a design. Therefore, the design specification of a certain oscillator property can be written as

$$\text{Design Spec.} = \text{Nominal Spec.} + \text{Technology margin} + \text{Safety margin.} \tag{4.1}$$

The numerical specification examples given in this chapter are nominal functional specifications, so we should keep in mind that process spread and possibly a safety margin must be added to derive the functional specifications that should be the target of the oscillator design.

4.2 Frequency and tuning range

Frequency and tuning range are important specifications that can have a significant impact on the level of difficulty when designing an oscillator. In general, the higher the center frequency of an application, the more difficult it is to design an oscillator for this application. For some applications only the center frequency is specified, but most oscillators need to cover a band of interest around a center frequency.

Table 4.1 shows the center frequency, absolute tuning range (minimum and maximum frequency) and relative tuning range[1] of a number of standards. For the first ten standards, the center frequency can of course be calculated, but is omitted since it usually is not specified, as it is already included in the tuning range specification. An additional specification is of importance instead: tuning accuracy. In principle not all frequencies within the tuning ranges shown in Table 4.1, need to be covered by the oscillator. The first ten standards all use channels equidistantly spaced on a certain frequency grid. For example, the Global System for Mobile communication (GSM) has a channel spacing of 200 kHz, whereas DECT uses 1.728 MHz. Therefore, theoretically, we could make an oscillator for standards with a frequency grid, which only produces discrete frequencies equal to the channel frequencies defined in the standard. However, in practice it is easier to implement a continuous tunable oscillator and realize the tuning accuracy with a PLL [3].

The relative tuning range requirements of Table 4.1 roughly divide the standards into two categories: wide range and narrow range standards (or applications). FM radio, television (TV), terrestrial Digital Video Broadcasting (DVB-T) and satellite TV are examples of the first category, whereas telecom standards such as GSM, Universal

[1] Maximum minus minimum frequency divided by the center frequency of the frequency band.

Standard	Center freq. [MHz]	Absolute tuning range [MHz]	Relative tuning range [%]
FM Radio front-end		87.5 – 108	21
TV receiver		41 – 960	184
DVB-T		470 – 860	59
GSM transmitter		890 – 915	2.8
GSM receiver		925 – 960	3.7
Satellite TV front-end		950 – 2150	77
DECT transceiver		1880 – 1920	2.1
UMTS transmitter (FDD)		1920 – 1980	3.1
UMTS receiver (FDD)		2110 – 2170	2.8
Bluetooth		2400 – 2483	3.4
SONET OC-3 (SDH STM-1)	155.52		
SONET OC-12 (SDH STM-4)	622.08		
SONET OC-48 (SDH STM-16	2488.32		
SONET OC-192 (SDH STM-64)	9953.28		

Table 4.1 Center frequency, absolute tuning range, and relative tuning range of several standards.

Mobile Telecommunications System (UMTS), DECT and Bluetooth are narrow range examples. A satellite TV receiver needs more than an octave of tuning range, whereas the required 40 MHz for a DECT receiver is almost negligible in comparison with the center frequency of roughly 2 GHz.

The four[2] optical standards listed in Table 4.1 have a fixed bit-rate, and only the required center frequency is denoted in the table. In practice, some tuning range may be desirable to cover standards, which are very close to the SONET/SDH bit-rate (e.g. Gigabit Ethernet or Fiber channel) or, for example, to make a flexible multi-rate system that covers multiple SONET/SDH bit-rates.

As for any specification, the tuning range specification must be met under worst case conditions. Therefore frequency deviations due to temperature changes, process spread and power supply variations should be added to the tuning range.

[2]Actually, eight standards. However the Synchronous Optical NETwork (SONET) and Synchronous Digital Hierarchy (SDH) standards become compatible at a bit-rate of 155.52 Mb/s.

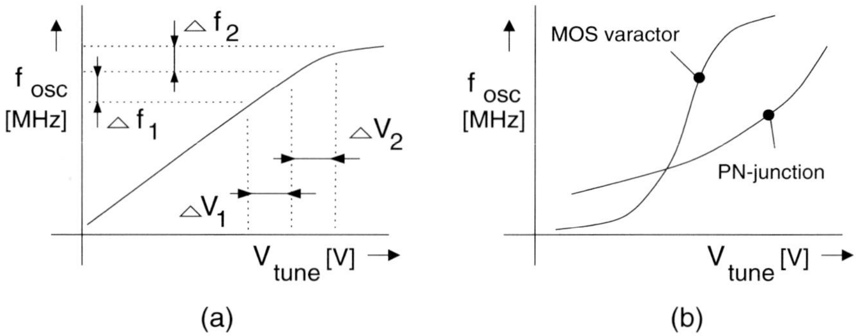

Figure 4.1 Typical tuning curve of a ring oscillator (a), and the tuning curve of an LC oscillator with MOS-varactor and PN-junction varactor (b).

For a ring oscillator, the process spread can easily be 20% to 40%. In an LC oscillator the (planar) inductor tolerance usually is very good (e.g. 1 to 5%) because of the accurate lithography of the IC process. However, the varactor and fixed capacitances can vary up to 20% in value, and the center frequency can therefore change by more than 10%. A more general discussion on temperature range and process spread can be found in Section 4.10.

4.2.1 Tuning constant and linearity

A VCO (or CCO) has a tuning constant K_{vco} that is specified in Hz/V (or K_{cco} in Hz/A). This constant is sometimes referred to as VCO gain too. For example, if the tuning range is 2.4 GHz to 2.483 GHz (Bluetooth standard) and the available tuning voltage range is 2.7 V, the tuning constant is about 30.7 MHz/V. This constant will become larger if the tuning range is increased, to cover process spread for example. If the supply voltage decreases, in the case of a redesign using a newer CMOS technology for example, the tuning-voltage range will also decrease and thus K_{vco} will increase.

> The phrase "tuning constant" is actually rather deceptive, since it rarely is a constant.

In practice, the tuning characteristic of an oscillator is a nonlinear function. Figure 4.1(a) shows the typical tuning curve of the two-integrator oscillator from Figure 2.17. For a large part of the tuning range, K_{vco} is equal to $\Delta f_1/\Delta V_1$. However, at the end of the tuning range, where parasitics start to dominate, K_{vco} decreases ($\Delta f_2/\Delta V_2 < \Delta f_1/\Delta V_1$).

Figure 4.1(b) shows the typical tuning characteristics of an oscillator tuned by a MOS varactor and a PN-junction varactor, respectively. The derivatives of the tuning curves vary significantly, especially for the MOS-varactor. So, instead of specifying the gain constant using one figure, the deviation from this nominal value is usually specified as well, for example: $K_{vco} \pm 30\%$.

4.3 Phase noise to carrier ratio

In Chapter 2, important non-ideal oscillator properties were introduced including phase noise sidebands. For convenience, the formula describing an oscillator with phase noise sidebands, which was given in Chapter 2, is repeated below,

$$v_{out}(t) = V_{peak}\cos(2\pi f_{osc}t + \theta_{pn}(t)). \tag{4.2}$$

The variable $\theta_{pn}(t)$ is a stochastic variable and its one-sided and double-sided power spectral density[3] are $S_{\theta_{pn}}(f)$ and $S^{ds}_{\theta_{pn}}(f)$, respectively.

It is difficult to measure $S_{\theta_{pn}}(f)$ directly. Normally, $S_{v_{out}}(f)$ is measured, being the power spectral density of v_{out}. The power density $S_{v_{out}}(f)$ versus frequency can be directly read from the analyzer display. Based on this measurement, $S_{\theta_{pn}}(f)$ of an oscillator (or, in general, of a tuning system) is characterized by the single-sideband (SSB) phase noise to carrier ratio $\mathcal{L}$[4] at an offset frequency f_m from the carrier frequency f_{osc}, which is defined as

$$\mathcal{L}(f_m) = \frac{P_{SSB}(f_{osc}+f_m)}{P_{signal}} \simeq \frac{S_{v_{out}}(f_{osc}+f_m)}{V_{peak}^2/2}, \tag{4.3}$$

where P_{SSB} and $S_{v_{out}}$ are measured in 1 Hz bandwidth.

Figure 4.2 illustrates the definition of $\mathcal{L}(f_m)$. As shown Figure 4.2, P_{signal} is the result of an integration of $S_{v_{out}}(f)$ around f_{osc}. Provided that the integration interval is large enough to capture about 99 %, P_{signal} is a good approximation of total signal power. In practice the carrier power $P_{carrier} = V_{peak}^2/2$ (the power arbitrarily chosen here in 1 Ω, of (4.2) with $\theta_{pn}(t) = 0$) is used as an approximation of P_{signal}. Clearly a high P_{signal}, and thus a large V_{peak} is desired for a small $\mathcal{L}(f_m)$. Commonly, $10\log(\mathcal{L}(f_m))$ is specified and its units are dBc/Hz, indicating that the phase noise is measured relative to the carrier and in an 1 Hz bandwidth[5]. Alternatively, the phase noise of an oscillator can be characterized by the Carrier to phase Noise Ratio CNR(f_m), which is simply $-10\log(\mathcal{L}(f_m))$.

A real oscillator spectrum has phase noise sidebands with three major regions, which can be approximated by power law curves (plotted on a log-log scale). These regions are shown in Figure 4.3. The largest part of an oscillator sideband is the

[3]Defined as the Fourier transform of the autocorrelation function of $\theta_{pn}(t)$.

[4]Pronounced "script l".

[5]In this book, $\mathcal{L}(f_m)$ figures are given in dBc/Hz, without explicitly writing $10\log(\mathcal{L}(f_m))$, when the phase noise in dBc/Hz is calculated (this would formally be more correct but does not improve readability).

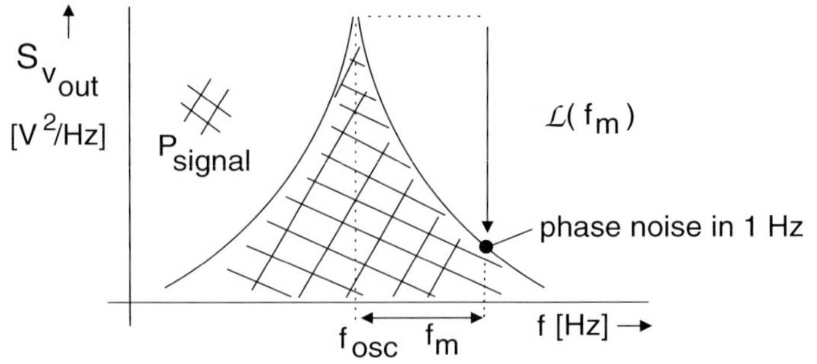

Figure 4.2 $\mathcal{L}(f_m)$ definition.

region with a slope of -6 dB/octave, and results from noise shaping of white noise sources in the oscillator (discussed on p. 60), and other, nonlinear, noise generation mechanisms, which will be explained in Chapter 6. This slope starts at the $1/f$ noise corner frequency $f_{1/f}$, and continues until it hits the white noise floor. The white noise floor often is the result of white noise generated by circuits, such as buffers, connected to the oscillator output. If devices in an oscillator, or at the tuning input of an oscillator, generate $1/f$ noise, this will result in an oscillator sideband region with a slope of -9 dB/octave that continues to $f_{1/f}$, at which point the phase noise with the $1/f^2$ slope starts dominating.

Figure 4.3 points out that oscillator phase noise in principle can not be specified by a single figure and it is best to plot $\mathcal{L}(f_m)$ versus f_m on a log-log scale for a complete specification. However, often $\mathcal{L}(f_m)$ is specified with only one figure, assuming a $1/f^2$ sideband slope. Under this assumption $\mathcal{L}(f_m)$ can be extrapolated to any offset frequency. For example, if $\mathcal{L}(10\,\text{kHz}) = -80\,\text{dBc/Hz}$ then it follows that $\mathcal{L}(20\,\text{kHz}) = -86\,\text{dBc/Hz}$ and $\mathcal{L}(2\,\text{MHz}) = -126\,\text{dBc/Hz}$.

As mentioned, $\mathcal{L}(f_m)$ can be measured directly using a spectrum analyzer. The spectrum analyzer measures the power spectrum of the oscillator, which not only includes the phase variations but amplitude variations too. However, the phase noise component is dominant in many practical oscillators, because the amplitude component is reduced by limiting mechanisms. In the literature, $\mathcal{L}(f_m)$ therefore is commonly used with the assumption that it only includes the phase noise component. This assumption is adopted in this book. There are several methods to unambiguously measure the phase noise power spectral density $S_{\theta_{pn}}(f_m)$ versus offset frequency f_m of an oscillator [90]. These are utilized in dedicated commercial phase noise measurement equipment.

An important relation between $S_{\theta_{pn}}(f_m)$ and $\mathcal{L}(f_m)$ can be derived, if we model the phase noise sidebands as the result of a baseband phase modulation process by sine waves. This model and underlying assumptions based on long-term statistical values are extensively discussed in [21]. In this model the phase noise sidebands are represented by numerous sine waves, each at a different frequency offset f_m, represented

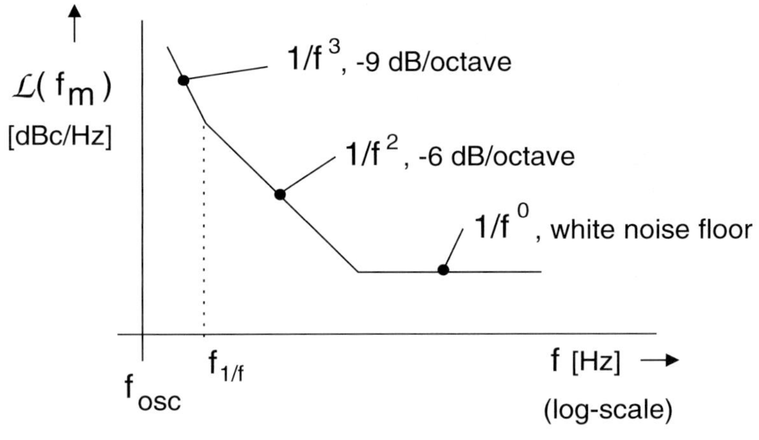

Figure 4.3 Power law approximation of a phase noise sideband.

as

$$\theta_{pn}(t) = \sqrt{2}\theta_{rms}(f_m) \cdot \sin(\,2\pi f_m t), \tag{4.4}$$

where the angular phase deviation $\theta^2_{rms}(f_m)$ (in rad^2/Hz) has the same average power in an 1 Hz bandwidth at offset frequency f_m as the phase noise [3]. Therefore, by definition, $\theta^2_{rms}(f_m) = S_{\theta_{pn}}(f_m){=}\,2S^{ds}_{\theta_{pn}}(f_m)$. Provided $\theta_{rms}(f_m) \ll 1$ (small modulation index or narrow band FM condition), modulation theory [21, 91] shows that $S_{v_{out}}(f)$ can be written as

$$S_{v_{out}}(f) = \frac{V^2_{peak}}{2}\left(\delta(f - f_{osc}) + S^{ds}_{\theta_{pn}}(f - f_{osc})\right). \tag{4.5}$$

Hence, under the previously mentioned assumption that only the phase noise component is present in $\mathcal{L}(f_m)$, and that total signal power is accurately enough approximated with that of an ideal sine wave, $\mathcal{L}(f_m)$ can be expressed in terms of $S_{\theta_{pn}}(f_m)$ by applying the phase noise to carrier ratio definition in (4.3),

$$\mathcal{L}(f_m) = \frac{\frac{V^2_{peak}}{2} S^{ds}_{\theta_{pn}}(f_m)}{\frac{V^2_{peak}}{2}} = \frac{S_{\theta_{pn}}(f_m)}{2} = \frac{\theta^2_{rms}(f_m)}{2}. \tag{4.6}$$

Equation (4.6) shows that $S_{\theta_{pn}}(f_m)$ of an oscillator (or in general $S_{\theta_{pn}}(f_m)$ of a signal source) can be determined via $\mathcal{L}(f_m)$ measurement, if a limiter is placed in front of a spectrum analyzer. As noted, (4.6) is valid for $\theta_{rms} \ll 1$. A useful rule of thumb is that $\mathcal{L}(f_m)$ should be less than -23 dBc/Hz: in that case $\theta_{rms}(f_m) < 0.1$ and (4.6) can be applied with negligible error.

Standard	$\mathcal{L}(f_m)$ [dBc/Hz]	f_m [MHz]	$\mathcal{L}(2\,\text{MHz})$ [dBc/Hz]
FM Radio front-end	-105	0.01	-151
TV receiver	-95	0.1	-121
DVB-T	-90	0.01	-136
GSM receiver	-127	0.6	-137
GSM transmitter	-162	20	-142
Satellite TV front-end	-78	0.01	-124
DECT transceiver	-110	1	-116
UMTS transmitter (FDD)	-147	20	-127
UMTS receiver (FDD)	-130	8	-118
Bluetooth	-110	1	-116

Table 4.2 $\mathcal{L}(f_m)$ requirements for several standards. In the last column, $\mathcal{L}(f_m)$ extrapolated to $f_m = 2\,\text{MHz}$ is listed, assuming a $1/f^2$ phase noise sideband slope.

An indication of the $\mathcal{L}(f_m)$ specification for several standards is given in Table 4.2. Exact values are not easy to specify because factors like production margin, whether a good $\mathcal{L}(f_m)$ is important as a selling feature, and the specification of the sub-systems around the oscillator all influence the phase noise requirements. However, the figures are in accordance with what is normally specified for the listed standards.

In the third column of Table 4.2, $\mathcal{L}(f_m)$ extrapolated to $f_m = 2\,\text{MHz}$ is shown for comparison as well. It is important to note that $\mathcal{L}(f_m)$ is proportional to f_{osc}^2, as will be derived in Chapter 6. Hence, although $\mathcal{L}(2\,\text{MHz})$ of FM radio looks tougher than the GSM transmitter specification, this specification has to be achieved around 100 MHz, which is a lot easier than meeting GSM transmitter requirements around 900 MHz.

Phase noise deteriorates transceiver performance. In other words, transceiver performance requirements will lead to a certain $\mathcal{L}(f_m)$ specification. In the following subsections, the effect of phase noise on transceiver performance will be discussed, and how the required phase noise to carrier ratio can be calculated.

4.3.1 Reciprocal mixing

Reciprocal mixing is illustrated in Figure 4.4. The desired channel at f_{RF} has an amplitude A_1. A strong adjacent channel is present at an offset f_m with amplitude A_2. The oscillator with amplitude A_3 has phase noise sidebands with an amplitude of A_4 at f_m. To simplify the discussion on reciprocal mixing, the phase noise sideband is

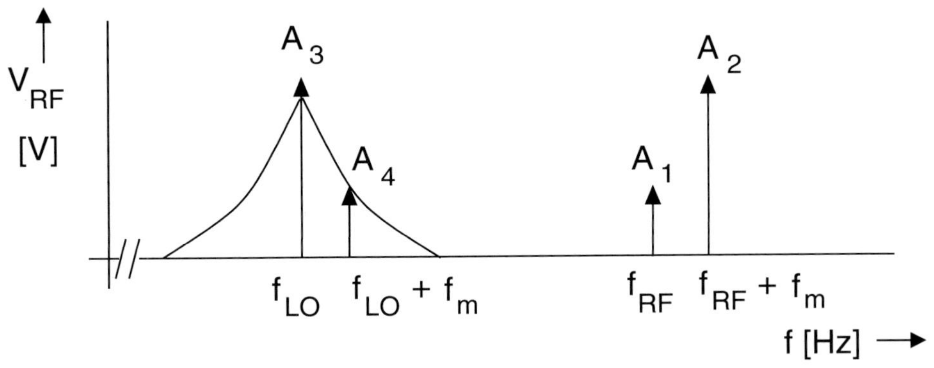

Figure 4.4 Reciprocal mixing: the desired signal A_1 and adjacent channel A_2 are both converted to the same frequency due to the phase noise sideband of the LO.

replaced by a single tone interferer at $f_{LO} + f_m$. After mixing, it is not only the desired signal with amplitude $\frac{1}{2}A_3A_1$ that is converted to $f_{RF} - f_{LO}$, but an unwanted signal with amplitude $\frac{1}{2}A_4A_2$ is also converted to this frequency. This lowers the signal-to-noise ratio (SNR) and hence A_4 should be sufficiently small to achieve a certain Bit Error Rate (BER) or SNR.

> One important mechanism that deteriorates the SNR of a desired channel is reciprocal mixing.

In a GSM system, the in-band interferer levels are specified, thus allowing the maximum tolerable $\mathcal{L}(f_m)$ to be calculated. For example, in the frequency band from 600 kHz to 800 kHz the interferer level can be -43 dBm, under minimum desired-signal conditions. More information is needed to calculate the phase noise specification of the oscillator: the reference sensitivity (-105 dBm), minimum signal-to-noise ratio for the demodulator (12 dB) and effective noise bandwidth (180 kHz). The unwanted signal $\frac{1}{2}A_4A_2$ at the wanted frequency should be no larger than -105 dBm minus 12 dB. This means that compared to the -43 dBm interferer, the power difference is 74 dB in a 180 kHz bandwidth. Therefore, the value for $\mathcal{L}(600\,\text{kHz})$ should be less than -74 dB minus $10\log(180 \cdot 10^3)$, which is -126.6 dBc/Hz (the value, rounded off, in Table 4.2).

4.3.2 Signal to noise degradation of FM signals

In frequency and phase modulated signals, information is carried in the phase of the carrier signal. When this signal is down-converted in a FM receiver using an oscillator, the oscillator phase noise is superimposed on the wanted signal and degrades the SNR.

For example, consider an FM modulated signal $v_{fm}(t)$ which is written as

$$v_{fm}(t) = V_{peak} \cos\left(2\pi(f_{carrier} + \Delta f \cos(2\pi f_m t))t\right), \tag{4.7}$$

with $f_{carrier}$ equal to 100 MHz, de peak frequency deviation is Δf is 50 kHz and modulation frequency f_m is 10 kHz. Note that (4.7) is identical to the equation of an FM signal given earlier on p. 15, but written in a different form.

The relationship between de frequency deviation $\delta f(f_m)$ of an oscillator at an offset f_m and $\theta_{rms}(f_m)$ can be derived using modulation theory [21]:

$$\delta f = \theta_{rms}(f_m) f_m. \tag{4.8}$$

We assume that the stochastic phase variable $\theta_{rms}(f_m)$ has an inverse dependency on f_m. In order words, we assume that the oscillator used for down-conversion of the signal in (4.7), has a $1/f^2$ phase noise sideband slope. In this case, as (4.8) makes clear, the resulting frequency noise $\delta f(f_m)$ will be white, independent of f_m.

Substitution of (4.8) in (4.6) relates the rms frequency deviation δf resulting from phase noise and $\mathcal{L}(f_m)$ in a noise bandwidth NBW and can be written as [21]

$$\mathcal{L}(f_m) = \frac{(\delta f)^2}{2 f_m^2 NBW}. \tag{4.9}$$

If the mentioned FM receiver requires a SNR of 80 dB after frequency conversion, a maximum δf of 3.5 Hz rms is allowed ($(50/\sqrt{2})$ kHz rms / $10^{(80/20)}$). This results in $\mathcal{L}(10\,\text{kHz})$ being equal to -109 dBc/Hz using (4.9), if the NBW of the FM receiver is 5 kHz. Note that we have assumed in this example that the signal-to-noise reduction due to reciprocal mixing is negligible.

4.3.3 Spurious emission

A transmitter of a telecom transceiver will emit power at unwanted frequencies in addition to the desired output signal, due to the phase noise of the oscillator used. This power at unwanted frequencies may enter the receive band RX and result in received noise. For this reason, maximum permitted emissions are specified in telecom standards. For example, in the type-approval template of a GSM system, the spurious emission at 20 MHz offset (GSM RX-band) from the 915 MHz carrier (worst case location in TX-band) must be -79 dBm down in the 100 kHz bandwidth relative to the +33 dBm carrier. This means that $\mathcal{L}(20\text{MHz})$ must be lower than -162 dBc/Hz[6].

[6] Filtering in the power amplifier and in the duplex filter of a transceiver may ease the oscillator specification.

4.4 Jitter

The time domain equivalent of phase noise is jitter, sometimes also referred to as timing jitter. Ideally, the time difference between zero-crossings of an oscillator output signal is constant. In practice this spacing is variable due to the stochastic nature of $\theta_{pn}(t)$ in (4.2). A graphical illustration of jitter was already presented in Figure 2.2 on p. 16. Assuming an oscillator with an oscillation period that has a Gaussian distribution, the mean of the distribution will be $f_{osc} = 1/\tau_{avg}$. If we defined τ_n as the period of cycle n, the cycle-to-cycle jitter of an oscillator can be defined as [92],

$$\sigma^2_{c-to-c} = \lim_{N \to \infty} \left(\frac{1}{N} \sum_{n=1}^{N} (\tau_n - \tau_{avg})^2 \right). \tag{4.10}$$

Cycle-to-cycle jitter measures the variance of each period to the average period. Its square root is the standard deviation of the above mentioned distribution. This definition, also referred to as cycle jitter or jitter in one clock cycle, is mostly used when jitter is specified as a single number in the unit pico-seconds.

In optical transceivers, jitter is often specified and measured in Unit Intervals (UI), where one UI corresponds to a phase deviation equal to one signal period [93]. The unit UI is used for jitter specification in the optical standards SONET and SDH. These standards have various hierarchical levels with different bit-rates. Since UIs are independent of bit-rate, the unit allows comparison at different hierarchical levels in the optical network.

The jitter in UI can be found by integration of $\mathcal{L}(f_m)$ over a offset frequency (f_m) range of interest. From (4.6), it follows that the jitter in units $\mathrm{UI_{rms}}$ can be calculated in a frequency band starting from $f_m = f_L$ up to $f_m = f_H$ with the expression

$$Jitter\,[UI] = \frac{1}{2\pi} \sqrt{\int_{f_L}^{f_H} \theta^2_{rms}(f_m)df_m} = \frac{1}{2\pi} \sqrt{\int_{f_L}^{f_H} 2\mathcal{L}(f_m)df_m}. \tag{4.11}$$

Table 4.3 gives an overview of the amount of jitter for four bit-rates, in the frequency band of interest f_L to f_H, which is allowed at the output of a SONET/SDH network element in absence of applied jitter. This specification is called the jitter generation requirement. Assuming that the network element is a complete transceiver, it should be noted that the amount of jitter is specified for the total transceiver, hence individual components in the transceiver must generate less. Normally, the jitter produced by individual components, such as the clock conversion block and the laser driver in an optical transceiver, is uncorrelated and should be added power-wise.

Application of (4.11) is illustrated with Figure 4.5. The output spectrum of a SONET OC-3 clock-conversion PLL is shown, which includes a VCO running at 622 MHz [94]. Total $\mathcal{L}(f_m)$ is plotted, as well as the individual contributions of the PLL

Standard	f_L [kHz]	f_H [MHz]	Jitter [$\mathrm{UI_{rms}}$]	Jitter [$\mathrm{UI_{p-p}}$]
SONET OC-3 (STM-1)	12	1.3	< 0.01	< 0.10
SONET OC-12 (STM-4)	12	5	< 0.01	< 0.10
SONET OC-48 (STM-16)	12	20	< 0.01	< 0.10
SONET OC-192 (STM-64)	50	80	< 0.01	< 0.10

Table 4.3 SONET (SDH) transceiver requirements for jitter generation.

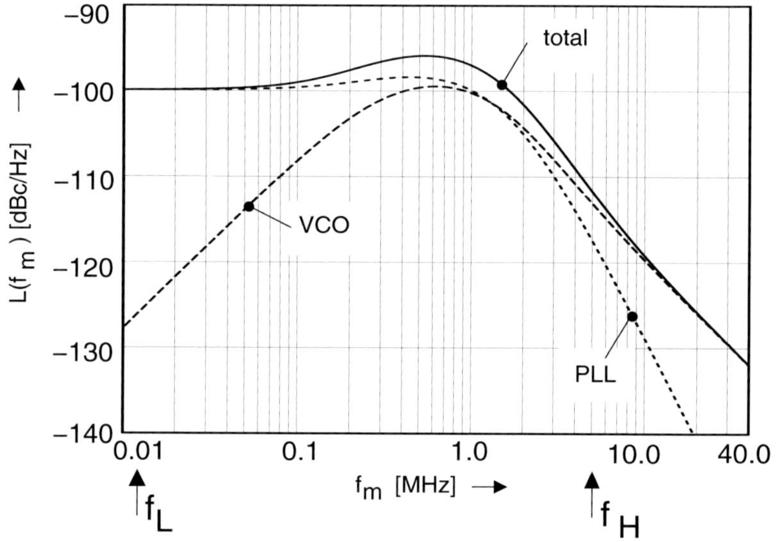

Figure 4.5 $\mathcal{L}(f_m)$ versus f_m of a clock-conversion PLL for SONET OC-3 transmitters.

components and the VCO. The loop bandwidth is about 1 MHz and outside the loop bandwidth, the VCO effectively is free-running. The generated jitter can be calculated by applying (4.11), using the integration limits specified in Table 4.3. Integration of the area under the total PLL noise over the frequency range f_L to f_H, and normalization to one period calculates to a jitter of 4.7 $\mathrm{mUI_{rms}}$ [94]. Therefore, the jitter generation specification for SONET OC-3 is met by this PLL design (see Table 4.3). The jitter in Table 4.3 is specified in $\mathrm{UI_{rms}}$ but also in peak-to-peak unit interval $\mathrm{UI_{p-p}}$.

α_{BER}	BER
6.2	10^{-3}
9.5	10^{-6}
14.1	10^{-12}

Table 4.4 Conversion parameter α_{BER} for three BER levels.

Since the jitter has a stochastic nature, peak jitter values have no meaning without specifying the probability of the peak value occurring.

If jitter has a Gaussian distribution it is impossible to specify a peak-to-peak jitter range that bounds the jitter 100% of the time, since the tails of the Gaussian distribution extend to infinity [95]. Under the assumption that the jitter has a Gaussian distribution (which is a valid assumption in many practical cases), and assuming that samples that fall outside the specified peak-to-peak range will cause errors, the units UI_{rms} can be converted into UI_{p-p} at a given Bit Error Rate (BER) [95], using the formula

$$Jitter_{UI_{p-p}} = \alpha_{BER} \cdot Jitter_{UI_{rms}}, \tag{4.12}$$

in which α_{BER} is determined by

$$BER = \frac{1}{2} erfc(\sqrt{2}\alpha_{BER}), \tag{4.13}$$

where $erfc$ is the complementary error function that is tabulated in many math and communication references (e.g. in [96]). The parameter α_{BER} and related BER is tabulated for three BER levels in Table 4.4. In literature $\alpha_{BER} = 6$ often is used, thus assuming that a BER level of approximately 10^{-3} is acceptable. The SONET/SDH standard uses $\alpha_{BER} = 10$ (see Table 4.3), which corresponds with a BER level close to 10^{-7}.

4.5 Waveform

In practice, every oscillator generates some power at other frequencies than the fundamental frequency desired. Due to nonlinearities in the oscillator, harmonics of the fundamental frequency will be generated, or in the case of modulation by spurious signals (see Section 4.9) at other frequencies.

These harmonics are specified in dB relative to the carrier (in dBc). In many cases, nonlinear operation of an oscillator improves the phase noise, but also generates harmonics. When an oscillator is used to drive mixers, a square wave oscillator output waveform may improve the switching of the mixer and thus the mixer noise figure. However, unwanted channel power or noise power located at harmonics may fall into the desired channel when the oscillator-mixer combination is used for frequency conversion. Care must be taken to ensure that the conversion products of the oscillator signal are sufficiently low, relative to the desired channel that was mixed with the fundamental. To some extent second-order harmonics can be suppressed (typically 30–40 dB) by a balanced design.

4.6 Carrier amplitude and power

The definition for phase noise to carrier ratio (see (4.3)) makes it clear that for minimizing $\mathcal{L}(f_m)$, the oscillator output level V_{peak} should be as high as possible. This can also be beneficial for the circuits that are cascaded to the oscillator. For example, if the LO-stage of a double-balanced mixer is driven by a large oscillator output swing, this stage switches, which generally improves the noise figure of this mixer type. The same is true when the oscillator circuit is driving a divider. If the input signal is too small, the divider will have poor phase noise performance.

> Any oscillator can be designed to provide a certain output voltage or output power to a specified load impedance, but in all practical cases it is better to insert a buffer between the oscillator and cascaded subsystems in a system.

The following reasons make a buffer cascaded after the oscillator useful:

- A buffer reduces the output level requirements and output load requirements of the oscillator. These requirements can contradict with other specifications, such as tuning range or phase noise to carrier ratio. Hence an output buffer orthogonalizes the oscillator design problem to some extent, while it takes care of delivering sufficient power to a specified load.

- A buffer provides isolation for the oscillator to load variations and attenuation of signals at the output of the buffer.

  - Signals at the output of the buffer, which can modulate the oscillator, cause problems such as self-mixing in zero-IF receivers [97]. Furthermore, without sufficient isolation between an oscillator and the circuit it is driving, an oscillator may be pulled away from its original frequency, by power from this circuit injected into the oscillator. By having sufficient isolation between an oscillator and circuits, which use the LO signal, frequency pulling can be avoided. A typical value for the isolation provided by one buffer stage is 20 dB.

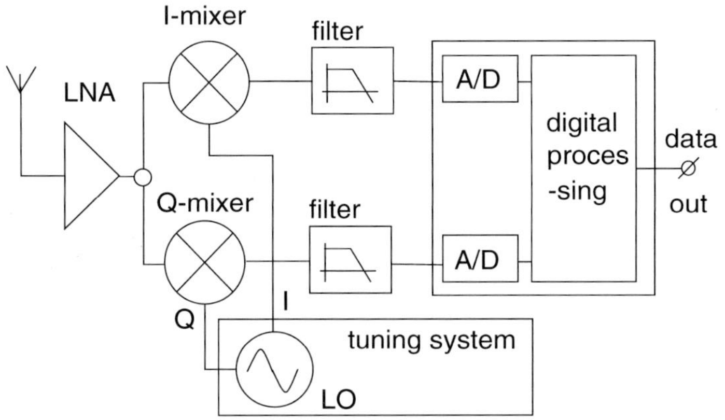

Figure 4.6 Simplified zero-IF architecture.

– Isolation from load variations is important for measurements. Without a buffer between measurement equipment and the oscillator being tested, it is very difficult to characterize the oscillator because the measurement equipment and cables may detune or corrupt the spectral purity of the oscillator.

Although, generally speaking, a high oscillator output level is desired, a large voltage swing can give rise to problems. A very large voltage swing may saturate transistors in the next stage and cause duty-cycle variation or corrupt the oscillator waveform in other ways. In this case the output level of the oscillator should be reduced or (for example capacitively) attenuated to adapt to the input voltage range of the cascaded stage.

4.7 Phase and amplitude matching

Many receiver architectures that allow a high degree of integration require quadrature signals. An example is the zero-IF architecture, see Figure 4.6. In this architecture, the RF signal is mixed with an I/Q oscillator signal and the resulting baseband signals are further processed in the in-phase and quadrature branches. Quadrature errors (both in phase and amplitude) in the branches or in the quadrature relation of the I/Q oscillator reduce the image rejection ratio (IRR) of a receiver front-end.

Consider the simplified zero-IF receiver from Figure 4.6, for example. The in-phase signal provided by the tuning system can be written as $\cos(\omega_{osc}t)$. When allocating all imperfections to the quadrature signal, this signal can be expressed as $(1+A_e)\sin(\omega_{osc}t+\phi_e)$. The relative amplitude error is A_e in this expression, and the

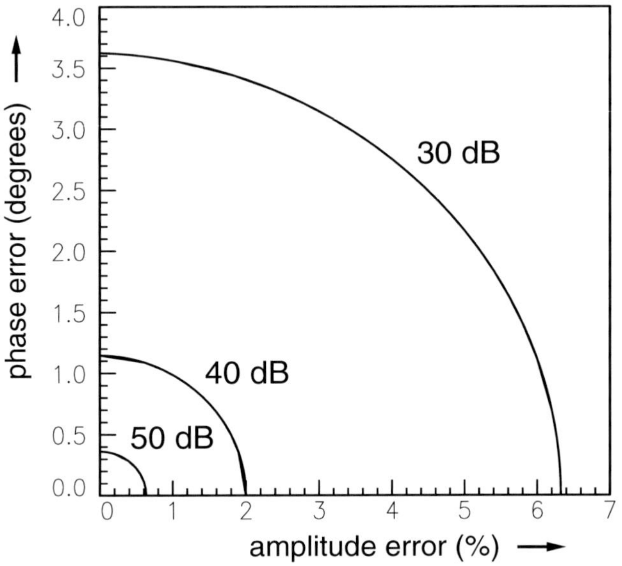

Figure 4.7 IRR resulting from imperfect quadrature signals of the LO in a zero-IF architecture.

phase error ϕ_e is in radians. The resulting IRR in dB due to non-zero A_e and ϕ_e is

$$\mathrm{IRR} = 10\log\left(\frac{4}{((1+A_e)\cos\phi_e - 1)^2 + ((1+A_e)\sin\phi_e)^2}\right) \tag{4.14}$$

Figure 4.7 shows the allowable amplitude error and phase error for 50, 40 and 30 dB IRR. Equation (4.14) can be approximated by

$$\mathrm{IRR} \approx 10\log\left(\frac{4}{A_e^2 + \phi_e^2}\right), \tag{4.15}$$

for small A_e and small ϕ_e.

This approximation can also be used to describe the IRR of image-reject architectures, such as the Hartley architecture [98], as a result of imperfect quadrature signals.

Accurate quadrature signals can be generated by, for example, using multi-stage oscillators (ring or LC) with an even number of stages. Other methods are discussed in Chapter 6.

One example is the two-integrator oscillator from Figure 2.17. Such an oscillator provides "correct-by-construction" I/Q signals, under perfect matching conditions. The symmetry in the topology of the oscillator dictates a 90° phase shift per section. Practical I/Q matching is limited by the device matching and the symmetry of the layout.

4.8 Power dissipation and supply voltage

An oscillator normally is part of a larger system with a restricted power budget. Low power design is important, especially for portable applications. For telecom transceivers, for example, talk and standby time are very important selling features. Hence the stringent phase noise specification for the telecom standards in Table 4.2 must be achieved at very low power levels (typically only several tens of mWs). Minimum power dissipation also is important for applications that are connected to the mains, since the whole system has to fit into a cheap package with a certain thermal resistance. In such a case, low power means the possibility of a higher degree of integration. The power budget of an oscillator is usually specified in milli-watts, or by the available current given by a supply voltage. Additional current is needed to realize output buffers, which form the interface between the oscillator and cascaded blocks such as mixers and dividers. The attainable $\mathcal{L}(f_m)$ levels in an oscillator are directly related to the power dissipation. Several phase noise models that will be discussed in Chapter 6 will explain how $\mathcal{L}(f_m)$ scales with power.

The maximum supply voltage can be dictated by the application or by the technology (e.g. breakdown voltages in CMOS). Obviously, a low supply voltage helps to bring down the dissipation of an oscillator. However, a low supply voltage may also result in a lower tuning range, for example because the varactor voltage in an LC oscillator can only be varied between 0 V and the supply voltage. To make things worse, a lower supply voltage can result in a lower maximum allowable carrier voltage swing, and thus a poorer $\mathcal{L}(f_m)$.

4.9 Supply pushing

A VCO (or CCO) has at least one tuning input and when the tuning voltage (or current) is varied, the frequency of the oscillator varies. Obviously, this is desired behavior and the tuning input has a certain designed tuning constant in Hz/V.

> Each practical oscillator has many unwanted tuning inputs.

A notorious, and generally undesired, tuning input is the power supply terminal of an oscillator. Noise or spurious signals on the power supply line modulate the oscillator and causes phase noise or spurious components, respectively. This phenomenon is called supply pushing.

Like the tuning constant of a VCO, supply pushing of an oscillator is specified in Hz/V. The supply pushing mechanism is illustrated in Figure 4.8. Due to a finite power supply rejection ratio (PSRR) to an arbitrary modulation point in the oscillator

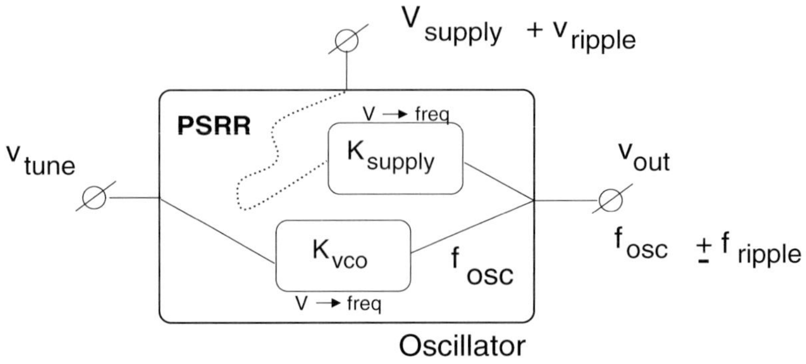

Figure 4.8 Schematic representation of supply pushing. Due to a finite PSRR, a modulation point in the oscillator is modulated by v_{ripple} and generates f_{ripple}.

circuit[7], with tuning sensitivity K_{supply}, the ripple voltage on the supply generates a spurious signal f_{ripple}.

Once the supply pushing constant K_{supply} is known, it is possible to calculate the amount of noise or ripple allowed on the power supply line given a $\mathcal{L}(f_m)$ requirement. Substitution of $\delta f = K_{supply} v_{noise}(f_m)$ and $NBW = 1\,\mathrm{Hz}$ in (4.9), and rearranging of (4.9) to make $v_{noise}(f_m)$ the dependent variable, leads to

$$v_{noise}(f_m) = \frac{\sqrt{2\mathcal{L}(f_m)}f_m}{K_{supply}}, \tag{4.16}$$

in which K_{supply} is specified in $\mathrm{Hz/V}$ and supply line noise voltage $v_{noise}(f_m)$ in $\mathrm{V_{rms}}/\sqrt{\mathrm{Hz}}$. The assumptions made when deriving (4.9) also place restrictions on (4.16) and require that $(K_{supply}v_{noise}(f_m))/f_m \ll 1$ (narrow band FM condition). For example, if $K_{supply} = 3\,\mathrm{MHz/V}$, and the target for $\mathcal{L}(2\,\mathrm{MHz}) = -100\,\mathrm{dBc/Hz}$, the noise voltage $v_{noise}(2\,\mathrm{MHz})$ on the supply line of the oscillator should not exceed $9.4\mu\mathrm{V_{rms}}$. Obviously, $v_{noise}(f_m)$ can be replaced by a ripple voltage $v_{ripple}(f_m)$, which causes (deterministic) spurious signals in the oscillator output spectrum.

4.10 Supply voltage variation, temperature range and process spread

An oscillator design meeting the target specification under nominal conditions is far from ready. All specifications have to be met over a certain temperature range and

[7]For example, the node voltage of a junction capacitor or varactor.

be robust against a certain degree of process spread. These specification aspects are discussed below.

4.10.1 Supply voltage variation

An oscillator design should function correctly within a certain supply voltage range, For example, the supply voltage can be 3 V nominal, but with a tolerance of $\pm 10\%$. The battery voltage of portable equipment will drop slowly to certain minimum voltage during its lifetime. Therefore all circuits in the system, including the oscillator, should be able to cope with the lowest specified voltage. If the nominal supply voltage already is low, e.g. 1 V, then a variation of -10% is a huge amount. Mains connected equipment will have a power supply subject to tolerances itself, and can age. Hence an oscillator design must also be able to cope with these supply variations.

4.10.2 Temperature range

Practical operating temperatures will not be identical to the nice constant junction temperature of 25°C or 27 °C used in simulation. For in-house consumer applications, a junction temperature from 0°C to 80°C is usually specified. The oscillator should meet all other specifications for any temperature in this range. Professional systems such as optical transceivers and systems operating under extreme conditions, such as the LNB of satellite dish, often require a junction temperature range of -40°C to 125°C. The difference between the ambient temperature and the junction temperature of an IC depends on the thermal resistance of the used IC package. The relation between these two temperatures is given by

$$T_{junction} = T_{ambient} + R_{thermal} \cdot P_{dissipated}, \tag{4.17}$$

in which $P_{dissipated}$ is the DC power dissipated in the IC in W and $R_{thermal}$ the thermal resistance of a package in K/W. The thermal resistance of a 8-pin Small Outline (SO) package with a cavity size of 2 mm^2 is 160 K/W, for example.

4.10.3 Process spread

Under nominal processing conditions a resistor designed to be 1 kΩ will be very close to this value. However, the parameters of a resistor and of any other device used in an oscillator is subject to statistical variations caused by variables during IC manufacture, which can be considered stochastic for a large ensemble of chips. This is called process spread. The process spread of a device X normally has a Gaussian distribution, symmetrically distributed around a nominal value with a standard deviation σ_X [99]. To have an acceptable yield, that is the percentage of oscillators which meet the target specification, design for 4 σ_X is common. For example, if the standard deviation $\sigma_{resistor}$ of the 1 kΩ resistor is 5 %, the oscillator in which the resistor is used should meet the specifications for resistors values in the range of 800 Ω to 1200 kΩ.

Component mismatch, or device mismatch is the parameter spread between devices of a similar type, designed to be equal. The process spread of two or more devices that are designed to be equal[8] will match closely. The spread between these devices is much smaller than the process spread and usually better than 1%. Whenever symmetry is required for the optimum performance of an oscillator (for example, for I/Q matching), good component matching is extremely important.

4.11 Technology and chip area

The technology plays an important role in the performance and cost of an oscillator. For example, the inductor quality is much better on high-ohmic substrates compared to low-ohmic substrates. It is therefore much easier to realize high-performance LC oscillators on high-ohmic substrates. The availability of good varactors also is of importance for sufficient tuning range and low phase noise. Technology parameters such as f_T, f_{max} and transistor parasitics like the collector (or drain) capacitances give an indication of how big the influence of parasitics will be on the LC or ring oscillator design, at a certain target oscillation frequency.

> If chip area minimization is important, the use of a large number of large planar coils should be avoided. Ring oscillators usually are much more compact than LC oscillators, but also are much noisier (as a rough rule of thumb, approx. Q^2 times) at comparable power dissipation levels.

4.12 Summary

In this chapter an overview has been given of the most important oscillator specifications. Deriving a clear set of specifications for the oscillator in a transceiver is an important first task in the design process, and requires a good understanding of the system in which the oscillator will be incorporated. The discussed specification items and their units are summarized in Table 4.5.

[8] That means they should be placed close to each other and have the same size and orientation.

Specification	Units
Frequency and tuning range	Hz (rad/s)
Tuning constant	Hz/V (Hz/A)
Tuning linearity	%
Phase noise to carrier ratio	dBc/Hz
Jitter	UI_{rms} (UI_{p-p}, s)
Power dissipation	W
Supply voltage	V
Carrier amplitude	V
Output load	Ω
Harmonics	dBc
Spurious signals	dBc
Supply pushing	Hz/V
Phase matching	°
Amplitude matching	% (dB)
Supply voltage variation	%
Temperature range	°C
Process spread & device matching	%
Technology	Bipolar, CMOS, BICMOS, ...
Chip area	μm^2

Table 4.5 Summary of important oscillator specifications.

5

Elementary properties

THE elementary properties of LC and ring oscillators are studied in this chapter. These properties mostly concern wanted, ideal, oscillator behavior, although we also will encounter properties that can be unwanted. Take for example frequency shifts due to harmonics.

> This chapter explores the designable space of LC and ring oscillators focusing on elementary oscillator properties.

Chapter 6, will study the practical, desired and unwanted properties of oscillators. The exploration in this chapter and Chapter 6 forms the basis for the definition of FOMs in Chapter 7, and provides the oscillator designer with valuable design insights and circuit ideas. Arguably, the elementary properties of LC and ring oscillators are of little interest to an oscillator designer, who needs to realize an oscillator design given all practical constraints imposed by design resources, especially by the IC technology. However, a separation between elementary and practical properties helps to point out the fundamental differences between the oscillator types in integrated transceivers. The basic equations that govern oscillator operation are more transparent, as the abundance of practical issues is left over to be discussed in Chapter 6.

In this “noiseless chapter”, frequency and phase, tuning, the waveform and carrier amplitude and power of LC and ring oscillators are studied in Sections 5.1 to 5.4. A summary of this chapter is given in Section 5.5.

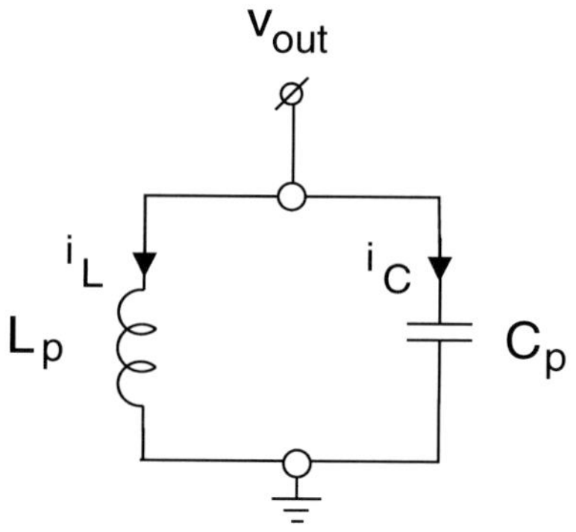

Figure 5.1 A behavioral model of the ideal LC oscillator.

5.1 Frequency and phase

A fundamental property of an oscillator is its frequency. An oscillator can have more than one output that provides this frequency with the same phase or with a certain fixed phase difference.

5.1.1 LC oscillators

First, the frequency of single-phase LC oscillators will be discussed. Second, the frequency and output phases of multi-phase LC oscillators are described.

Single-phase LC oscillators

Figure 5.1 shows the behavioral model of an ideal single-phase oscillator. No losses are present and therefore there is no need for an element to replenish lost energy, such as a transconductance.

> The magnetic energy $W_{L_p} = 1/2 \cdot L_p I_L^2$ (in which I_L is the peak value of $i_L(t)$) stored in the inductor L_p is each period transfered to the capacitor and stored as electrical energy $W_{C_p} = 1/2 \cdot C_p V_L^2$, and vice versa.

Kirchoff's current law applied on this circuit leads to the differential equation

$$\frac{d^2 v_{out}(t)}{dt^2} + \frac{v_{out}(t)}{L_p C_p} = 0, \tag{5.1}$$

and its solution, which was already encountered in Chapter 2

$$v_{out}(t) = V_{carrier} \cos(\, \omega_{LC}\, t + \phi_0), \tag{5.2}$$

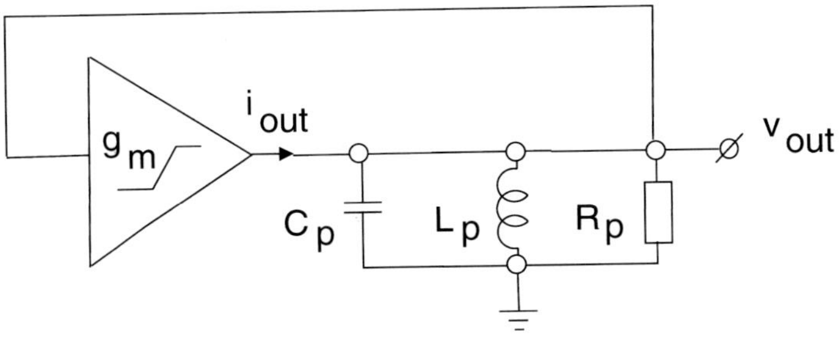

Figure 5.2 A behavioral model an ideal single-phase LC oscillator with a nonlinear transconductor and load R_p.

in which amplitude $V_{carrier}$ and initial phase ϕ_0 depend on the initial conditions. The frequency ω_{LC} is identical to the frequency of the LC oscillator with losses in Figure 2.7 on p. 24,

$$\omega_{LC} = \frac{1}{\sqrt{L_p C_p}} \tag{5.3}$$

Since there are no losses in this model, the parallel LC circuit is the electrical equivalent of a "perpetuum mobile"[1].

In Chapter 2, we discussed the ideal properties of a "black box" oscillator in Section 2.1 and concluded that harmonics can be a desired property of an oscillator. It therefore is of interest to investigate the effect of harmonics generated by the oscillator on the frequency. These harmonics can originate from a linear building block in an oscillator, such as a delay element, or (which is more common) can be generated by nonlinear building blocks in the oscillator. The model in Figure 5.1 contains no nonlinear element (or a delay element for example), no harmonics are present and the oscillation frequency will be always exactly ω_{LC}. In order to investigate the frequency deviation due to harmonics, the single-phase LC behavioral model in Figure 5.2 is introduced. In this oscillator model, a noiseless load resistor R_p is present, to which we want to provide energy. This energy is replenished by a nonlinear transconductor which has a small-signal gain g_m.

The frequency deviation due to the harmonics generated by the nonlinear transconductor, $\Delta\omega_{LC}$ can be assessed using the method of reactive power balance of harmonics [9]. Parameter $\Delta\omega_{LC}$ is defined as the frequency shift with respect to the nominal oscillation frequency ω_{LC} (see (5.3)). We will see that this shift always is negative. The reactive power balance of harmonics applied to the oscillator from Figure 5.2 states that for a time-invariant nonlinear transconductor with a transfer characteristic

[1]Of the third kind: elimination of fraction. A perpetuum mobile of the first kind creates energy.

$I_{out} = f(V_{in})$, the following integral[2] is valid [9, 100]:

$$\oint i_{out} dv_{in} = 0. \tag{5.4}$$

Evaluation of this integral (see [9] for a detailed derivation), for the oscillator from Figure 5.2, yields the following for the frequency deviation $\Delta\omega_{LC}$:

$$\Delta\omega_{LC} = \frac{-1}{2Q_p^2} \sum_{n=2}^{\infty} \frac{n^2}{n^2-1} h_{i(n)}^2 \cdot \omega_{LC}, \tag{5.5}$$

in which the harmonic current content $h_{i(n)}$ is defined as

$$h_{i(n)} = \frac{I_{out(n)}}{I_{out(1)}}, \tag{5.6}$$

where $I_{out(1)}$ is the transconductor's output current amplitude of the fundamental ($n = 1$). Hence $h_{i(1)}$ is equal to unity. Since the maximum (in terms of total power) in harmonic content is reached when i_{out} is a square wave, the maximum frequency deviation can be calculated [100]. Substitution of the amplitude of the harmonics of a square wave in (5.5) results in (5.7)

The maximum frequency deviation due to transconductor nonlinearities is (see also Figure 5.4),

$$\Delta\omega_{LC,max} \approx \frac{-1}{8Q_p^2}\omega_{LC}. \tag{5.7}$$

Note that the frequency deviation always is negative, meaning that the value always is lower than that predicted by linear analysis. The physical explanation for this frequency reduction is provided in [9]. For the oscillator under discussion, the energy stored in the capacitor at resonance is larger than the energy stored in the inductor, when harmonics are present in current i_{out}. To restore the energy balance between W_{C_p} and W_{L_p}, the oscillation frequency must decrease somewhat with respect to ω_{LC}.

To illustrate the frequency deviation predicted by (5.7) with the model in Figure 5.2, the limiting characteristic of the nonlinear transconductor needs to be specified. In the majority of oscillators in integrated transceivers, the amplitude stabilization mechanism is the self-limiting mechanism. Keeping that in mind as well as the fact that the most basic building block in integrated circuit design, a differential amplifier, has a limiting characteristic, a generic choice for the transconductor's transfer characteristic is[3]

$$I_{out}(V_{in}) = \begin{cases} I_{lim} = g_m V_{lim} & \text{if } V_{in} \geq V_{lim}, \\ g_m V_{in} & \text{if } -V_{lim} < V_{in} < V_{lim}, \\ -I_{lim} = -g_m V_{lim} & \text{if } V_{in} \leq -V_{lim}. \end{cases} \tag{5.8}$$

[2] The integral describes the area circumscribed in the I-V characteristic of the transconductor in one oscillation period.

[3] All simulations involving nonlinear transconductors in this chapter are performed using this transfer function.

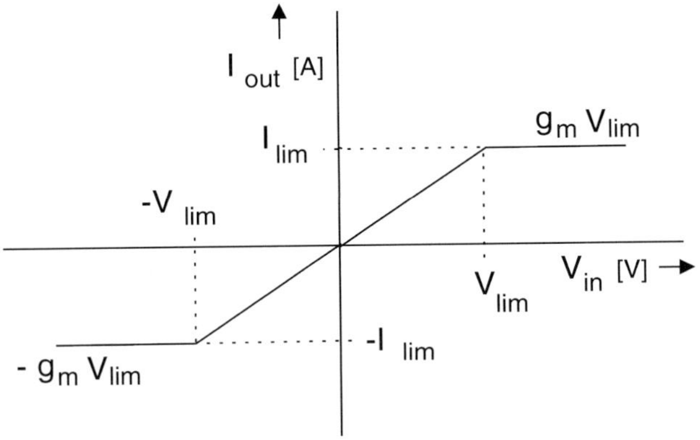

Figure 5.3 An ideal transfer characteristic of a limiting transconductance.

In Appendix C, the limiting $I_{out}(V_{in})$-transfer characteristic of a bipolar and CMOS differential pair is approximated by the piece-wise linear $I_{out}(V_{in})$-curve of the ideal limiter characteristic described by (5.8). A graphical representation of the limiting transfer characteristic is shown in Figure 5.3.

The simulation results of the nonlinear oscillator in Figure 5.2 implemented with a transconductor with the transfer function denoted in (5.8) are shown in Figure 5.4. The oscillation frequency is set to 1 GHz and Q_p to 4. According to (5.7), the frequency deviation should be about 7.8 MHz as a maximum. For increasing open-loop gain $\alpha_{ol} = g_m R_p$, V_{in} (which is identical to the oscillator's output voltage V_{out}) in (5.8) will increase, be larger than V_{lim} most of the time, and the transconductor's output I_{out} current will eventually be a square wave (with maximum harmonic content). Indeed, Figure 5.4 shows that for a very high open loop-gain ($\alpha_{ol} = 50$), the negative frequency shift limit of 7.8 MHz is almost reached.

Multi-phase LC oscillators

Many transceiver architectures require a signal source that provides multiple phases of a certain output frequency. There are many ways to generate multi-phase signals[4]. As we will see in the next section, multi-phase signals are an inherent feature of ring oscillators. Alternatively, for example, a multi-phase LC oscillator can be used to generate a multi-phase output signal.

An elementary behavioral model of multi-phase LC oscillator is shown in Figure 5.5. The model in Figure 5.5 consists of N identical stages. In each stage a phase shifter is present, followed by a transconductor with transconductance g_m that is connected to a parallel LCR resonator (C_p, L_p and R_p). The phase shifter provides a phase shift θ between input and output voltage. This phase shifter is essential for multi-phase outputs in this model, but will also prove most useful in Chapter 6 for $\mathcal{L}(f_m)$

[4]Nine possible implementations for I/Q signal generation are presented in Appendix D.

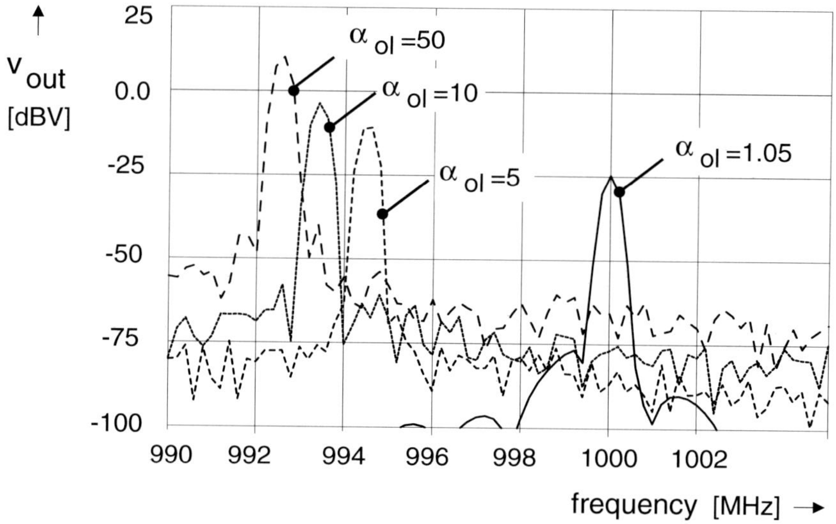

Figure 5.4 Simulations of the frequency deviation of a 1 GHz ideal nonlinear LC oscillator for open-loop gain values α_{ol} equal to 1.05, 5, 10 and 50. The quality factor Q_p is set to 4.

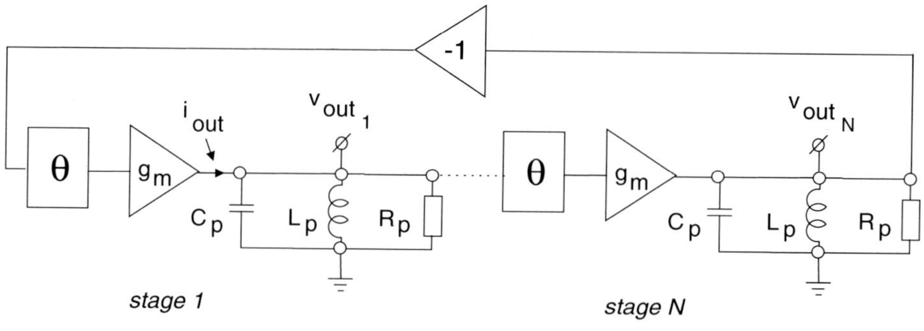

Figure 5.5 A behavioral model of an ideal N-stage LC oscillator.

minimization. Omission of the inverter and the phase shifters in Figure 5.5 leads to a behavioral model of a multi-stage single-phase LC oscillator. This oscillator has no particular advantage over a single-stage single-phase oscillator as will become clear in the next chapter, and will not be discussed here.

The oscillation frequency of the multi-phase LC oscillator can be derived by applying the phase condition for oscillation. Since the inverter provides a 180° phase shift, each stage must provide a phase shift of $\pm 180°/N$ to obey the phase condition for oscillation. Hence, the phase condition for oscillation can be written as

$$\theta + \phi_{res} = \pm 180^\circ/N, \tag{5.9}$$

in which ϕ_{res} is the resonator phase shift. Equation (5.9) shows what happens if $\theta \neq \pm 180^\circ/N$. For example, suppose that $N = 4$, $\theta = 40^\circ$ and all four stages in the multi-phase LC oscillator are perfectly matched. To obey the phase condition for oscillation, the loop will cause ϕ_{res} to be equal to 5°. The phase shift of a parallel LCR resonator can be expressed in terms of Q_p and v, which were defined on p. 24

$$\phi_{res} = -\arctan(vQ_p). \tag{5.10}$$

The oscillation frequency of the N-stage LC oscillator can be calculated by taking $\omega = \omega_{NLC}$ in v and solving (5.10) for ω_{NLC}:

$$\omega_{NLC} = \frac{-\tan(\phi_{res}) + \sqrt{4\,Q_p^2 + \tan^2(\phi_{res})}}{2Q_p} \cdot \omega_{LC}. \tag{5.11}$$

Equation (5.11) shows that for $\phi_{res} = 0^\circ$ the frequency ω_{NLC} of the multi-phase LC oscillator becomes equal to that of a single-phase LC oscillator.

The minimum number of identical stages of the oscillator model to have multi-phase output signals is two. For a two-stage LC oscillator, output voltages v_{out_1} and v_{out_2} will be in quadrature, and $\phi_{res} + \theta = \pm 90^\circ$. The I/Q relation is based on the phase condition for oscillation and symmetry (assuming perfect matching) in this case, and we designate this by the term: "correct-by-construction". When the signal-processing functions in a transceiver require more than two output phases, more LC stages can be used.

5.1.2 Ring oscillators

In this section the frequency and phase of the ideal two-integrator oscillator will be discussed, followed by a derivation of the frequency of an N-stage ring oscillator with $N \geq 3$.

Two-integrator oscillators

A perhaps classic example of a two-stage ring oscillator is the two-integrator oscillator. A behavioral model of this oscillator is shown in Figure 5.6. Two ideal integrators are cascaded, and the output of the second stage is inverted and fed back to the input of the first stage. Other two-stage ring oscillators are conceivable. For example, replacement of the capacitors by inductors in Figure 5.6 yields a "two-differentiator" oscillator. Because the two-integrator oscillator is most often encountered in transceivers, we will limit the discussion of ideal two-stage oscillators to this oscillator. The two-integrator is a ring oscillator: like the N-stage ring oscillator with $N \geq 3$ that will be

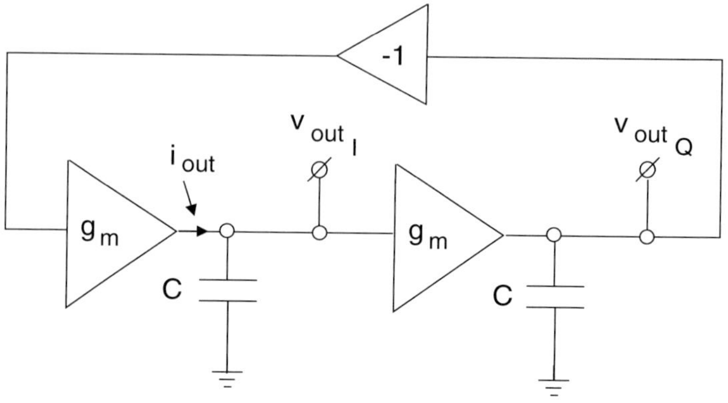

Figure 5.6 A behavioral model of an ideal two-stage ring oscillator; the two-integrator oscillator.

discussed on p. 97, it consists of identical stages connected in a loop. However, a distinguishing property of the two-integrator oscillator is the perfect integration in each stage. Another difference between the two-integrator oscillator and ring oscillators with more stages is the oscillation condition that determines the frequency. We will see that this is the gain condition in the case of the two-integrator oscillator, and the phase condition for ring oscillators with more stages. The above distinctive properties of the two-integrator oscillator justify a separate discussion of this oscillator and ring oscillators with more stages.

Since each stage of the two-integrator oscillator provides a $-90°$ phase shift for any frequency, the phase condition for oscillation is always fulfilled. In combination with perfect matching of g_m's and C's, that also means that output voltages v_{out_I} and v_{out_Q} are in quadrature (and the I/Q relation is correct-by-construction). The amplitude condition for steady-state oscillation determines the frequency of a two-integrator oscillator. Since the magnitude of the open-loop transfer function of this oscillator is

$$|H(j\omega)| = \frac{g_m^2}{\omega^2 C^2}, \tag{5.12}$$

the oscillation frequency can be derived by solving $|H(j\omega)| = 1$ for $\omega = \omega_{two-int}$, which yields

$$\omega_{two-int} = \frac{g_m}{C}. \tag{5.13}$$

The frequency $\omega_{two-int}$ is derived using a linear model and thus is the small-signal oscillation frequency of a two-integrator oscillator. If the transconductors in Figure 5.6 are nonlinear, the nonlinear operation of the oscillator, or in other words generation of harmonics, will influence the oscillation frequency.

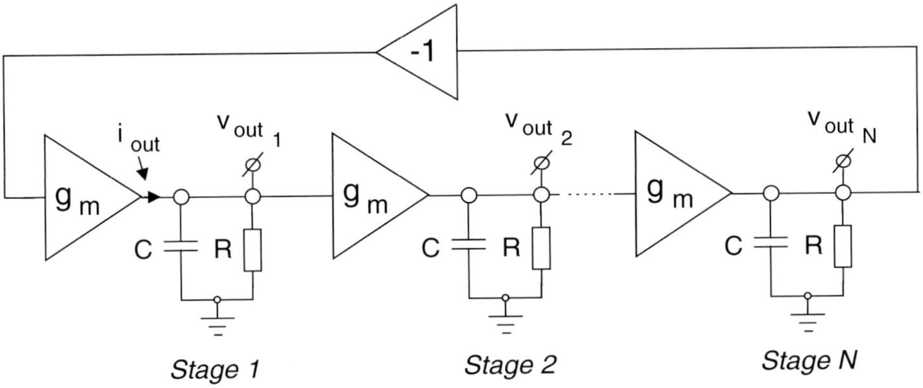

Figure 5.7 An behavioral model of an ideal N-stage ($N \geq 3$) ring oscillator.

The large-signal frequency of a ring oscillator can be expressed in terms of the sum of the charging and discharging propagation delays[5] of the capacitance in a stage, τ_c and τ_d, respectively. If τ_c and τ_d are both equal to τ_{delay}, the large-signal frequency of the two-integrator oscillator is

$$\omega_{two-int,large-signal} = \frac{2\pi}{4\tau_{delay}}. \tag{5.14}$$

When the linear transconductors in Figure 5.6 are replaced by nonlinear ones having the transfer characteristic as described in (5.8), τ_{delay} is easily derived when V_{out_I} (and thus V_{out_Q}) is large compared to V_{lim}. In this case, the transconductors in the two-integrator oscillator are fully switching and are charging and de-charging the capacitor C with a constant current I_{lim} each half period. If $V_{out} = V_{out_I} = V_{out_Q}$ represents the peak voltage on an output terminal of the two-integrator oscillator, τ_{delay} equals $(V_{out}C)/I_{lim}$ and therefore (5.14) becomes for a fully switching two-integrator oscillator

$$\omega_{two-int,large-signal} = \frac{2\pi I_{lim}}{4CV_{out}}. \tag{5.15}$$

N-stage ring oscillators

A behavioral model of an N-stage ring oscillator is shown in Figure 5.7. The resistor in the RC-network in each stage is noiseless in this ideal model. Because of its presence, the phase shift in each stage will always be less than $-90°$, and the minimum of

[5]The charging or discharging propagation delay is defined as the delay-time between the 50% points (50% of the peak-to-peak waveform height) of the input and output waveform of a stage [101].

oscillator stages is three ($N \geq 3$). So at least three phases are provided by this ring oscillator, and in the case of perfect matching the phase relation of the output signals is correct-by-construction. For an even number of stages quadrature signals will be available.

The small-signal oscillation frequency of an N-stage ring oscillator is determined by the phase condition for oscillation. The argument of the open-loop transfer function calculates to

$$\arg(H(j\omega)) = -N \arctan(\omega RC). \tag{5.16}$$

Solving $\arg(H(j\omega)) = -\pi$ for $\omega = \omega_{Nring}$, leads to the oscillation frequency for an N-stage ring oscillator

$$\omega_{Nring} = \frac{\tan(\pi/N)}{RC}. \tag{5.17}$$

Analogous to the discussion of the two-integrator, the actual frequency of an N-stage ring oscillator will deviate from ω_{Nring}, if the transconductors in Figure 5.7 are replaced by ones with a nonlinear transfer characteristic. Assuming again that $\tau_c = \tau_d = \tau_{delay}$, the large-signal frequency of an N-stage ring oscillator can be written as

$$\omega_{Nring,large-signal} = \frac{2\pi}{2N\tau_{delay}}. \tag{5.18}$$

The value τ_{delay} and thus the frequency for an ideal switching N-stage ring oscillator is derived in Appendix E. As shown in this appendix, τ_{delay} can be approximated for large N by $RC\ln(2)$. Accordingly, for a switching N-stage ring oscillator (5.18) approximately becomes

$$\omega_{Nring,large-signal} \approx \frac{2\pi}{2NRC\ln(2)}. \tag{5.19}$$

For $N = 3$ the error made when using (5.19) is large, namely more than 30 %. However, this error for $N = 4$ already is less than 13% and for $N = 6$ it is less than 3%. Figure 5.8 shows the large signal frequency of a perfect switching N-stage ring oscillator divided by the small-signal frequency of a ring oscillator (see (5.17)). The plotted values are listed in Table E.1 on p. 269 of Appendix E. It turns out that the large signal frequency of an N-stage ring oscillator always is higher than its small-signal frequency. For $N = 3$ frequency $\omega_{Nring,large-signal}$ is a factor 1.26 higher. This factor approximates to 1.44 for $N \to \infty$.

5.2 Tuning

Once the frequency of an oscillator is determined, the tuning possibilities can be determined by investigating the oscillator components that determine the oscillation frequency. The tuning possibilities for LC and ring oscillators are studied in this section.

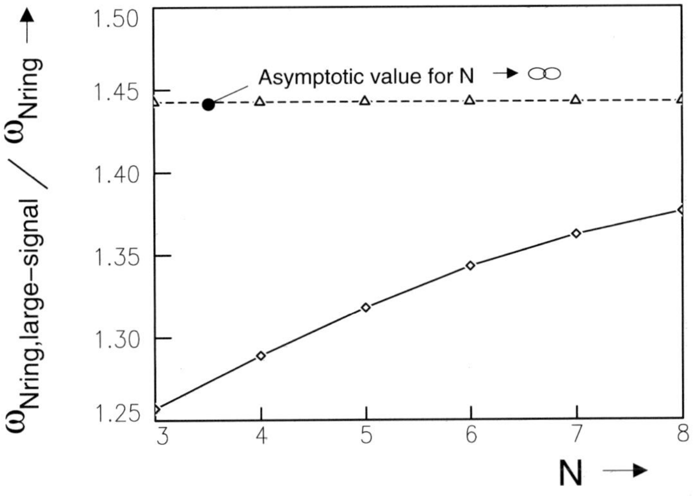

Figure 5.8 $\omega_{Nring,large-signal}/\omega_{Nring}$ versus the number of ring oscillator stages N. The asymptotic value for $N \to \infty$ is ≈ 1.44.

5.2.1 LC oscillators

Tuning methods for single-phase and multi-phase LC oscillators are discussed in the following two subsections.

Single-phase LC oscillators

The frequency of a single-phase LC oscillator is equal to $\omega_{LC} = (\sqrt{L_p C_p})^{-1}$. Therefore tuning of the oscillation frequency can be accomplished by either varying the inductor in Figure 5.1 or the capacitor. In practice, the inductor is not easily varied continuously in integrated circuit design, but in theory it makes no difference whether L_p or C_p or both are varied, to set a certain frequency. The most common method of varying ω_{LC} is to vary capacitance C_p. For example, C_p can be a varactor that is voltage controlled: $C_p(V_{tune})$. The minimum frequency of the tuning range wanted, f_{min}, and the maximum frequency of the tuning range, f_{max}, specify the required capacitance variation $\alpha_{varactor}$

$$\alpha_{varactor} = \frac{C_{p_{max}}}{C_{p_{min}}} = \left(\frac{f_{max}}{f_{min}}\right)^2. \tag{5.20}$$

For example, looking at Table 4.1 on p. 69, a single-phase LC oscillator in a satellite TV front-end would require a varactor with $\alpha_{varactor} = 5.1$, whereas $\alpha_{varactor}$ only needs to be 1.06 nominally in a UMTS receiver.

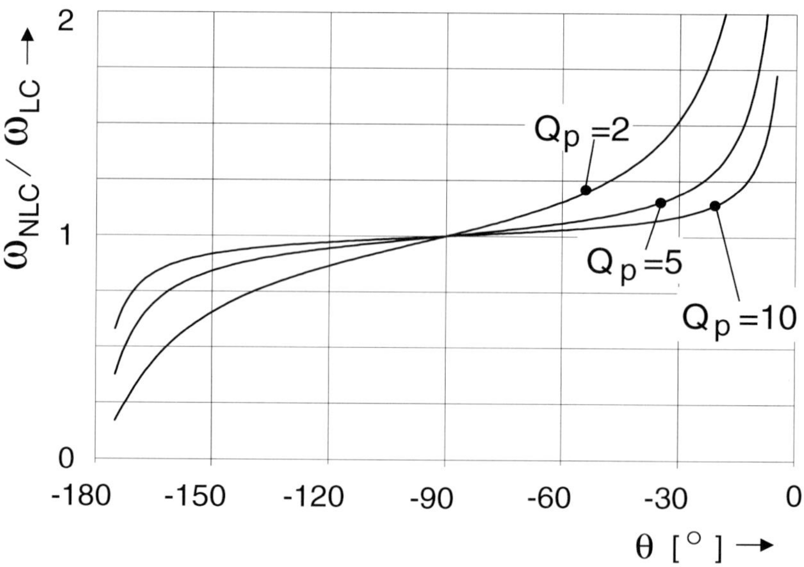

Figure 5.9 The normalized frequency ω_{NLC}/ω_{LC} of a two-stage LC oscillator versus phase-shift θ, for Q_p is 2, 5 and 10.

Multi-phase LC oscillators

A multi-phase LC oscillator can be tuned in exactly the same manner as a single-phase LC oscillator, as (5.11) makes clear. The oscillation frequency ω_{NLC} is proportional to ω_{LC}. If we substitute $\phi_{res} = \pm 180°/N - \theta$ (derived from (5.9)) in (5.11), ω_{NLC} becomes

$$\omega_{NLC} = \frac{-\tan(\pm 180°/N - \theta) + \sqrt{4Q_p^2 + \tan^2(\pm 180°/N - \theta)}}{2Q_p} \cdot \omega_{LC}. \quad (5.21)$$

This means that in addition to changing the inductors and/or capacitors in a multi-phase LC oscillator, phase shift θ can be utilized to vary the oscillation frequency[6]. Jumping ahead to Chapter 6, it is important to note that we will see that this tuning method leads to increased phase noise in multi-phase LC oscillators.

Figure 5.9 shows the normalized frequency ω_{NLC}/ω_{LC} of a two-stage (I/Q) LC oscillator versus θ. At a phase shift $\theta = -90°$ the resonator phase shift is 0°, and therefore the normalized frequency is unity. Effectively, by varying θ, the oscillation

[6] A single-phase LC oscillator can also be tuned in this way, if a controllable phase-shifter is incorporated in the feedback loop.

frequency moves across the resonator phase characteristic, which is normally plotted as phase versus frequency. Hence, the frequency variation achieved with this tuning method is the largest for low values of Q_p, since for low values the slope of the resonator phase characteristic is less steep compared to high values of Q_p.

5.2.2 Ring oscillators

Theoretically, both the two-integrator oscillator and the N-stage ring oscillator as modeled in Figure 5.6 and Figure 5.7 can be tuned using two methods. These methods are discussed in the following two subsections.

Two-integrator oscillators

In two-integrator oscillators, tuning of the oscillation frequency can be achieved by varying g_m or C (see (5.13)). The most common method is variation of transconductance g_m, because implementations of the transconductors in Figure 5.6 (e.g. a bipolar or a CMOS differential pair) can, in general, be varied over a much larger range than tunable implementations of C, as will become apparent in Chapter 6.

N-stage ring oscillators

For the ideal N-stage ring oscillator, capacitive and resistive tuning can be used to vary the oscillation frequency. Theoretically, variation of C or R are both completely equivalent tuning methods; the stage delay in the ring oscillator is varied. However, for the same reason as was mentioned for the two-integrator oscillator, variation of C is less common compared to resistive tuning.

5.3 Waveform

All oscillators under discussion, the single phase LC oscillator (Figure 5.2), the multiphase LC oscillator (Figure 5.5), the two-integrator oscillator (Figure 5.6) and the N-stage ring oscillator (Figure 5.7) can theoretically produce ideal sinusoidal waveforms. In this case the oscillators work completely in linear mode. This requires a proper AGC control to be present that takes care of meeting the gain condition for oscillation, since the self-limiting amplitude stabilization mechanism generates harmonics. On the other hand, when all transconductors in the four oscillator types have the limiting transfer characteristic defined in (5.8), and the output voltage is much larger than V_{lim}, these oscillators work in strongly nonlinear mode. In strongly nonlinear mode the transconductors in the oscillators are fully switching and deliver $-I_{lim}$ or I_{lim} with a 50% duty cycle.

The waveforms of the oscillators under discussion, when operating in linear and in strongly nonlinear (switching) mode, are highlighted in this section. Terminal subscripts (that distinguish oscillator stages) are omitted for clarity, since the voltages and currents have identical amplitudes in every oscillator stage, for multi-stage oscillators. The notations $V_{out(1)}$ and $I_{out(n)}$ denote the voltage amplitude of the fundamental and the amplitude of the nth harmonic of the output current of transconductor, respectively.

5.3.1 LC oscillators

LC oscillators are known for their frequency stability. In Section 5.1, we showed that the quality factor Q_p of an LC resonator is an important parameter, which determines the amount of frequency deviation when an LC oscillator works in strongly nonlinear mode. This quality factor also largely determines the waveform generated by LC oscillators, as will be derived in this section.

Single-phase LC oscillators

The waveform produced by an ideal single-phase LC oscillator working in the linear region is a perfect sine wave. Consider the oscillator model from Figure 5.2. If the transconductor's transfer function is modeled by (5.8), and its output voltage V_{out} is less than V_{lim}, the oscillator operates in linear mode. As mentioned in the introduction to this section on waveforms, a proper AGC control is needed in this case which, for example, controls g_m, such that the average $\alpha_{ol} = g_m R_p$ equals unity. The amplitude at the fundamental frequency ω_{LC} in linear mode is equal to

$$V_{out(1),LC-linear} = I_{out} R_p, \tag{5.22}$$

in which I_{out} is the amplitude of the transconductor's output current.

Analogous to the definition in (5.6), we introduce the harmonic voltage content $h_{v(n)}$, which is defined as

$$h_{v(n)} = \frac{V_{out(n)}}{V_{out(1)}}, \tag{5.23}$$

with $V_{out(1)}$ the voltage amplitude of the fundamental ($n = 1$).

Hence $h_{v(1)}$ is equal to unity. Obviously,

$$h_{v(n),linear} = 0, \tag{5.24}$$

for $n > 1$, for the single-phase LC oscillator working in linear mode.

When the transconductance g_m in Figure 5.2 is increased (and therefore I_{lim} is increased, see (5.8)), V_{out} will eventually become much larger than V_{lim}. The transconductor is now fully switching and its output current alternates between $-I_{lim}$ and I_{lim}. Also, α_{ol} will be much larger than one, and self-limiting can be used as an amplitude stabilization mechanism. The output current of the transconductor, when the oscillator works in this strongly nonlinear mode, is a square wave with Fourier series:

$$i_{out} = \frac{4I_{lim}}{\pi}\left(\sin(\omega_{osc}t) + \frac{\sin(3\omega_{osc}t)}{3} + \frac{\sin(5\omega_{osc}t)}{5} + \ldots\right), \tag{5.25}$$

assuming a zero starting phase. This current is injected into the resonator, and therefore the amplitude of the fundamental and harmonics of the resulting voltage across the resonator, can be written as follows

$$V_{out(n)} = I_{out(n)} \frac{R_p}{\sqrt{1 + (\nu^2 Q_p^2)}}, \tag{5.26}$$

and since for harmonics $\omega = n\omega_{LC}$, ν is equal to

$$\nu = \frac{\omega}{\omega_{LC}} - \frac{\omega_{LC}}{\omega} = \frac{n^2 - 1}{n}. \tag{5.27}$$

From (5.26) it follows that, the amplitude of the fundamental ($n = 1$) of a switching single-phase LC oscillator at frequency ω_{LC} is

$$V_{out(1),switching} = \frac{4}{\pi} I_{lim} R_p, \tag{5.28}$$

because ν is zero for $n = 1$. For large n, the expression $(n^2 - 1)/n$ (see (5.27)) can be approximated by n, and the harmonic voltage content can be written as

$$h_{\nu(n),switching} \approx \frac{1}{n^2 Q_p}, \tag{5.29}$$

for $n = 3, 5, 7, \ldots$ (only odd harmonics).

Notice that a factor $1/n$ in (5.29) originates from the harmonic content of the transconductor's output current (see (5.25)), which decreases with this factor. All even harmonics are zero due to the symmetric characteristic of the transconductance.

Figure 5.10 illustrates the preceding discussion by showing two waveforms of an LC oscillator operating in linear mode, (a) and (c), and two waveforms generated by an LC oscillator operating in strongly nonlinear mode, (b) and (d). Figure 5.10(b) is a waveform of an LC oscillator with $Q_p = 2$. Distortion of the waveform is clearly visible. As predicted by (5.29), the waveform is much more sinusoidal for a higher Q_p, which is illustrated by Figure 5.10(d), which shows the waveform of a switching LC oscillator for $Q_p = 10$.

Multi-phase LC oscillators

The transconductors in the multi-phase LC oscillator model from Figure 5.5 work in linear mode if $V_{out} < V_{lim}$, but can also work in switching mode if $V_{out} \gg V_{lim}$,

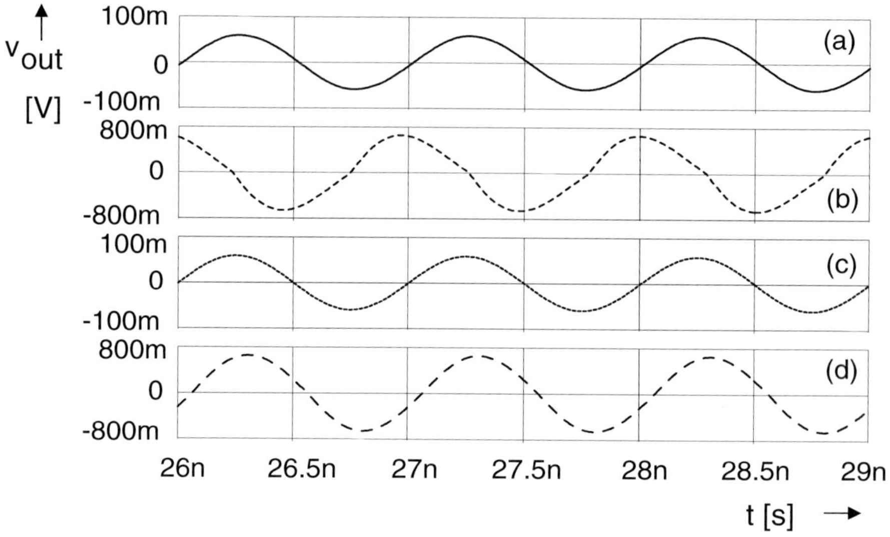

Figure 5.10 Four output waveforms a of single-phase LC oscillator operating in linear mode for $Q_p = 2$ (a), operating in strongly nonlinear mode (i.e.switching) for $Q_p = 2$ (b), linear mode for $Q_p = 10$ (c), and strongly nonlinear mode for $Q_p = 10$ (d).

assuming the N transconductances are nonlinear and modeled by (5.8). The expression for $V_{out(n)}$ can be written as

$$V_{out(n)} = I_{out(n)} \frac{R_p}{\sqrt{1 + ((\frac{n\omega_{NLC}}{\omega_{LC}} - \frac{\omega_{LC}}{n\omega_{NLC}})^2 Q_p^2)}}. \tag{5.30}$$

Substitution of $n = 1$ and ω_{NLC} (see (5.11)) in this expression, yields the following for linear operation of a multi-phase LC oscillator

$$V_{out(1)|_{linear}} = I_{out} R_p \cos(\phi_{res}), \tag{5.31}$$

and all harmonics are zero.

In strongly nonlinear mode, all N transconductors in Figure 5.5 are fully switching and the amplitude of the fundamental at frequency ω_{NLC} is equal to

$$V_{out(1),switching} = \frac{4}{\pi} I_{lim} R_p \cos(\phi_{res}). \tag{5.32}$$

The harmonic voltage content of a multi-stage LC oscillator can derived by dividing (5.30) by (5.32). This gives a rather complex expression that offers little insight. Fortunately, for large n and small resonator phase shift ϕ_{res}, the harmonic voltage content can be approximated by

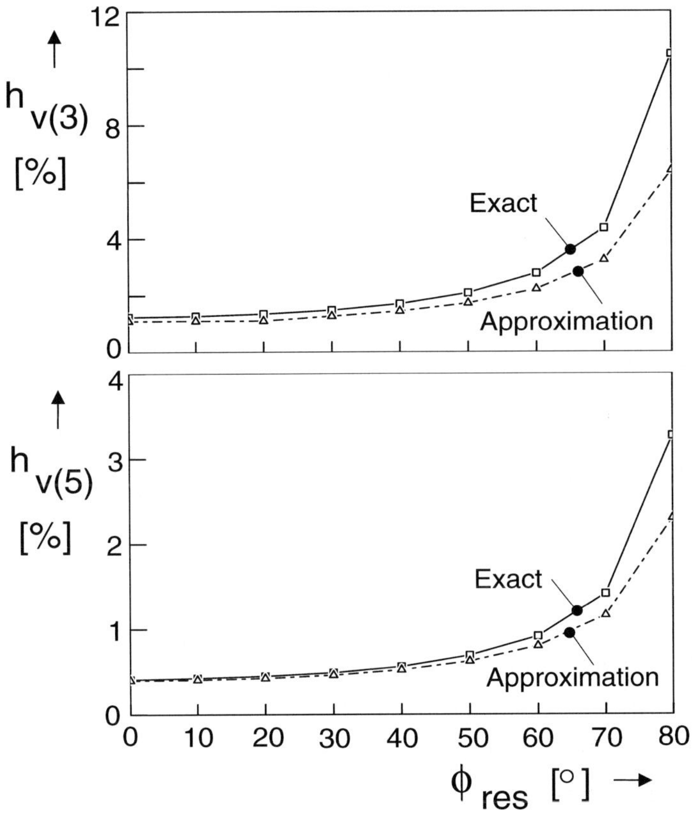

Figure 5.11 Approximated (see (5.33)) and exact harmonic voltage content $h_{v(3)}$, and $h_{v(5)}$ of a multi-stage LC oscillator versus resonator phase shift ϕ_{res}. Quality factor Q_p is set to 10.

$$h_{v(n),switching} \approx \frac{1}{n^2 Q_p \cos(\phi_{res})} \tag{5.33}$$

for $n = 3, 5, 7, \ldots$ (only odd harmonics).

The exact value for $h_{v(n)}$ and its approximation denoted in (5.33) are plotted versus ϕ_{res} in Figure 5.11, for n is 3 and 5. Figure 5.11 shows that (5.33) is quite accurate for small ϕ_{res} and, most importantly, that the harmonic signal content can increase drastically for large ϕ_{res}. In other words, zero resonator phase shift is optimum for minimum $h_{v(n)}$ of multi-phase LC oscillators.

5.3.2 Ring oscillators

The two-integrator oscillator as well as the N-stage ring oscillator theoretically can produce undistorted sine waves, when a perfect AGC control loop takes care of the conditions for steady-state oscillation. In this section, the waveforms of these oscillators will be discussed when operating in this linear mode, as well as the waveforms generated when operating in strongly nonlinear mode.

Two-integrator oscillators

In linear operation mode the two-integrator oscillator (see Figure 5.6) output voltage simply is

$$V_{out(1),linear} = I_{out}\frac{1}{\omega_{two-int}C} = I_{out}\frac{1}{g_m}, \tag{5.34}$$

and all harmonics are zero. When the two transconductors have a limiting characteristic as denoted (5.8) and are fully switching, the amplitude of the fundamental at frequency $\omega_{two-int}$ is equal to

$$V_{out(1),switching} = \frac{4}{\pi}I_{lim}\frac{1}{\omega_{two-int}C}. \tag{5.35}$$

The harmonic voltage content can be calculated by realizing that $V_{out(n)}$ is equal to $V_{out(1)}$ divided by $n\omega_{two-int}C$, and therefore is

$$h_{v(n),switching} = \frac{1}{n^2}, \tag{5.36}$$

for $n = 3, 5, 7, \ldots$ (only odd harmonics).

Figure 5.12 shows the waveforms generated by the two-integrator oscillator in linear and strongly nonlinear mode. A perfectly triangular waveform is generated when the transconductors in the oscillator are fully switching. Note that compared with (5.29), the harmonic voltage content of a two-integrator oscillator has one parameter less: the quality factor (which is effectively one).

***N*-stage ring oscillators**

In linear operation mode the N-stage ring oscillator modeled in Figure 5.6 will generate a sine wave that has an amplitude

$$V_{out(1),linear} = I_{out}\frac{R}{\sqrt{1+\tan^2(\pi/N)}} = I_{out}\frac{1}{g_m}, \tag{5.37}$$

and no harmonics are generated in this case. In the case the N transconductors in Figure 5.6 have a nonlinear characteristic described by (5.8), and are switching, the

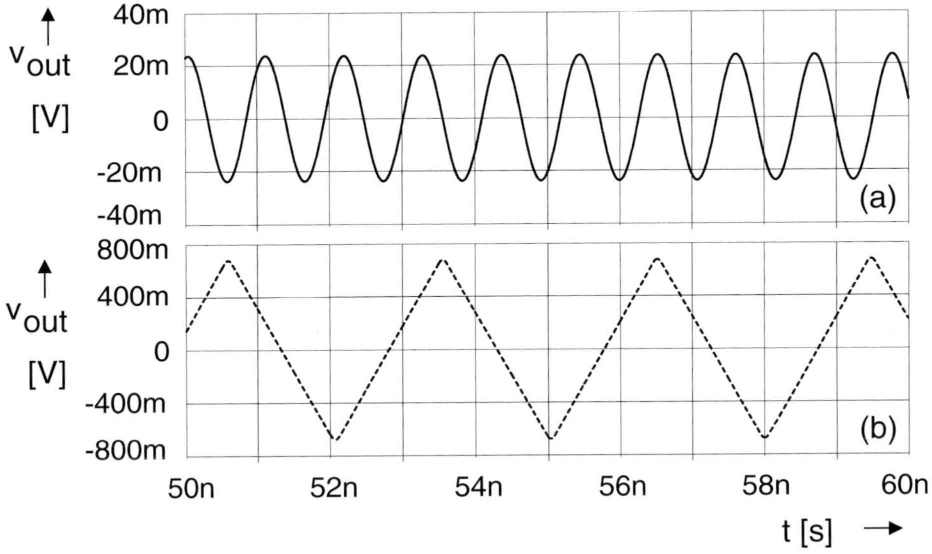

Figure 5.12 Two output waveforms of the two-integrator oscillator when the two transconductors operate in the linear region (a), and when they are switching (b).

amplitude of the fundamental at oscillation frequency ω_{Nring} is approximately

$$V_{out(1),switching} \approx \frac{4}{\pi} I_{lim} \frac{R}{\sqrt{1+(\frac{\pi}{N\ln(2)})^2}}, \tag{5.38}$$

for $N \geq 3$. For $N \gg 3$, $V_{out(1)}$ approximates $4/\pi \cdot I_{lim}R$. Equation (5.38) is derived by multiplication of the fundamental amplitude of the transconductor output current (a square wave) with the transfer function of a parallel RC circuit, with $\omega = \omega_{Nring}$. As shown in Appendix E, for $N = 5$ and larger, ω_{Nring} is approximated with quite good accuracy by (5.19). Following the same route as was followed for (5.38) for deriving $V_{out(n)}$ of a switching ring oscillator, and applying the definition for $h_{v(n)}$ yields

$$h_{v(n),switching} \approx \frac{\sqrt{1+(\frac{\pi}{N\ln(2)})^2}}{n\sqrt{1+(\frac{n\pi}{N\ln(2)})^2}}, \tag{5.39}$$

for $n = 3,5,7,\ldots$ (only odd harmonics).

Especially for a large number of stages, the ring oscillator has a higher harmonic voltage content compared to the two-integrator oscillator.

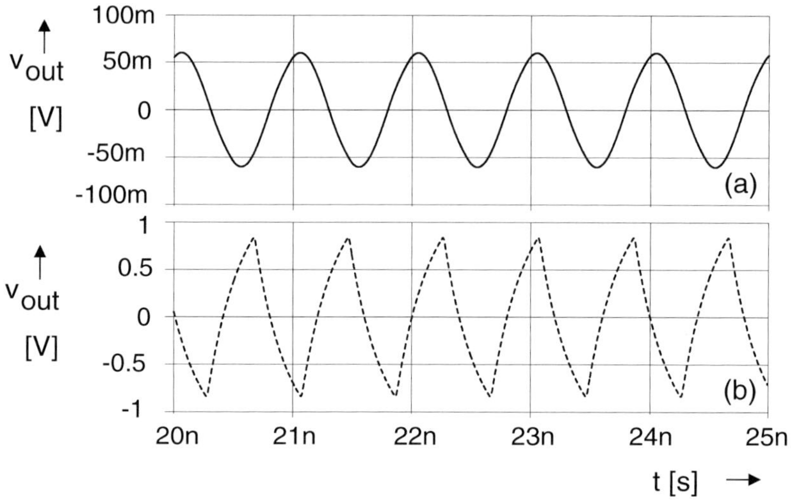

Figure 5.13 Two output waveforms of a three stage ring oscillator; when the transconductors operate in the linear region (a), and when they operate in the strongly nonlinear region (i.e. are switching) (b).

Figure 5.13 illustrates the preceding discussion by showing the sine wave produced by a three-stage ring oscillator in linear mode, and the exponential waveform of a switching three-stage ring oscillator.

5.4 Carrier amplitude and power

The carrier amplitudes of four oscillator types discussed in this chapter were already derived in Section 5.3, and are repeated in Table 5.1. Conversion of the peak current values in Table 5.1 to rms values, and squaring the obtained rms values and multiplication by the impedance seen by the transconductor in the (ideal) oscillator, gives the delivered carrier power at the fundamental frequency.

5.5 Summary

In this chapter, the elementary properties of single-phase LC oscillators, multi-phase LC oscillators, two-integrator oscillators and ring oscillators (with $N \geq 3$) have been discussed. All these oscillator types provide multi-phase output signals, except for the single-phase LC oscillator. An oscillator with an even number of stages provides

Oscillator	linear operation	strongly nonlinear operation
single-phase LC	$I_{out}R_p$	$\frac{4}{\pi}I_{lim}R_p$
multi-phase LC	$I_{out}R_p\cos(\phi_{res})$	$\frac{4}{\pi}I_{lim}R_p\cos(\phi_{res})$
two-integrator	$I_{out}\frac{1}{gm}$	$\frac{4}{\pi}I_{lim}\frac{1}{\omega_{two-int}C}$
N-stage ring	$I_{out}\frac{1}{gm}$	$\approx\frac{4}{\pi}I_{lim}\frac{R}{\sqrt{1+(\frac{\pi}{N\ln(2)})^2}}$

Table 5.1 Carrier amplitudes, in peak voltages, of the oscillator types under discussion for linear operation and strongly nonlinear (switching) operation.

correct-by-construction I/Q signals, if all oscillator stages are perfectly matched. The small-signal frequency and large-signal frequency for the four oscillator types have been derived. Interestingly, the large-signal frequency of LC oscillators is lower compared to the small-signal value, whereas it is higher for an ideal N-stage ring oscillator. This frequency deviation from the small-signal frequency is a function of the quality factor Q_p for LC oscillators, and can therefore be extremely small for high Q_p.

Tuning of LC oscillators can be accomplished by varying the inductance or capacitance of the resonator. In addition, a third tuning option exists. By varying the phase shift in an LC oscillator, the resonator phase shift becomes nonzero (which is necessary to meet the phase condition for oscillation), and the frequency will vary. Since the slope of the phase-frequency characteristic of an LC resonator is a function of Q_p, this frequency variation can be large for low Q_p values, but it is small for high Q_p values. The cost of this method is increased phase noise, as will become apparent in the next chapter. In the two-integrator oscillator and the N-stage ring oscillator, tuning is performed by varying transconductance g_m or the capacitance of a stage, and time constant RC, respectively.

All discussed oscillator types can generate perfect sine waves. LC oscillators working in strongly nonlinear mode, but having a high Q_p, still generate almost perfect sine waves. In LC oscillators, the degree of attenuation of harmonics is proportional to Q_p. When the two transconductors in a two-integrator oscillator are switching, the ideal two-integrator oscillator generates a triangular waveform. Under the same conditions, the N-stage ring oscillator produces an exponential "saw-tooth-like" waveform that has the highest harmonic content of the four oscillator types.

When the transconductors in the discussed oscillator types work in their linear region (linear mode operation), a proper AGC-control must be present, for stable oscillation (i.e. the gain condition for steady-state oscillation must be met). If the AGC-set level results in an output current I_{out}, the peak output voltage of an oscillator simply is this current multiplied by the impedance at the oscillation frequency, connected to the output of the transconductor. In strongly non-linear operation, the transconductors in an oscillator switch between the maximum output current levels $-I_{lim}$ and I_{lim}. Obviously, harmonics are generated and self-limiting can be used for amplitude stabilization. The peak amplitude in this case is $4/\pi \cdot I_{lim}$ times the impedance at the oscillation frequency, seen at the transconductor's output.

6

Practical properties

IN PRACTICE, oscillator properties are non-ideal and much more complex compared to the elementary properties discussed in the previous chapter. If this had not been the case, this book would have been a lot shorter, and the life of an oscillator designer probably boring. Parasitics reduce the maximum achievable frequency, parasitics reduce the tuning range, non-ideal varactors make the tuning characteristic nonlinear, noise corrupts the spectral purity, mismatch gives rise to I/Q mismatch, etcetera, etcetera.

> In this chapter, practical properties of LC oscillators and ring oscillators are described, both on a behavioral and on a circuit level.

Not all properties are discussed both qualitatively and quantitatively as is shown in the overview of this chapter, see Table 6.1. The properties of oscillators in transceivers which, in general, present the greatest challenge to the oscillator designer are discussed both qualitatively and quantitatively. Other properties, such as I/Q matching, tuning of ring oscillators and practical oscillator waveforms are addressed only qualitatively. Sections 6.1 and 6.2 describe frequency and phase, and tuning range, respectively. The discussion on phase noise to carrier ratio is split into two parts. In Section 6.3, the linear time invariant (LTI) system level modeling from Section 3.4.1 is used to derive insightful $\mathcal{L}(f_m)$ expressions for oscillators, in which all devices work in the linear region (linear mode). In general, low phase noise oscillators work in (strongly) nonlinear mode and several phase noise generation mechanisms, such as up and down-conversion, add to the total phase noise. On the other hand, modulation of noise sources in a practical oscillator can improve $\mathcal{L}(f_m)$. These linear time-variant

Practical property	Qualitatively discussed	Quantitatively discussed	Section	Page
Frequency of			6.1	113
single-phase LC oscillators	$\surd$	$\surd$	6.1.1	113
multi-phase LC oscillators	$\surd$	$\surd$	6.1.2	118
two-integrator oscillator	$\surd$	$\surd$	6.1.3	123
N-stage ring oscillators	$\surd$	$\surd$	6.1.4	128
Phase, I/Q-matching of			6.1	113
multi-phase LC and ring oscillators	$\surd$		6.1	113
Tuning of			6.2	132
LC oscillators	$\surd$	$\surd$	6.2.1	133
ring oscillators	$\surd$		6.2.2	152
LTI $\mathcal{L}(f_m)$ modeling of			6.3	155
LC oscillators	$\surd$	$\surd$	6.3.1	156
ring oscillators	$\surd$	$\surd$	6.3.2	165
LTV and nonlin. $\mathcal{L}(f_m)$ modeling of			6.4	169
LC oscillators	$\surd$	$\surd$	6.4	169
ring oscillators	$\surd$	$\surd$	6.4	169
Waveform of			6.5	177
LC oscillators	$\surd$		6.5	177
ring oscillators	$\surd$		6.5	177
Carrier amplitude and power of			6.6	178
LC oscillators	$\surd$		6.6	178
ring oscillators	$\surd$		6.6	178
Power dissipation of			6.7	181
LC oscillators	$\surd$	$\surd$	6.7	181
ring oscillators	$\surd$	$\surd$	6.7	181

Table 6.1 Overview of the practical properties covered only qualitatively or both qualitatively and quantitatively in this chapter.

(LTV) and nonlinear issues concerning $\mathcal{L}(f_m)$ modeling will be discussed in Section 6.4. Section 6.5 up to and including Section 6.7 cover waveform, carrier amplitude and carrier power, power dissipation and supply voltage, respectively. This chapter is concluded by a summary in Section 6.8.

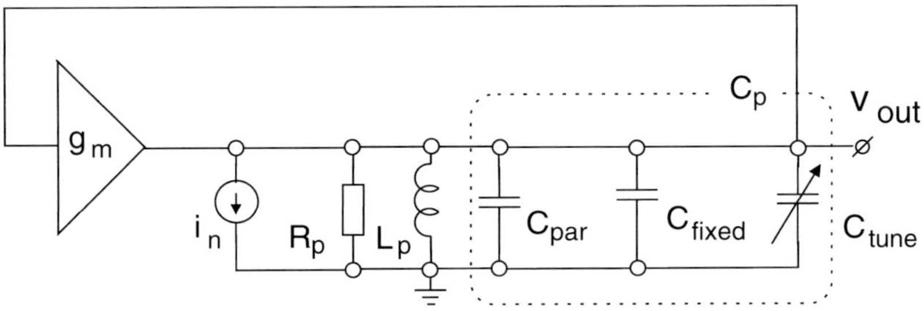

Figure 6.1 LC oscillator model tunable with varactor C_{tune} and with parasitic tank capacitance C_{par}.

6.1 Frequency and phase

The oscillation frequency of single-phase and multi-phase LC and ring oscillators is highlighted in this section. For both types of oscillators, circuit implementations of oscillator topologies often used in transceivers are discussed to illustrate the influence of parasitic components on the frequency. Phase matching is briefly discussed too.

6.1.1 Single-phase LC oscillators

Figure 6.1 shows the linear LC oscillator model, which will be used to investigate the frequency of a single-phase LC oscillator. Compared to the ideal oscillator from Figure 5.2, R_p is introduced as a parasitic element (not a load resistor any more), additional parasitic capacitance is introduced and noise source i_n is added. Varactor C_{tune}, fixed capacitance C_{fixed} and parasitic capacitance C_{par} form the total tank capacitance C_p. Transconductance g_m is needed to compensate the losses of the resonator modeled by R_p. All noise sources are lumped together and modeled by noise current source i_n, which will be of interest later on in Section 6.3.

In Figure 6.1, all losses are lumped in resistor R_p, which simplifies calculations and leads to the quality factor Q_p of the resonator as defined in (2.13) on p. 24. This includes the series resistance of practical coils[1] and capacitors as shown in Figure 6.2(a).

Using the component quality factor of an inductor $Q_{ls} = \omega L_s / R_{ls}$ and of a capacitor $Q_{cs} = 1/(\omega C_s R_{cs})$, R_p, L_p and C_p can be expressed with the circuit elements from Figure 6.2(a). The parallel resistor R_p can be written as

$$R_p = \frac{R_{cs}\left(1+Q_{cs}^2\right) R_{ls}\left(1+Q_{ls}^2\right)}{R_{cs}\left(1+Q_{cs}^2\right)+R_{ls}\left(1+Q_{ls}^2\right)}, \tag{6.1}$$

[1]The model of the practical coil in Figure 6.2 is a simplified model. In a somewhat more realistic model a capacitor is present in parallel with the series connection of R_{ls} and L_s, that determines the self-resonance frequency of the coil. We assume that this capacitor is included in C_{par} of Figure 6.1. See [102] for a physically based model of planar spiral inductors which includes losses due to self resonance and the substrate.

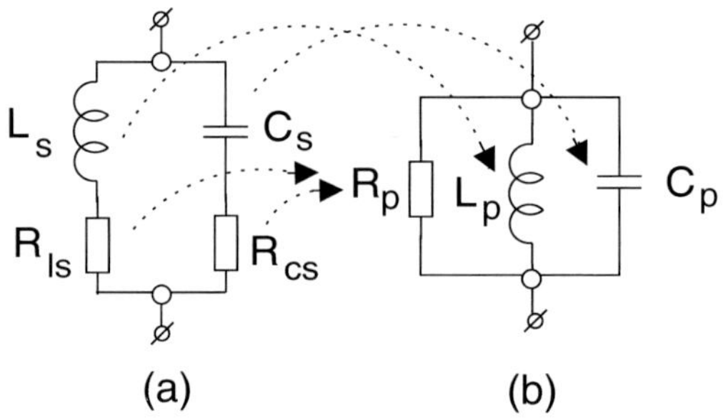

Figure 6.2 A simple model of a practical resonator (a) and the resonator model with losses lumped in R_p (b).

and

$$L_p = L_s \frac{1+Q_{ls}^2}{Q_{ls}^2}, \tag{6.2}$$

and finally

$$C_p = C_s \frac{Q_{cs}^2}{1+Q_{cs}^2}. \tag{6.3}$$

Note that for high component quality factors Q_{ls} and Q_{cs}, L_s and C_s are accurately represented by L_p and C_p, respectively.

The oscillation frequency of the behavioral model in Figure 6.1 is equal to

$$\omega_{LC} = \frac{1}{\sqrt{L_P (C_{tune} + C_{fixed} + C_{par})}}. \tag{6.4}$$

Both L_p and C_p (the sum of all three capacitances) can be used to tune the oscillator. Inductor L_p is usually fixed. Capacitor C_{fixed} is an intentionally added capacitor (which often is zero in practice because it reduces the tuning range, see Section 6.2.1) to improve the quality factor of C_{tune}, for example. Fixed capacitance normally has a better quality factor than a varactor. If this capacitor and C_{tune} are set at zero, the maximum angular oscillation frequency $1/\sqrt{L_p C_{par}}$ is reached. Capacitance C_{par} consists of several contributions: the active oscillator part, parasitic capacitance in the resonator (for example, from the inductor), capacitance of cascaded circuits (loading) and inter-connect capacitance. Clearly, L_p and C_{par} must be made small for a high maximum frequency (assuming the starting condition for oscillation is met).

The frequency of an oscillator with the practical tank circuit –without additional parasitic capacitance– shown in Figure 6.2(a) is equal to

$$\omega_{LC_{prac}} = \frac{\sqrt{L_s - C_s R_{ls}^2}}{\sqrt{C_s L_s}\sqrt{L_s - C_s R_{cs}^2}}, \tag{6.5}$$

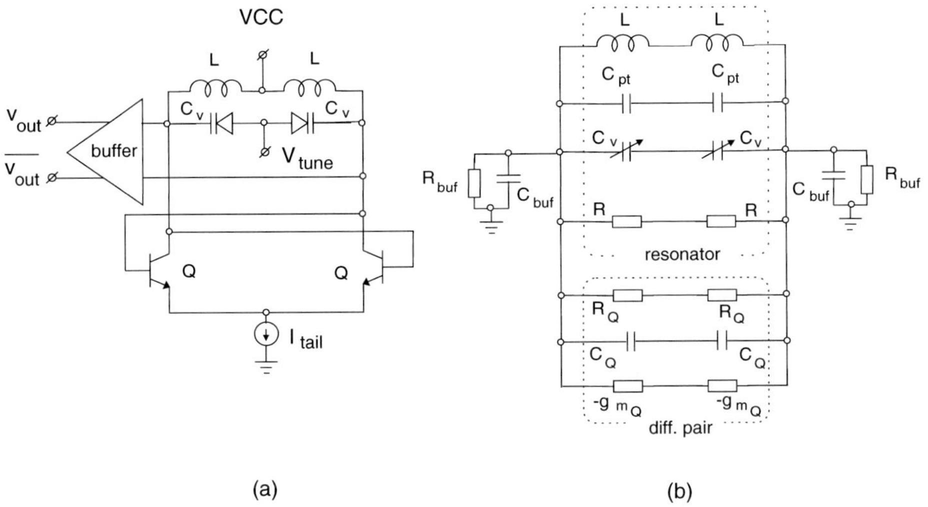

Figure 6.3 Bipolar LC oscillator implementation (a) and its equivalent small signal differential mode circuit (b).

which can be rewritten in terms of Q_{ls} and Q_{cs} as

$$\omega_{LC_{prac}} = \frac{1}{\sqrt{L_s C_s}} \sqrt{\frac{1 - \frac{1}{Q_{ls}^2}}{1 - \frac{1}{Q_{cs}^2}}}. \tag{6.6}$$

Equation (6.6) shows that low quality factors cause a frequency shift of the nominal oscillation frequency $1/\sqrt{C_s L_s}$ for $Q_{cs} \neq Q_{ls}$. Energy losses in the inductive branch cause the frequency to decrease, whereas losses in the capacitive branch cause an increase of the frequency [9]. However, for moderate (say >4) to high quality factors, $\omega_{LC_{prac}}$ is accurately described by $1/\sqrt{C_s L_s} \approx \omega_{LC}$.

Single-phase LC oscillator circuit examples

Figure 6.3(a) presents a cross-coupled pair LC oscillator topology, implemented in bipolar technology. The cross-coupled pair realizes a negative resistance, which compensates for the losses of the tank circuit and other losses such as the cascaded buffer. Therefore, negative resistance modeling discussed in Section 2.4.2 on p. 29 can be used for this oscillator topology. In Figure 6.3(b), its equivalent small-signal model is shown, which will be used to assess the oscillation frequency.

In principle, the maximum oscillation frequency of an LC oscillator circuit is equal to the f_{MAX} of an IC technology used for implementation[2]. In Appendix F, f_{MAX}

[2]The frequency where the maximum available power gain becomes unity.

expressions for bipolar and MOS technology are given. The literature shows that this frequency can indeed be reached (e.g. 50 GHz for 0.25μm CMOS technology [103]). However, practical applications of interest (see for example Table 4.1 on p. 69) operate at much lower frequencies, and f_{MAX} seldom is a limiting factor. Nevertheless, a higher f_{MAX} means lower parasitics, which can result in a greater tuning range (see Section 6.2).

The equivalent model in Figure 6.3(b) consists of three parts: the differential pair, the resonator, and the loading of the buffer. To start with the first part, the half-circuit of the differential pair is modeled with three elements: resistance R_Q, capacitance C_Q and transconductance $-g_{m_Q}$. The value of these elements can be derived by using a simplified small-signal transistor model, such as the model introduced in Appendix F. If the base resistance of this model (Figure F.1 on p. 272) is neglected, R_Q, C_Q and $-g_{m_Q}$ in Figure 6.3(b) can be calculated as[3]

$$R_Q = R_\pi, \tag{6.7}$$

$$C_Q = C_\pi + 4C_\mu + C_{js}, \tag{6.8}$$

$$-g_{m_Q} = -g_{m_{bipo}}. \tag{6.9}$$

As pointed out in Appendix F, adaptation of above equations to a MOS implementation is straightforward. Thus, $R_Q = \infty$, $C_Q = C_{gs} + 4C_{gd} + C_{db}$ and $-gm_Q = -g_{m_{MOS}}$, if the gate resistance is neglected. The half-circuit of the resonator is modeled with inductor L, varactor C_v, loss resistor R and parasitic capacitance C_{pt}. The latter models all parasitic resonator capacitance, which includes interconnect. Loading of the oscillator in Figure 6.3(a) by cascaded circuits (e.g. a buffer, mixer or divider) is modeled with resistor R_{buf} and capacitor C_{buf}

Equation (6.4) can be used to calculate the oscillation frequency of the model in Figure 6.3(b) with

$$C_{tune} = \frac{1}{2}C_v, \tag{6.10}$$

$$C_{par} = \frac{1}{2}(C_\pi + 4C_\mu + C_{js} + C_{pt} + C_{buf}), \tag{6.11}$$

$$L_p = 2L. \tag{6.12}$$

The frequency predicted by (6.4) shifts somewhat due to transistor nonlinearities, as we have seen in Section 5.1.1 on p. 90. This frequency shift is a function of the quality factor of the resonator (see (5.7) on p. 92) and the open loop gain (see Figure

[3]Bipolar and MOS transistor parameters are defined in the Glossary and in Appendix F. Therefore, introduction of these parameters is omitted here.

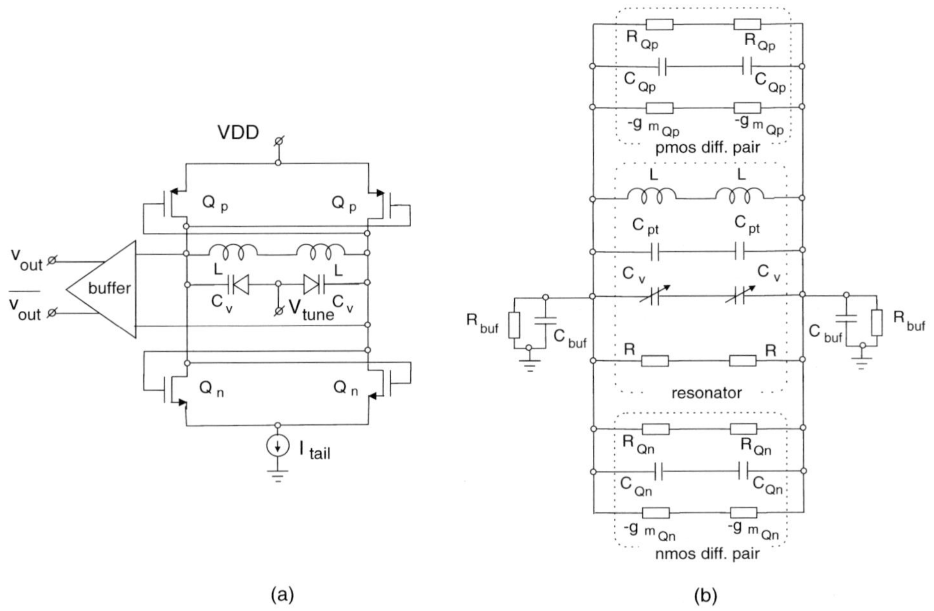

Figure 6.4 Complementary MOS LC oscillator implementation (a) and its equivalent small-signal differential mode circuit (b).

5.4 on p. 94). Figure 6.3(b) reveals that the resonator, with unloaded quality factor $Q_p = R/(\omega L)$ is loaded by resistances R_Q and R_{buf}. The loaded resonator Q can be written as

$$Q_{P_{loaded}} = \frac{(R_Q \parallel R_{buf})}{(R_Q \parallel R_{buf}) + R} Q_p. \tag{6.13}$$

Loaded quality factors should be used for practical oscillators when assessing the frequency shift with (5.7). Note however that for practical open loop gain values (e.g. 3) and loaded quality factors (e.g. 4 or higher) the frequency shift due to transistor non-linearities is less than 0.5% (see p. 94), and of no concern in general. In Appendix A, the importance of a high resonator quality factor for low-noise low-power oscillators was stressed. Equation (6.13) shows that for a high loaded quality factor not only Q_p must be as high as possible, but the loading of the resonator must be minimized as well.

Figure 6.4(a) shows a complementary MOS LC oscillator [76]. The equivalent model of this oscillator in Figure 6.4(b) directly reveals that compared to a simple cross-coupled pair more parasitics are present. The oscillation frequency of this oscillator is given by (6.4) with

$$C_{tune} = \frac{1}{2} C_v, \tag{6.14}$$

$$C_{par} = \frac{1}{2}(C_{gs_n} + 4C_{gd_n} + C_{db_n} + C_{gs_p} + 4C_{gd_p} + C_{db_p} + C_{pt} + C_{buf}) \tag{6.15}$$

$$L_p = 2L, \tag{6.16}$$

if the gate resistance of the NMOS and PMOS transistors are neglected (and some other parasitics, see Appendix F for the model used). The higher parasitics of the complementary MOS implementation are a disadvantage that can reduce the tuning range as we will see, but the architecture also has advantages. In Section 6.4.2 we will see that the symmetry of the topology (if $g_{m_{Qn}}$ and $g_{m_{Qp}}$ are optimized for optimum waveform symmetry) results in a low $f_{1/f}$ corner of the phase noise [104]. Furthermore, the total differential transconductance $1/2 \cdot (g_{m_{Qn}} + g_{m_{Qp}})$ is doubled and so is the carrier amplitude compared to the oscillator in Figure 6.3(a), if $g_{m_{Qn}} = g_{m_{Qp}}$ [105].

In Appendix G, the performance of a number of single-phase LC oscillator designs from literature is listed.

6.1.2 Multi-phase LC oscillators

As pointed out earlier, in Chapter 5, many modern transceiver architectures require quadrature signals. Some architectures may even require more than two phases.

There are many ways of generating quadrature signals. Nine implementation examples are listed in table form in Appendix D.

Moreover, the properties of each implementation are described qualitatively in this appendix. Even-stage ring oscillators are often used, because they are easy to integrate, have a compact chip area and a wide tuning range. However, as will be shown in Section 6.3, LC oscillators have superior phase noise performance compared to ring oscillators. Another common method for I/Q signal generation is the use of a single-phase LC oscillator running at twice the desired oscillation frequency with 50% duty cycle, followed by a divider. Here, duty-cycle errors are a limiting factor for the I/Q matching. For this reason, if accurate multi-phase signals are needed in combination with a very small $\mathcal{L}(f_m)$, the use of multi-phase LC oscillators can be the best option for a particular transceiver design.

Figure 6.5 shows the model that will be used to assess the frequency of practical multi-phase LC oscillators, and later on to calculate $\mathcal{L}(f_m)$ of this linear model. Compared to the ideal multi-phase LC oscillator model in Figure 5.5, transconductance

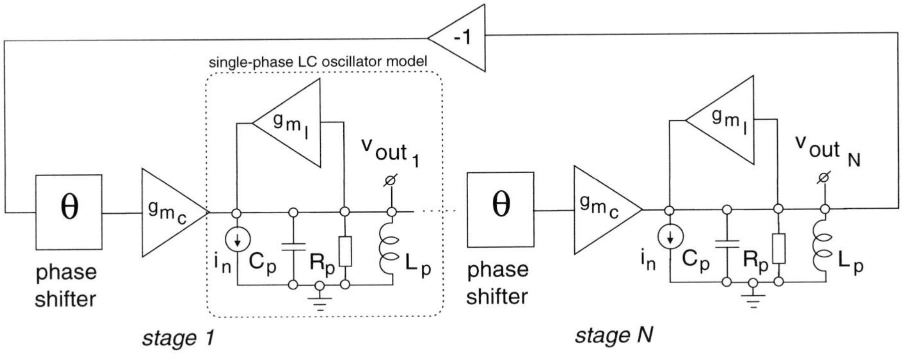

Figure 6.5 Multi-phase LC oscillator.

g_{m_l} is added which compensates the losses modeled by R_p. It also provides an additional control, which in practice is commonly used to set the oscillation level. Note, however, that the presence of transconductances g_{m_l} is not mandatory to meet the oscillation conditions, provided phase shift θ is nonzero. Furthermore, noise source i_n is added and represents the noise generated by one stage. Basically, the model is formed by N single-phase LC oscillators in a loop (see Figure 6.1), connected by a cascade of a phase shifter with phase shift θ and transconductance g_{m_c}. Therefore, the discussion in the previous section on practical resonator properties also is of interest and applicable to each stage in multi-phase LC oscillators.

> Qualitatively Figure 6.6 shows how the ratio between g_{m_l} and g_{m_c} influences the degree of coupling between the LC oscillators stages. If the coupling is too weak, multi-oscillations can occur [106].

The principle of operation of the oscillator model in Figure 6.5 is explained by Figure 6.7(a), for the case $N = 2$ and $\theta = 0$. Adaptation for $N > 2$ is straightforward and therefore omitted. For $N = 2$, the model reduces to a quadrature LC oscillator with two outputs $v_{out_1} = v_{out_I}$ and $v_{out_N} = v_{out_Q}$. The output currents of transconductances g_{m_c} and g_{m_l} are referred to as i_c and i_l, respectively.

Assuming a quadrature oscillation mode exists, output voltages $\vec{v}_{out_I}$ and $\vec{v}_{out_Q}$ will be in quadrature; see Figure 6.7(a). Output voltage $\vec{v}_{out_Q}$ may lead or lag $\vec{v}_{out_I}$ by 90°, hence the phasors $\vec{v}_{out_Q,lead}$ and $\vec{v}_{out_Q,lag}$ are both drawn. The phasors (both lead and lag) of the currents in the quadrature stage, $\vec{i}_l = g_{m_l} \cdot \vec{v}_{out_Q}$ and $\vec{i}_c = g_{m_c} \cdot \vec{v}_{out_I}$ are shown in Figure 6.7(a) as well[4]. Based on symmetry, the currents in the in-phase stage will

[4] In a high frequency I/Q oscillator, parasitic shifts will cause $\vec{i}_c$ and $\vec{i}_l$ to lag $\vec{v}_{out_I}$ and $\vec{v}_{out_Q}$, respectively. Moreover, the quadrature relation between the currents may not be present in a practical circuit, due to non-identical parasitic phase shift and other second order effects. However, this does not alter the principle of operation, and for brevity an in-depth discussion of these non-idealities is omitted.

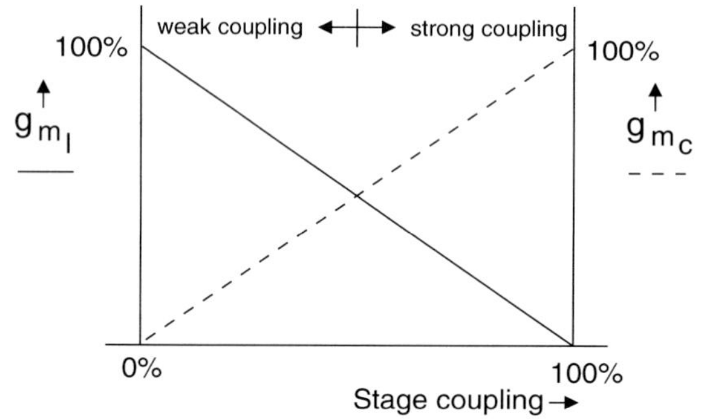

Figure 6.6 Depending on the value of transconductances g_{m_l} and g_{m_c}, the coupling between the LC oscillator stages can be weak or strong.

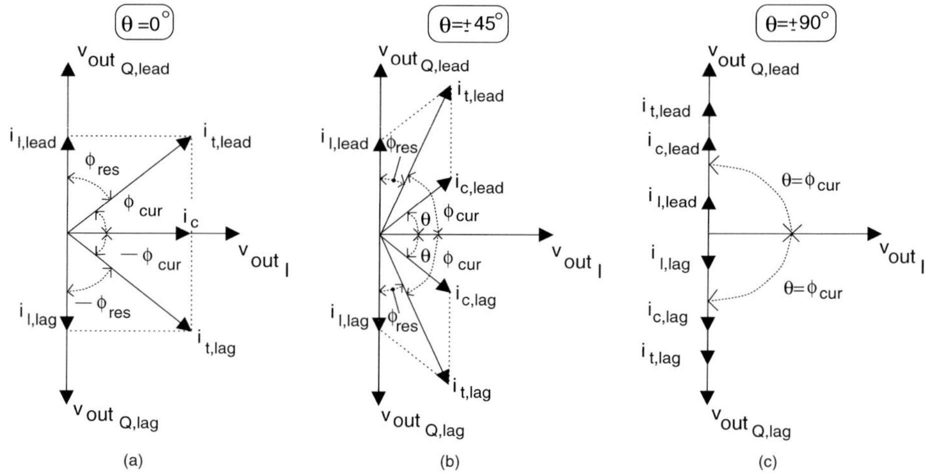

Figure 6.7 Combined phasor diagram of the output voltages $\vec{v}_{out_I}$ and $\vec{v}_{out_Q}$, and the currents in the quadrature stage, $\vec{i}_l$ and $\vec{i}_c$, for $\theta = 0, \pm 45°, \pm 90°$.

be identical in amplitude and phase relation to the currents in the quadrature stage, and we can restrict our analysis to one stage. The sum of $\vec{i}_c$ and $\vec{i}_l$ can be written as $\vec{i}_t = i_{tot}e^{j\phi_{cur}}$ with $i_{tot} = |\vec{i}_l + \vec{i}_c|$ and $\tan(\phi_{cur}) = |\vec{i}_l|/|\vec{i}_c|$ (in case $\vec{i}_t$ and $\vec{v}_{out_Q}$ are leading $\vec{v}_{out_I}$). The current $\vec{i}_t$ is injected into the resonator, which provides a phase shift ϕ_{res}. For the behavioral model in Figure 6.5, it can be shown that $|\vec{v}_{out_Q}| = i_{tot}R_p\cos(\phi_{res})$.

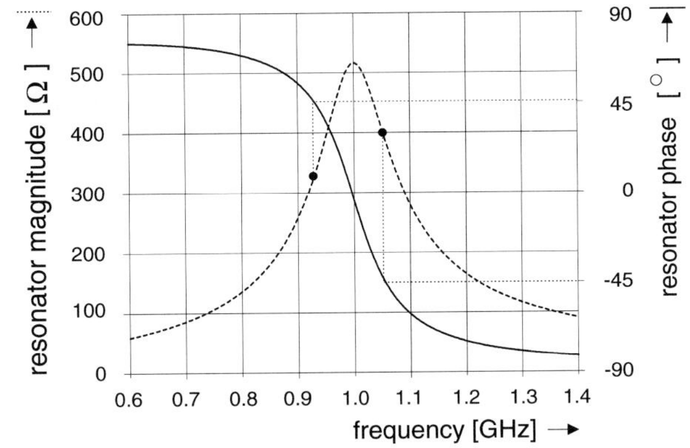

Figure 6.8 Magnitude and phase of the practical resonator from Figure 6.2 with $L_s = 10\text{nH}$, $C_s = 2.5\text{pF}$, $Q_{ls} = 10$ and $Q_{cs} = 40$ at 1 GHz.

> As illustrated in Figure 6.7(a), $\phi_{cur} + \phi_{res}$ must be either $+90°$ or $-90°$ for quadrature operation. In practice this ambiguity of several solutions for quadrature oscillation is solved, because a practical resonator, unlike the theoretical parallel resonator in Figure 6.5, is asymmetrical.

For a practical resonator there can be two solutions (or more, see [107]), where the oscillation conditions for quadrature operation are met. This is illustrated in Figure 6.8: the loop gain is for $-45°$ resonator phase shift higher than for $+45°$. This asymmetry provides a unique solution to the oscillation conditions, where the phase condition for quadrature oscillation is met and the loop gain is the highest [107].

From the preceding discussion it is clear that for $\theta = 0$ the resonator phase shift ϕ_{res} is nonzero. For example, if $\vec{i}_c$ and $\vec{i}_l$ have the same amplitude, $\phi_{cur} = \pm 45°$ and ϕ_{res} will also be equal to $\pm 45°$. When θ is nonzero, due to parasitic phase shifts or deliberately design, ϕ_{res} reduces when θ is increased. Compare Figure 6.7(a) and Figure 6.7(b) for example. If phase shift θ is made $\pm 90°$, $\vec{i}_c$ and $\vec{i}_l$ both will have the same phase and ϕ_{cur} will be $\pm 90°$ too. This is illustrated in Figure 6.7(c); notice that $\phi_{res} = 0$ in this case. We will see in Section 6.3 that $\phi_{res} = \pm 0°$ is the optimum operating point for maximum quality factor and minimum $\mathcal{L}(f_m)$ of multi-phase LC oscillators.

The frequency ω_{NLC} of a multi-phase LC oscillator was already derived in the previous chapter and denoted in (5.11) on p. 95. The resonator phase shift $|\phi_{res}|$ in (5.11) can be expressed as $180/N - |\phi_{cur}|$ for the model in Figure 6.5, with ϕ_{cur} equal to

$$\phi_{cur} = \theta + \arctan\left(\frac{|\vec{i}_l| \sin\varphi}{|\vec{i}_c| + |\vec{i}_l| \cos\varphi}\right), \tag{6.17}$$

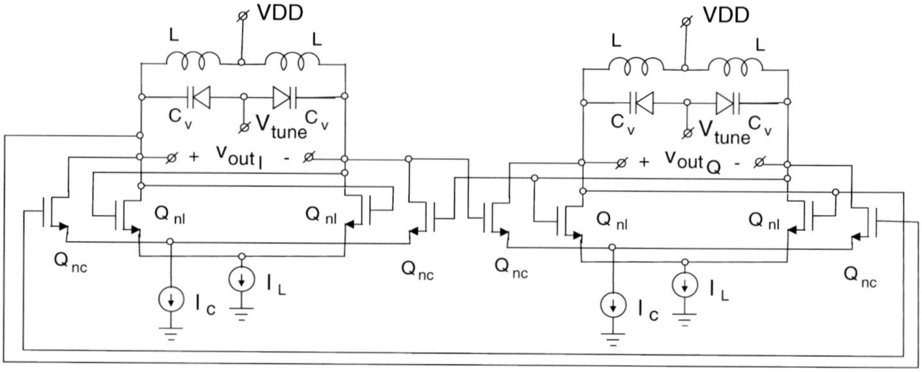

Figure 6.9 MOS circuit implementation of an I/Q LC oscillator.

which follows from the "cosine-rule" and where φ is the angle between $\vec{i}_l$ and $\vec{i}_c$ (ideally, without parasitic phase shift in the transconductances, $\varphi = 180/N - \theta$) [108].

The matching between the phases of a multi-phase oscillator is limited by device matching and layout symmetry. A quantitative analysis of the I/Q errors is beyond the scope of this work, and statistical analysis can provide the designer with this information. Note that calibration techniques can be applied to improve I/Q matching of oscillators (see Appendix D).

Multi-phase oscillator circuit example

Figure 6.9 shows an I/Q LC oscillator implementation [184]. Each stage is implemented with a cross-coupled LC stage (in which transistors Q_{nl} implement the transconductance g_{m_l} in Figure 6.5), and the stages are coupled with transistors Q_{nc}. These coupling transistors implement transconductance g_{m_c} in the behavioral model shown in Figure 6.5 and also add phase shift (modeled by θ in Figure 6.5).

A stage of a differential multi-phase LC oscillator can be implemented by any differential single-phase LC oscillator. For example, the two single-phase LC oscillator circuit topologies discussed a few pages ago, both can be used in combination with any resonator implementation. In addition, discussions on parasitics (the previous section) and tuning of single-phase LC oscillators (Section 6.2), are similar for multi-phase LC oscillators. Obviously, in addition to the parasitics of each LC stage, the coupling transistors Q_{nc} in Figure 6.9 add some loading, parasitic capacitance and parasitic phase shift.

More than two LC stages can be coupled in the same manner as given in Figure 6.9, to realize an N-stage LC oscillator [185]. More stages result in more power dissipation and more chip area. On the other hand, a number of advantages are obtained. Phase noise performance generally improves with more stages (see Section 6.3) at the

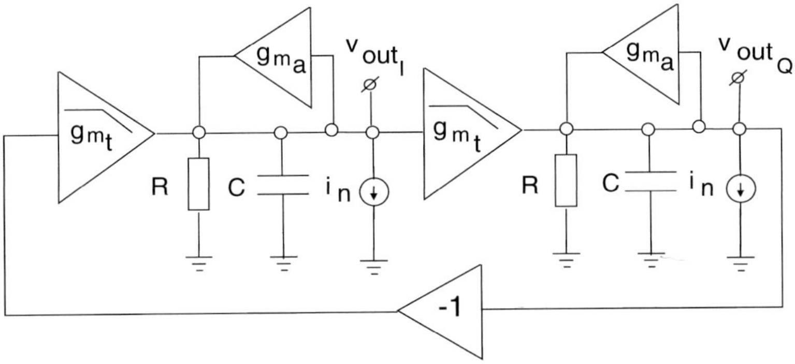

Figure 6.10 Behavioral model of a two-integrator oscillator.

expense of more power dissipation[5]. In addition, N low-phase-noise output signals are available from an N-stage LC oscillator.

> An example of an alternative coupling method is to place transistors Q_{nc} in series with the transistors Q_{nl} in Figure 6.9, instead of in parallel with transistors Q_{nl}, and have the gates of transistors Q_{nl} cross-coupled connected to the drains of transistors Q_{nc} [186].

The tail current of the cross-coupled pairs is reused in this way. At the same time the parasitics and load on the resonator is reduced.

> In Appendix G, a number of I/Q oscillator references are listed together with a summary of the performance.

6.1.3 The two-integrator oscillator

A behavioral model of a practical two-integrator oscillator is shown in Figure 6.10. Compared to the ideal two-integrator model from Figure 5.6 on p. 96, four non-ideal aspects are added. In each stage a resistor R is present, which models all losses (e.g. collector or drain resistors of an implementation and the loading of the next stage). To compensate for these losses, transconductance g_{m_a} is present, which realizes a negative resistance in parallel with R. If transconductances g_{m_t} were to have zero phase shift, g_{m_a} would be equal to $1/R$ to have perfect integration in each stage. However, as indicated by the low-pass curve in the symbols of transconductances[6] g_{m_t}, these

[5] However with respect to phase noise performance normalized for power, the multi-phase LC oscillator has, no advantage compared to a single-phase LC oscillator: see p. 163.

[6] See Appendix B, for behavioral modeling building blocks.

are modeled with a single-pole model: $g_{m_t}/(1+j\omega\tau_t)$. In many practical cases, ring oscillator stage implementations satisfy dominant pole behavior [109, 110], and the reduced order of a single-pole model can be used advantageously for meaningful estimates and insightful formulas. Transconductances g_{m_a} are intentionally modeled without a pole. Normally these transconductances are much smaller than g_{m_t}, as the parasitics and noise g_{m_a} produces are inevitable (R needs to be partly compensated) but unwanted (it introduces extra noise). All noise sources in a stage are combined in noise current source i_n. Note that if a differential implementation is assumed, the inversion in the behavioral model simply is a cross-coupling of wires.

The modulus and argument of the open loop transfer of one two-integrator stage from Figure 6.10 are,

$$|H_{ol}| = \frac{g_{m_t}R}{\sqrt{(-1+g_{m_a}R)^2+C^2R^2\omega^2}\sqrt{1+\tau_t^2\omega^2}}, \tag{6.18}$$

and

$$\arg(H_{ol}) = -\arctan\left(\frac{(CR+\tau_t-g_{m_a}R\tau_t)\omega}{-1+g_{m_a}R+CR\tau_t\omega^2}\right). \tag{6.19}$$

For $\tau_t = 0$ and $g_{m_a} = 1/R$, (6.18) and (6.19) reduce to the modules and argument of the ideal two-integrator oscillator on p. 96. In other words, the phase shift per stage will be $-90°$ for all frequencies, and the frequency of the two-integrator oscillator becomes $\omega_{two-int} = g_{m_t}/C$ (see (5.13) on p. 96).

For nonzero τ_t, the two-integrator oscillator oscillates at a frequency ω_{osc} provided that the phase shift at this frequency is $-90°$ and the stage gain is exactly unity. This means the denominator of (6.19) must be zero. Solving $-1+g_{m_a}R+CR\tau_t\omega_{osc}^2 = 0$ for g_{m_a} yields

$$g_{m_a} = \frac{1-CR\tau_t\omega_{osc}^2}{R}. \tag{6.20}$$

The value for transconductance g_{m_t} at which the stage gain is unity, can be expressed in terms of g_{m_a} and ω_{osc} by solving $|H_{ol}| = 1$:

$$g_{m_t} = \frac{\sqrt{((-1+g_{m_a}R)^2+C^2R^2\omega_{osc}^2)(1+\tau_t^2\omega_{osc}^2)}}{R}. \tag{6.21}$$

Equations (6.20) and (6.21) can be used to derive ω_{osc} of the two-integrator oscillator in Figure 6.10[7], but this expression is rather complicated, whereas (6.20) and (6.21) provide some insight. Equation (6.21) shows that the stage gain for high frequencies is dominated by the time-constants RC and τ_t and becomes almost independent of the setting for g_{m_a}. In other words, the amplitude condition for oscillation is set by g_{m_t}, practically independent of g_{m_a}. Equation (6.20) shows that the g_{m_a} needs to be reduced when the oscillation frequency is increased.

[7] Working in linear mode.

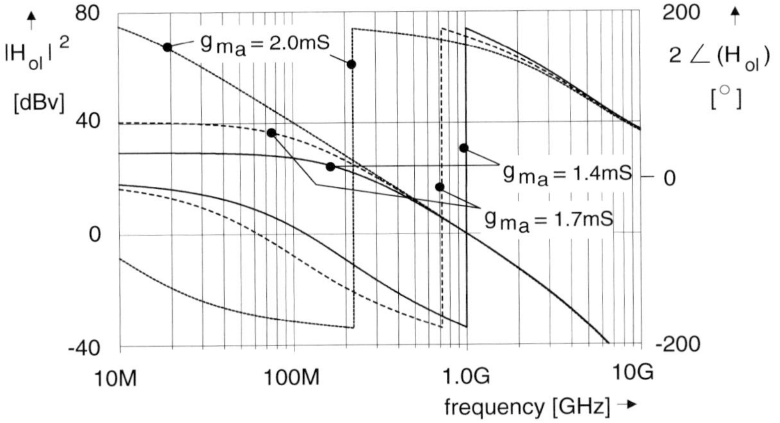

Figure 6.11 Squared modules and argument of two stages of the open-loop transfer function of the two-integrator oscillator model in Figure 6.10, for three values of g_{m_a}.

> The phase shift of transconductance g_{m_t} due to time constant τ_t increases when the oscillation frequency increases, and therefore time constant $(R \,||\, -1/g_{m_a})C$ must be reduced to meet the phase condition for oscillation.

Figure 6.11 illustrates the discussion of (6.20) and (6.21). The open-loop gain and phase of the two stages in Figure 6.10 is shown for three g_{m_a} settings. The frequency was set to 1 GHz, τ_t to 30 ps, R to 500 Ω and C to 500 fF. For $g_{m_a} = 1.4$mS, the phase condition for oscillation is exactly met at 1 GHz. Figure 6.11 shows that higher g_{m_a} values produce a significant phase shift and the loop gain for lower frequencies increases. However the 0 dB crossing of the loop gain is virtually unchanged. In a practical two-integrator oscillator, g_{m_t} is commonly used for frequency control and g_{m_a} for amplitude control.

> For linear operation of the two-integrator oscillator (and any other oscillator), an AGC control is needed that adjusts g_{m_a} such that the phase condition for oscillation is met at the 0 dB crossing of the loop gain. An efficient implementation of an AGC loop for the two-integrator oscillator was shown in Figure 2.17 on p. 35.

Two-integrator oscillator circuit example

A bipolar circuit implementation of a two-integrator oscillator is presented in Figure 6.12 [52]. Transconductance g_{m_t} from Figure 6.10 is implemented with a differential

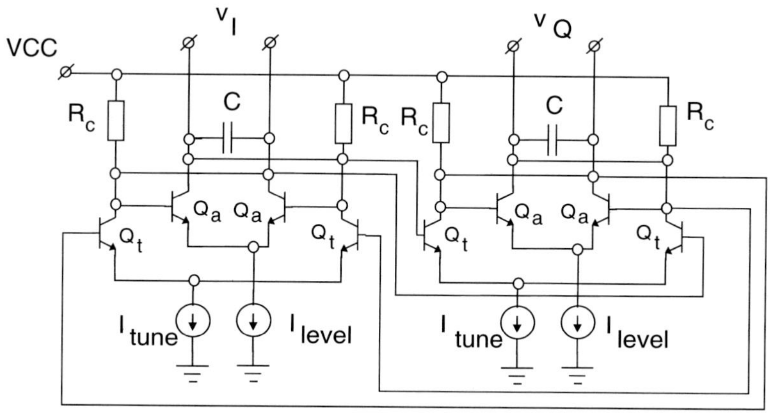

Figure 6.12 Bipolar circuit implementation of a two-integrator oscillator.

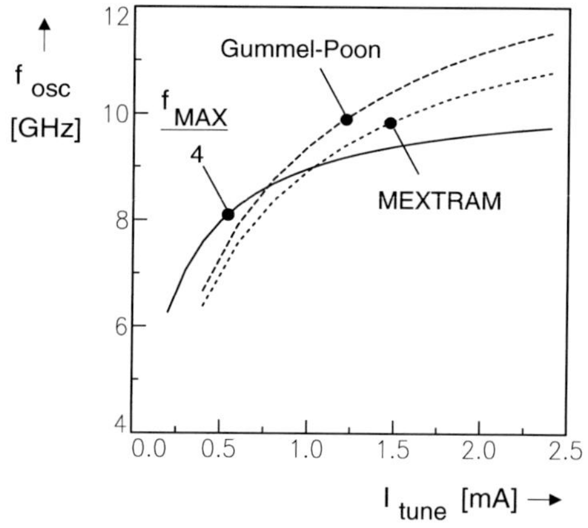

Figure 6.13 Simulated oscillation frequency of the oscillator from Figure 6.12 with Gummel-Poon and MEXTRAM transistor models. The frequency $f_{MAX}/4$ of the implementation technology, a 30 GHz f_T double-poly BiCMOS process, is also shown.

pair, whereas g_{m_a} is realized with a cross-coupled differential pair. In an 11 GHz f_T BiCMOS process, this implementation easily achieves an oscillation frequency ranging from 0.9 to 2.2 GHz (with C omitted: the integration capacitance only consists of device parasitics and interconnect) [52]. For this particular design the transistors that implemented g_{m_t} were four times larger sized than the transistors which implemented g_{m_a}. A detailed two-integrator design example is given in Chapter 9.

For N-stage ring oscillators, a reasonable estimate of the oscillation frequency can be written as [109]

$$f_{osc} \approx \frac{f_{MAX}}{2N}, \tag{6.22}$$

which becomes $f_{MAX}/4$ for the two-integrator oscillator. For f_{MAX} calculation, the (bipolar) f_{MAX} formula in Appendix F on p. 273 can be used. Note that an IC technology usually has one f_{MAX} figure specified, which usually is valid for the best transistor under optimal bias conditions. The transistor and bias conditions (e.g. base-collector voltage) may be quite different for the transistor used in an oscillator design. Therefore, the f_{MAX} should be calculated using the transistor and bias conditions representative of the oscillator design. The oscillation frequency estimator $f_{MAX}/4$, together with transient simulation results of the oscillation frequency using a Gummel-Poon model and a MEXTRAM transistor model, are shown in Figure 6.13 for a bipolar two-integrator oscillator design. The design was simulated in a 30 GHz f_T double-poly BiCMOS process, with a circuit implementation identical to the one shown in Figure 6.12 but with capacitors C omitted. The size of the transistors Q_t were four times the size of Q_a. Tuning current I_{tune} was varied from 200 μA to 2.4 mA. Resistor R_c was set at $(1V)/I_{tune}$, resulting in a voltage swing of 200 mV peak at the maximum frequency. Transistor Q_t has its peak f_T (approximately 26 GHz at a collector-base voltage of 0 V) at a collector current of 1.8 mA. Assuming the transistors are fully switching, I_{tune} should not exceed the peak f_T current, since transistor operation beyond peak f_T current normally is not well characterized. Some improvement of the maximum f_{osc} beyond $I_{tune} = 1.8$ mA is visible in Figure 6.13, because transistors Q_t are not fully switching with an external base voltage of 200 mV around 10 GHz. At peak f_T current, the error of oscillation frequency estimator (6.22) with respect to the simulation results using the MEXTRAM model is less than 12%.

Figure 6.14 shows the circuit diagram of a V/I-converter, which was designed for a 622 MHz version of the oscillator in Figure 6.12 [94]. The oscillator with V/I-converter is part of a clock-conversion PLL for SONET OC-12 (SDH STM-4) optical transmitters, and implemented in an 18 GHz f_T BiCMOS technology. Since the loop filter of a PLL usually provides an output voltage, a V/I-converter is needed to make the two-integrator oscillator voltage controlled, instead of current controlled. As leakage current of a PLL loop filter causes reference breakthrough, this V/I converter needs a high-ohmic input [3]. This input is realized with a rail-to-rail CMOS buffer [111]. The output of the input buffer is mapped to a voltage ranging from stabilizer voltage V_{stabi} (2.4 V) to approximately 1.7 V. This provides room for the base-emitter voltage of Q_1, the voltage drop across R_3 and one saturation voltage of the current source. In this way, the rail-to-rail property of the CMOS input buffer is maintained. It is desirable to use as much of the available voltage range for tuning as possible, because this results in a lower tuning constant K_{vco}. A low K_{vco} means low sensitivity to PLL loop-filter noise[8] and a lower reference breakthrough [88]. Transistor Q_1 implements V/I-conversion as well as temperature compensation. The V/I-converter has a posi-

[8] See (4.16) on p. 84, with K_{supply} replaced by K_{vco}.

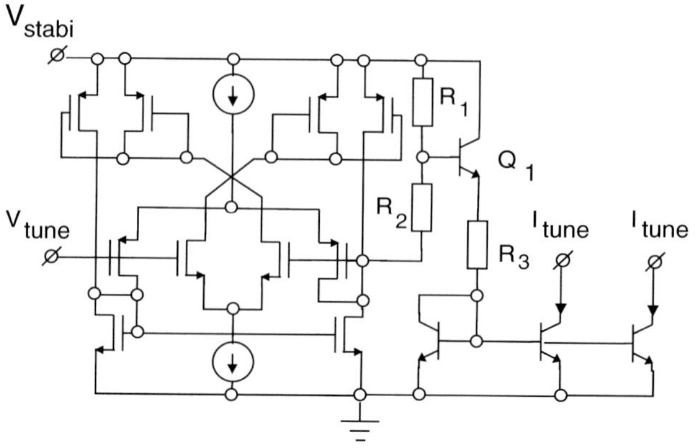

Figure 6.14 Example of a V/I-converter design with high-ohmic input and incorporated temperature compensation for a two-integrator oscillator.

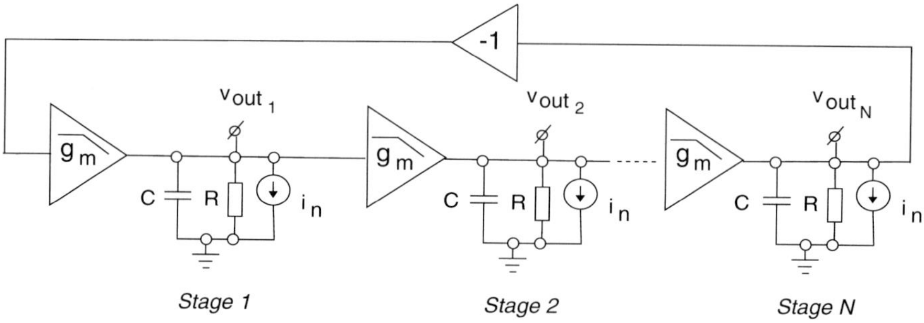

Figure 6.15 Multi-phase ring oscillator behavioral model.

tive temperature coefficient, due to Q_1, which compensates the oscillation frequency dependence on thermal voltage V_T.

> In Appendix H, two-integrator oscillator designs and other ring oscillator designs are listed to together with a performance summary

6.1.4 *N*-stage ring oscillators

Figure 6.15 shows the behavioral model of an N-stage ring oscillator, which will be used to analyze practical properties of ring oscillators with three stages or more. Compared to the ideal ring oscillator behavioral model, which was discussed on p. 97 of Chapter 5, several non-ideal properties are added. The transconductances are modeled with a single pole model, as indicated by the low-pass curve in the transconductances, i.e. the transconductance in a stage is modeled by $g_m/(1+j\omega\tau_1)$. All noise sources in a stage are combined in one noise current source i_n. Note that the resistors in an N-stage ring oscillator stage were already present in the ideal behavioral model on p. 97, whereas they were introduced as losses in this chapter for the two-integrator oscillator (which is ideally resistor-less, as discussed on p. 96).

The modulus and arguments of the transfer function of one stage are described by (6.18) and (6.19), respectively, with g_{m_a} set to zero, g_{m_t} replaced by g_m, and τ_t by τ_1. The oscillation frequency of the N-stage ring oscillator model in Figure 6.15 can be found by solving the phase condition for oscillation for $\omega = \omega_{osc}$,

$$\arctan\left(\frac{(RC+\tau_1)\omega_{osc}}{-1+RC\tau_1\omega_{osc}^2}\right) = \frac{\pi}{N}, \tag{6.23}$$

which leads to the oscillation frequency of the ring oscillator working in linear mode

$$\omega_{osc} = \frac{\cot(\frac{\pi}{N})(RC+\tau_1+\sqrt{(RC+\tau_1)^2+4RC\tau_1\tan^2(\frac{\pi}{N})})}{2RC\tau_1}. \tag{6.24}$$

Solving (6.23) with $\tau_1 = 0$, leads to the ideal oscillation frequency expression ω_{Nring} already encountered on p. 98.

Usually, a ring oscillator works in strongly nonlinear mode, and its oscillation frequency can be expressed as $1/(2N\tau_{delay})$, if the input rise time is equal to the output rise time of a stage. A simple estimation of τ_{delay} can be made if we model a ring oscillator stage as a high-gain, single-pole model [110],

$$H(j\omega) = \frac{A}{1+j\omega\tau}. \tag{6.25}$$

In a practical ring oscillator many higher-order poles are present, but these are neglected under the assumption that time-constant τ is dominant. In the case of Figure 6.15, $A = g_mR$ and $\tau = RC+\tau_1$. For $A \gg 1$ the 0 dB frequency of the ring oscillator loop gain is approximately A/τ. The phase margin at this frequency is $\pi - N\arctan(A)$, which must be negative for oscillation, hence

$$N > \frac{\pi}{\arctan(A)}. \tag{6.26}$$

With a typical loop gain ranging from $4\ldots10$, minimum N ranges from 2.4 to 2.1 for this model. In practice, higher order poles allow oscillation for $N < 3$. The propagation delay of a one-pole system with a linear ramp input can be approximated by $\tau_{delay} = 0.8\tau$ [110]. The factor 0.8 tends towards the factor $\ln(2)$ derived in Appendix E, if square wave signals are assumed, which is less realistic for small N and high frequencies. The frequency of an N-stage ring oscillator with dominant pole behavior can now be written as

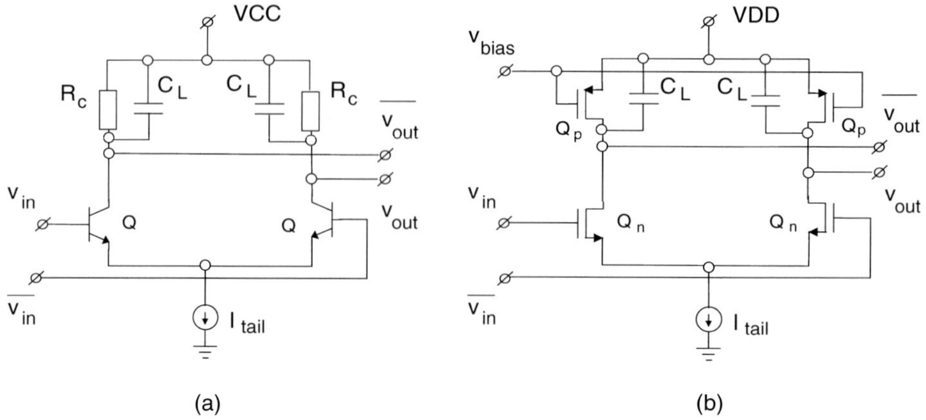

Figure 6.16 Bipolar CML ring oscillator stage (a) and a source coupled differential pair with PMOS loads (b).

$$f_{osc} \approx \frac{1}{2N0.8\tau}. \tag{6.27}$$

Expression (6.27) will be used in the following section to estimate the frequency of two ring-oscillator circuit implementations.

N-stage ring oscillator circuit examples

Two implementations of a ring oscillator stage are shown in Figure 6.16. In Figure 6.16(a) a bipolar current mode logic (CML) cell is presented. The simulated oscillation frequency of this stage, implemented in a 30 GHz f_T double-poly BiCMOS process, versus bias current I_{tail} is plotted in Figure 6.17. The oscillation frequency is plotted for N ranging from 3 to 8, and transistors Q in Figure 6.16(a) were identical to the ones used for the two-integrator oscillator simulation example. In addition to the results of the transient simulations, (6.22) is plotted in Figure 6.17. Whereas (6.22) provided reasonable oscillation frequency predictions for a two-integrator oscillator, it is quite pessimistic for maximum bias currents in its estimations, at least for the presented case. Its error for $N = 3$ at maximum frequency is 29 %, for example.

A quite accurate frequency estimation of the CML stage in terms of technology parameters and bias conditions can be found using (6.27). The transistor model used for τ calculation is discussed in Appendix F. One modification of this model is needed: for accurate τ calculations, capacitance C_μ is split up into an intrinsic part $X_\mu C_\mu$ and an extrinsic part $(1-X_\mu)C_\mu$ [112]. This is modeled in a similar manner in the Gummel-Poon transistor model[9]. This is done to take the distributed nature of C_μ into account.

[9]From the Gummel-Poon model parameter X_μ can be derived, as well as the other technology parameter values.

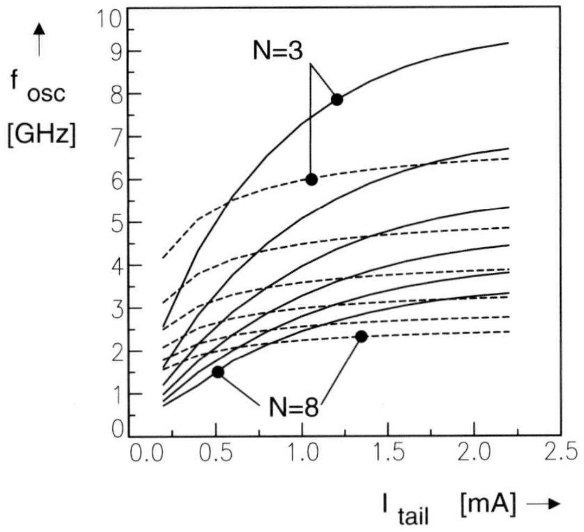

Figure 6.17 Simulated oscillation frequency of a CML ring oscillator (Figure 6.16(a)) for $N = 3\ldots8$. The frequency $f_{MAX}/(2N)$ of the implementation technology, a 30 GHz f_T double-poly BiCMOS process, is also shown (dashed lines).

The time constant of the dominant pole of the CML circuit in Figure 6.16(a) can be calculated with

$$\begin{aligned} \tau_{CML-stage} &\approx R_{base}C_\pi + R_{base}X_\mu C_\mu(1+g_{m_{bipo}}(R_c \,||\, R_{in})), \\ &+ (R_c \,||\, R_{in})(X_\mu C_\mu + C_{js}) + (R_c \,||\, R_{in})C_L, \end{aligned} \tag{6.28}$$

in which R_{in} and C_L can be approximated by [112],

$$R_{in} \approx R_\pi \tag{6.29}$$

$$C_L = C_{in_{CML-Stage}} \approx (1-X_\mu)C_\mu(1+g_{m_{bipo}}R_c). \tag{6.30}$$

Equation (6.28) is substituted in (6.27), evaluated, and plotted for N ranging from 3 to 8 in Figure 6.18. The technology and bias conditions were as discussed for Figure 6.17. The simulated oscillation frequency of the CML oscillator, which was already plotted in Figure 6.17, is re-plotted for comparison in Figure 6.18. As illustrated by this figure, the dominant pole model is a much better estimator for the oscillation frequency than (6.22). At the maximum frequency, 9.2 GHz ($N = 3$), the error is less than 5 %, and at the minimum frequency, 0.72 GHz ($N = 8$), the error is approximately 18%.

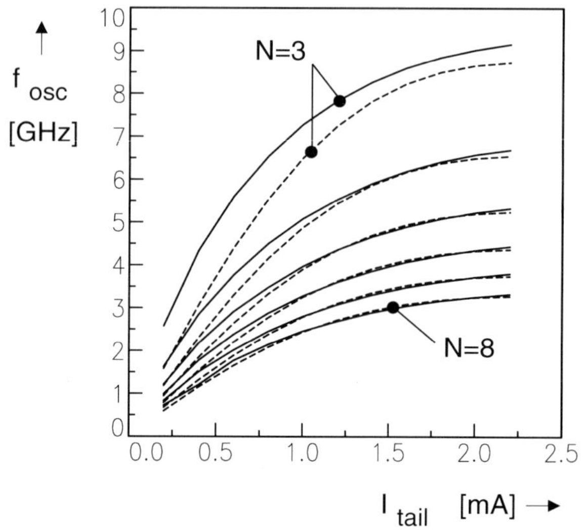

Figure 6.18 Simulated oscillation frequency of a CML ring oscillator (Figure 6.16(a)) for $N = 3 \ldots 8$, and a frequency estimation (dashed lines) calculated with (6.27), in which τ is described by (6.28).

In Figure 6.16(b) a MOS ring oscillator stage is shown. The PMOS loads of this circuit normally operate in the triode region. In Section 6.2.2 resistive tuning will be discussed, and one possible implementation of this tuning method involves PMOS loads in the linear region, with tuned gate voltage. The time constant of the dominant pole of the MOS oscillator stage is given by

$$\tau_{CMOS-stage} \approx R_{Q_p}(C_{gd_n} + C_{db_n} + C_{gd_p} + C_{db_p} + C_L), \qquad (6.31)$$

in which R_{Q_p} is the load resistance[10] of the PMOS in Figure 6.16(b), and C_L is the input capacitance of the next stage. Equation (6.31) was validated in a 0.8 μm process with a 3.3 V supply, 500 mV voltage swing, for $N = 4, 6, 8$ [110]. The maximum error compared to simulations was found to be below 24 %.

6.2 Tuning

In this section tuning of LC and ring oscillators will be discussed. Tuning of ring oscillators, which is usually less problematic in integrated transceivers than LC os-

[10] Which is equal to $1/(\mu_p C_{ox}(W/L)(V_{sg} - V_{TH}))$ in the triode region provided that $V_{sd} \ll (V_{sg} - V_{TH})$.

cillator tuning, will be discussed only briefly. But first an in depth discussion of LC oscillator tuning follows.

6.2.1 LC oscillators

In general, tuning of LC oscillators can be performed by varying L_p or C_{tune} as is evident from (6.4). Inductance L_p and C_{tune} can be varied in a continuous manner or in discrete steps. Furthermore, they can be implemented by passive elements or active circuits, which exhibit inductive or capacitive behavior in a frequency range of interest. In addition, the frequency of an LC oscillator can be varied by changing the (additional) resonator phase shift (see (5.11) on p. 95), or several oscillators that are switched on and off can be used. These tuning methods can be used in single-phase and multi-phase LC oscillators. Each method is highlighted below.

> Since the resonator Q should be maximized for low-power low-phase-noise LC oscillators (see Appendix A and Section 6.3), the influence of a specific tuning method on the quality factor is included in the discussion.

Continuous passive capacitive tuning

Capacitance C_p in Figure 6.1 consists of the varactor C_{tune}, intentionally added fixed capacitor C_{fixed} and parasitic capacitance C_{par}. To simplify the discussion, we assume that $C_{fixed} = 0$. The required maximum capacitance C_{max} to minimum capacitance C_{min} ratio of varactor C_{tune} can then be expressed as a function of maximum and minimum frequency (f_{max} and f_{min}, respectively), C_{par} and C_{min} as

$$\alpha_{varactor} = \frac{C_{max}}{C_{min}} = \left(\frac{f_{max}}{f_{min}}\right)^2 + \frac{C_{par}}{C_{min}}\left(\left(\frac{f_{max}}{f_{min}}\right)^2 - 1\right). \tag{6.32}$$

If C_{par} is zero, $\alpha_{varactor}$ is simply the squared ratio of f_{max} over f_{min}. However, if C_{par} is substantial compared to C_{min}, the required $\alpha_{varactor}$ to realize the required tuning range increases.

> Equation (6.32) shows that at high frequencies, where parasitics start to dominate, a higher $\alpha_{varactor}$ is needed to realize a certain tuning range compared to low frequencies. This trend is shown in Figure 6.19.

Figure 6.20 shows a resonator consisting of a series circuit with a varactor and a capacitor, a parallel capacitor and an inductor. As is evident from the preceding section, additional capacitance decreases the effective $\alpha_{varactor}$ and thus the tuning range. Nevertheless, in some cases it might be beneficial to add a capacitor in series

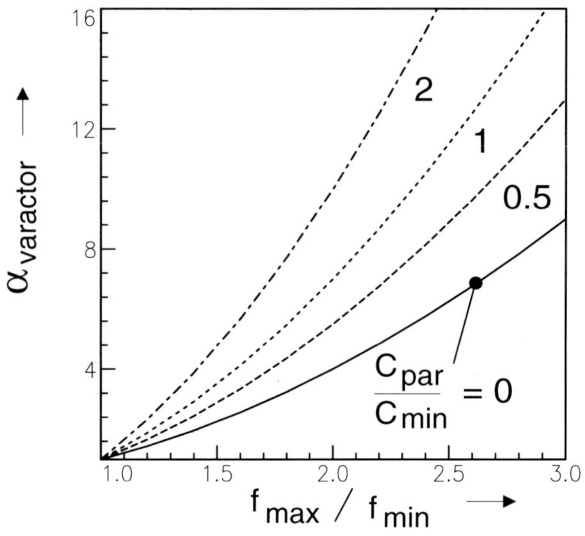

Figure 6.19 $\alpha_{varactor}$ versus f_{max}/f_{min} for four values of C_{par} over C_{min}.

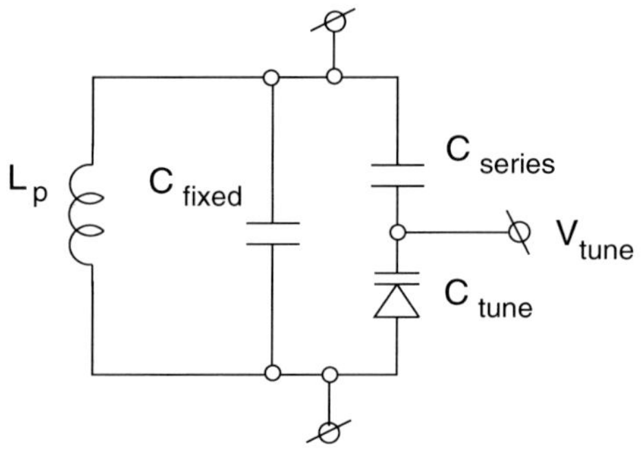

Figure 6.20 Resonator with varactor C_{tune} in series with C_{series}, and a parallel capacitor C_{fixed}.

with the varactor, and/or in parallel. First, as we shall see below, the overall quality factor of the compound capacitor, including the varactor, can have an increased quality factor with respect to the varactor quality factor $Q_{varactor}$. Second, it can be needed to isolate one terminal of the varactor from the rest of the circuit with respect to direct current, to supply the tuning voltage [113]. Note that this isolation might be required in a single-ended oscillator, whereas in a differential oscillator, the varactors can be connected back-to-back (see Figure 6.3(a), for example). Third, due to the capacitive voltage division, the oscillator voltage swing across the varactor is reduced and the

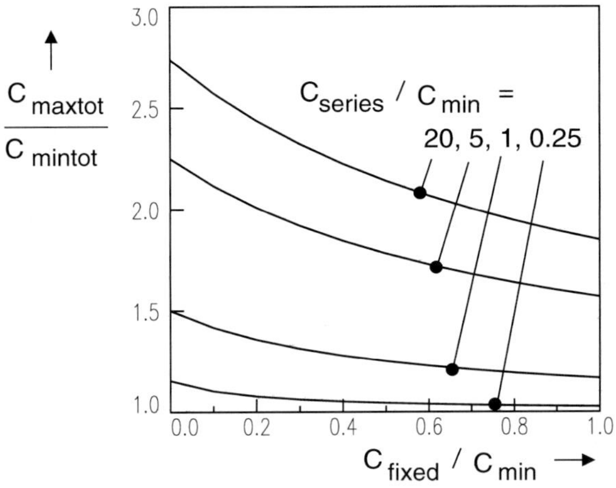

Figure 6.21 C_{maxtot}/C_{mintot} versus C_{fixed}/C_{min} for four values of C_{series}/C_{min} and $\alpha_{varactor} = 3$.

effective varactor ratio is somewhat increased[11]. Fourth, capacitive voltage division can be used to feedback only a certain ratio of the (large) voltage swing across the resonator to the active part, and avoid saturation. The reduction of varactor ratio $\alpha_{varactor}$ as a result of the capacitors C_{fixed} and C_{series} can be written as

$$\frac{C_{maxtot}}{C_{mintot}} = \alpha_{varactor} \left(\frac{\frac{C_{fixed}}{C_{max}} + \frac{C_{series}}{C_{max}+C_{series}}}{\frac{C_{fixed}}{C_{min}} + \frac{C_{series}}{C_{min}+C_{series}}} \right). \tag{6.34}$$

Figure 6.21 illustrates (6.34) for $\alpha_{varactor} = 3$. Note that if C_{series}/C_{min} is made sufficiently large (e.g. 20), the reduction of effective varactor ratio is small. For $C_{min} \to \infty$ and $C_{fixed} = 0$, the total capacitance ratio C_{maxtot}/C_{mintot} becomes equal to $\alpha_{varactor}$ (3).

By good approximation the quality factor of the compound varactor from Figure 6.20 is equal to

$$Q_{tot} \approx Q_{varactor} \frac{C_{series}(C_{series} + C_{tune}) + C_{fixed}(2C_{series} + C_{tune})}{C_{series}^2}, \tag{6.35}$$

if we we neglect the series resistances of C_{fixed} and C_{series}. For example, with $Q_{varactor} =$ 10 at 1 GHz, $C_{fixed}/C_{min} = 0$ and $C_{series}/C_{min} = 1$, Q_{tot} becomes approximately 20,

[11] For example, if the varactor C_{tune} is a PN-junction type with zero bias capacitance C_{j0} then its equivalent linear value is [114]

$$C_{tune} = \frac{C_{j0} V_j^{MJ}}{V_{max} - V_{min}} \left(\frac{(V_j - V_{min})^{1-MJ}}{1 - MJ} - \frac{(V_j - V_{max})^{1-MJ}}{1 - MJ} \right), \tag{6.33}$$

with V_j the built-in junction potential, MJ the junction grading coefficient and V_{min} and V_{max} the minimum and maximum voltage across the junction (negative voltages since the junction is reverse biased).

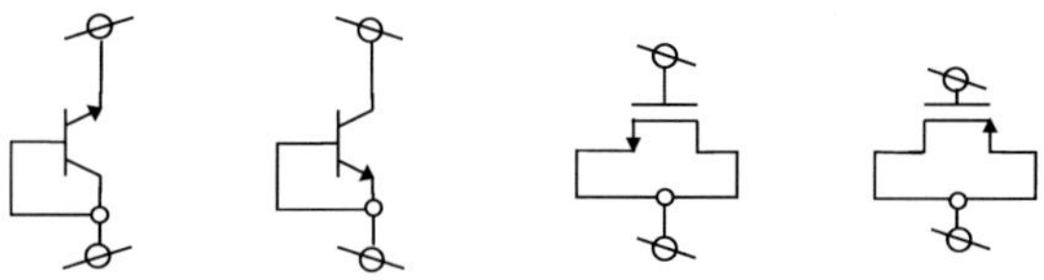

Figure 6.22 Four varactor implementations. The emitter-base junction of an NPN-transistor (a), the collector-base junction of an NPN-transistor (b), an NMOS-transistor with drain and source shorted (c), and its PMOS-transistor equivalent (d).

and with $C_{fixed}/C_{min} = 1$, Q_{tot} increases to 50. In Figure 6.21 we see that the effective varactor ratio is halved (reduced from 3 to 1.5) for $C_{fixed}/C_{min} = 0$ and $C_{series}/C_{min} = 1$. In other words, the quality factor enhancement can be significant, but the varactor ratio reduces very quickly.

> The quality factor of a varactor can be significantly enhanced by adding series or parallel fixed capacitance with a higher quality factor, at the cost of a (often significant) reduction in tuning range.

Given the limited varactor ratios available in IC technologies[12] and the challenging tuning range requirements (certainly, if we take process spread into account), there normally is little design freedom to exchange a reduced $\alpha_{varactor}$ for an improved quality factor.

Varactor implementation examples

Figure 6.22 shows four possible varactor implementations; two constructed with bipolar transistors[13], and two with MOS transistors. Either the emitter-base junction (Figure 6.22(a)) or the collector-base junction (Figure 6.22(b)) of an NPN-transistor can be used. The latter junction generally has the highest breakdown voltage[14], and can thus withstand a higher reverse voltage, which makes the collector-base junction a more robust choice[15].

[12] See varactor implementation examples below. Note that discrete varactor ratios can be as high as 20, but at the cost of very high tuning voltages (e.g. as high as 30 V) [113].

[13] PNP-type varactors are rarely used. Most bipolar and BiCMOS processes only have lateral structure PNP transistors (in some technologies high-speed vertical PNPs can be a process option) that are 50 to 100 times slower (and have more fixed parasitics capacitance) than NPNs.

[14] Typical values for B_{vbco} and B_{vbeo} in a 5 V, 11 GHz f_T bipolar process are 6.5 V and 4.5 V, respectively.

[15] Close to break-down the reverse leakage current will also increase. When the oscillator with junction-type varactors is incorporated in a PLL, this results in an increase of spurious signals in the PLL output spectrum (reference breakthrough) [3]. Leakage currents should normally be less than several tens of nano-amperes.

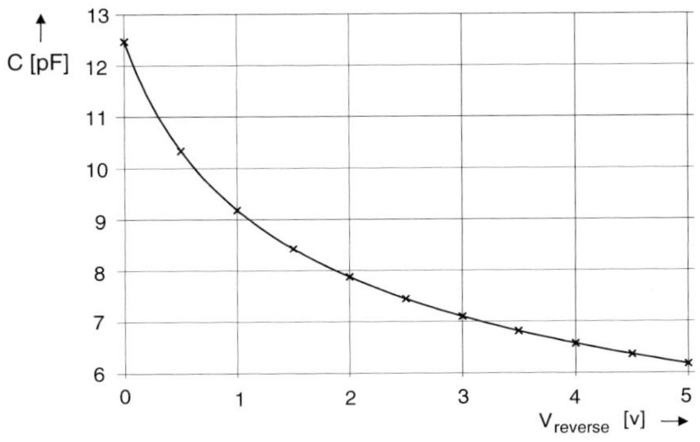

Figure 6.23 Simulated capacitance curve of a varactor constructed with the collector-base junction of 20 large NPN-transistors in an 11 GHz f_T bipolar process.

The capacitance of the varactor implementation from Figure 6.22(b) is approximated by the junction capacitor

$$C_{cb} = \frac{C_{jc}}{\left(1 + \frac{V_{cb}}{V_{jc}}\right)^{MJC}}, \tag{6.36}$$

in parallel with the collector-substrate junction capacitor C_{js} of the NPN-transistor (similar equation). In (6.36), C_{jc} (zero bias collector-base (cb) capacitance), V_{jc} (built-in cb-junction potential), and MJC (cb-junction grading coefficient) are Gummel-Poon transistor model parameters. Voltage V_{cb} is the voltage across the cb-junction.

Figure 6.23 shows a simulation of a cb-junction varactor constructed with 20 NPN transistors in parallel, in an 11 GHz technology. This varactor was investigated as an option for a reference oscillator for a satellite TV receiver (see Section 9.3 on p. 233). The ratio $\alpha_{varactor}$ is about two for this varactor, if the reverse voltage across the varactor is varied from 0 to 5 V. Two or less is a typical $\alpha_{varactor}$ value for varactors constructed with NPN transistors. Normally, the grading coefficients of the junctions of an integrated NPN transistor are in the range of 0.25[16] to 0.33, which gives low $\alpha_{varactor}$ values[17]. Moreover, two additional constraints can reduce the available tuning voltage range, limiting the effective capacitance variation. First, the supply voltage limits the available tuning range. This voltage can be lower than the allowable voltage before reliability reduction or before an increased leakage current occurs. Second, only a part of the supply voltage may be available for tuning. For example, a charge pump in a PLL, which generates the varactor tuning voltage via a loop filter, needs some headroom to operate properly. If the effective capacitance ratio of an NPN-type

[16] MJC is 0.25 for the simulated varactor and VJC is 0.7 V.

[17] Discrete varactors often have hyper-abrupt junctions with high grading coefficients (e.g. 0.5–2) and accordingly higher capacitance ratios.

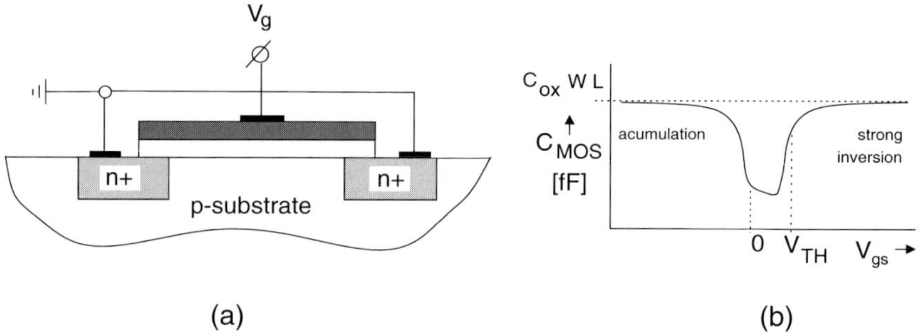

Figure 6.24 NMOS cross-section (a) and the capacitance characteristic versus tuning voltage of an NMOS varactor (b).

varactor is insufficient to meet the tuning range requirements, a MOS-type varactor can be a solution (see next section). Quality factors of PN-junction varactors[18] can be higher than 60 at 1 GHz: see Table 6.2 and Figure 9.5 on p. 220. The design of a monolithic SOA LC oscillator with optimized PN-junction varactors will be described in Section 9.1.

Both NMOS and PMOS transistors can be used as a varactor, if the source of the transistor is connected with the drain, as illustrated in Figure 6.22(c) and Figure 6.22(d), respectively. A simplified cross-section of an NMOS varactor is shown in Figure 6.24(a). The capacitance variation of this device with source and drain grounded, is shown in 6.24(b) [89]. For large negative V_{gs} voltages, the varactor works in the accumulation region, and its capacitance C_{MOS} is $C_{ox}WL$[19]. In this operating region holes are accumulated under the oxide interface. Another extreme situation occurs for $V_{gs} \gg V_{TH}$. The MOS varactor is now in strong inversion and a channel is formed. Again, C_{MOS} is maximum and equal to $C_{ox}WL$. Between accumulation and strong inversion C_{MOS} can be modeled as a series circuit of $C_{ox}WL$ and a depletion capacitance, which value is a function of V_{gs} [115]. The series connection of these two capacitance reaches its minimum around $V_{gs} = 0$. Figure 6.24(b) demonstrates that a MOS-varactor can be used in accumulation mode and inversion mode. Models describing C_{MOS} and the quality factor for both modes can be found in [116–118]. In Section 9.3, the design of an LC oscillator with PMOS varactor will be described.

Table 6.2 shows a number of measured performance parameters of MOS-varactors, reported in the literature. All MOS varactors in this table have an N-well[20], in con-

[18]In oscillator literature, the Q versus reverse bias voltage is often assumed to be constant or is not mentioned. However, the series resistance of PN-junction varactors, and of course their capacitance, are a function of the reverse voltage. In general, the series resistance is the highest at zero reverse voltage and decreases when the reverse bias voltage is increased (because the depletion region of the diode increases) [113].

[19]Neglecting overlap and fringing capacitances.

[20]One could call this structure a PMOS varactor or an NMOS varactor with N-well: we will refer to it

Ref.	type, mode	Control voltage swing	$\alpha_{varactor}$	Min. Q_{var}
[118]	PN-junction	2.4 V	1.83	69 at 1 GHz
[118]	PN-junction	2.4 V	1.61	130 at 1 GHz
[53]	PN-junction	5 V	1.7	26 at 1 GHz
[53]	PN-junction	5 V	1.6	60 at 1 GHz
[118]	MOS, Inv. ($L = 1\mu$m)	0.6 V	1.83	7 at 1 GHz
[118]	MOS, Inv. ($L = 0.5\mu$m)	0.8 V	1.56	36 at 1 GHz
[118]	MOS, Acc. ($L = 1\mu$m)	0.8 V	2.45	20 at 1 GHz
[118]	MOS, Acc. ($L = 0.5\mu$m)	1 V	1.76	95 at 1 GHz
[119]	MOS, Acc. ($L = 0.35\mu$m)	2 V	1.85	17 at 1.8 GHz
[116]	MOS, Acc. ($L = 0.5\mu$m)	0.75 V	3	23 at 1 GHz
[117]	MOS, Acc. ($L = 0.25\mu$m)	0.75 V	2.4	15 at 3.6 GHz

Table 6.2 Selection of in the literature reported measured PN-junction varactor and MOS-varactor performance.

trast with the drawing in Figure 6.24(a), which is realized in the p-substrate. The N-well eliminates the parasitic PN-junction at the source and the drain, and this yields a higher $\alpha_{varactor}$. The D/S contacts are formed by n+ regions in this varactor and it is working in accumulation or depletion only. The n-well also enables the use of the well-potential as a tuning voltage terminal.

In first order modeling, the quality factor of MOS-varactors is inverse proportional to channel length L [115]. Therefore, it improves with technology scaling. It is important to note that the layout of MOS-varactors has a big influence on $\alpha_{varactor}$ and the quality factor. Multiple gate-fingers should be used and the gate must be connected on both sides. In fact, all connections to the MOS-varactor (including the s/d connection) are critical and must be far smaller than 1Ω for common designs [118]. Parasitic capacitance, which limits maximum capacitance variation, can be reduced by deep or shallow trench isolation technology [117]. In comparison with junction-type varactors, MOS varactors have the advantage that the capacitance variation can be achieved within a small voltage swing; see Table 6.2. This is an important feature for low-voltage designs. Furthermore, unlike PN-junctions, there is no risk of signal

simply as MOS varactor. Note that a PMOS varactor with P-well is an alternative structure but it has a lower quality factor due to the lower mobility of the PMOS device [116].

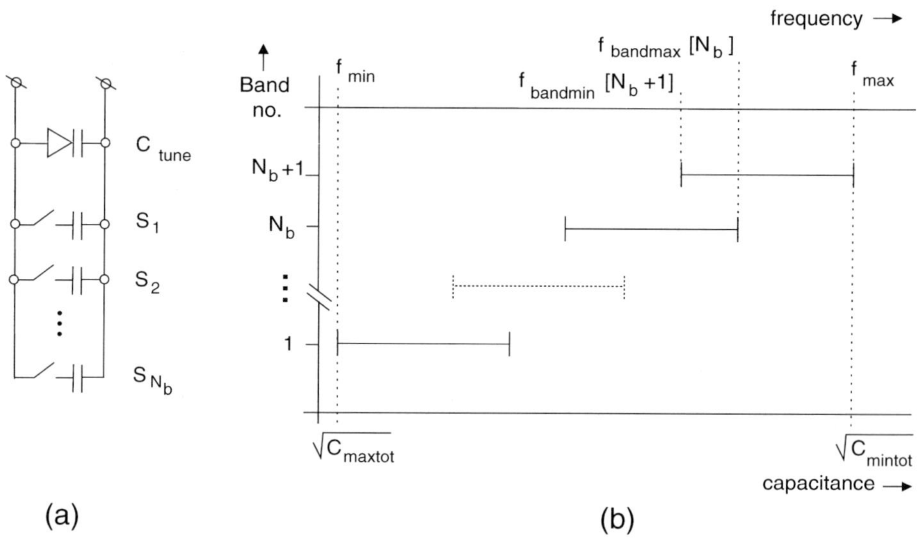

Figure 6.25 Band-switched capacitor array in parallel with a continuous varactor (a). Overlapping N_b+1 frequency bands (b).

clamping. The large MOS-varactor capacitance swing for a low tuning voltage swing can also be a big disadvantage. It means that tuning constant K_{VCO} will be high. This causes a high sensitivity to noise on the VCO tuning input, which in turn increases $\mathcal{L}(f_m)$ by means of FM modulation. In practice, a large voltage swing across a MOS-varactor causes an effective reduction of the tuning slope, and hence a lower tuning constant. A high tuning slope also makes an LC oscillator more susceptible to AM to PM conversion, as will be discussed in Section 6.4. PN-junction varactors (e.g. see Table 6.2) can therefore be preferable in a VCO design, provided that the power supply voltage is sufficiently high to accommodate the required control voltage swing.

> In Appendix G, the relative tuning range of a number of single-phase and multi-phase LC oscillators is listed. The majority of achieved relative tuning figures is in the range of 5% to 20%.

Discrete passive capacitive or inductive tuning: band-switching

Integrated varactors may not provide sufficient tuning range for an LC oscillator in an application. Band-switching can then be used to meet the tuning range requirements [120]. Figure 6.25 shows the principle of band-switching. A total of N_b switched capacitors divide the tuning range into N_b+1 bands. Varactor C_{tune} realizes continuous tuning within a band. Provided that the N_b+1 frequency bands overlap, all frequency within the range f_{min} to f_{max} can be reached.

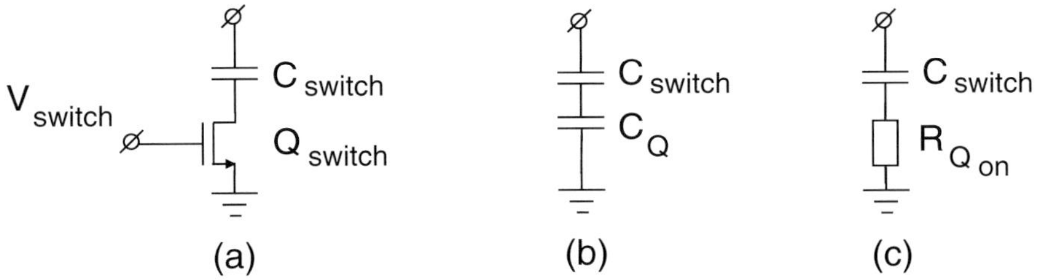

Figure 6.26 Typical switched capacitor circuit implementation (a), a simplified off-state model (b), and a simplified on-state model.

> Band-switching reduces the requirements of the continuous varactor. Furthermore, it reduces the oscillator's tuning constant K_{vco}, making it less sensitive to noise concerning its tuning input, which is beneficial to $\mathcal{L}(f_m)$ [187].

Therefore, even if a varactor capacitance ratio is sufficient to meet the tuning range requirements, band-switching can be needed to improve $\mathcal{L}(f_m)$. As shown in the previous section, especially MOS-varactors have a high tuning slope.

The total capacitance variation of a band-switched capacitor array plus the varactor can be expressed as

$$\frac{C_{maxtot}}{C_{mintot}} = \left(\frac{f_{max}}{f_{min}}\right)^2 = \frac{C_{min}\alpha_{varactor} + N_b C_{bandmin}\alpha_{band}}{C_{min} + N_b C_{bandmin}}, \tag{6.37}$$

with α_{band} capacitance ratio of the switched capacitor, C_{min} the minimum (continuous) varactor capacitance, and $C_{bandmin}$ the minimum capacitance of the switched capacitor (in off-state). To guarantee that the whole desired frequency band ranging from f_{min} to f_{max} is covered, the frequency bands need to overlap. Note that the switched capacitors S_1 to S_{N_b} are all assumed to be identical in Figure 6.25. If they are binary weighted, which is more efficient in terms of the number of needed control signals, the number of bands is equal to $2^{N_b} - 1$.

In Figure 6.26, simplified models for the off-state and on-state of a typical band-switch circuit implementation are shown. In the off-state, the minimum parasitic capacitance of the switch is an important design parameter. Capacitance $C_{bandmin}$ is equal to $(C_Q^{-1} + C_{switch}^{-1})^{-1}$. Therefore C_Q should be minimized for the minimum fixed capacitance introduced by the band-switches, since this reduces the effective tuning range of the varactor. In the on-state, the quality factor of a band-switch is of primary importance, and in the first order it is equal to

$$Q_b = \frac{1}{\omega C_{switch} R_{Q_{on}}} \approx \frac{1}{\omega C_{switch}(\mu_n C_{ox} W/L(V_{switch} - V_{TH}))^{-1}}. \tag{6.38}$$

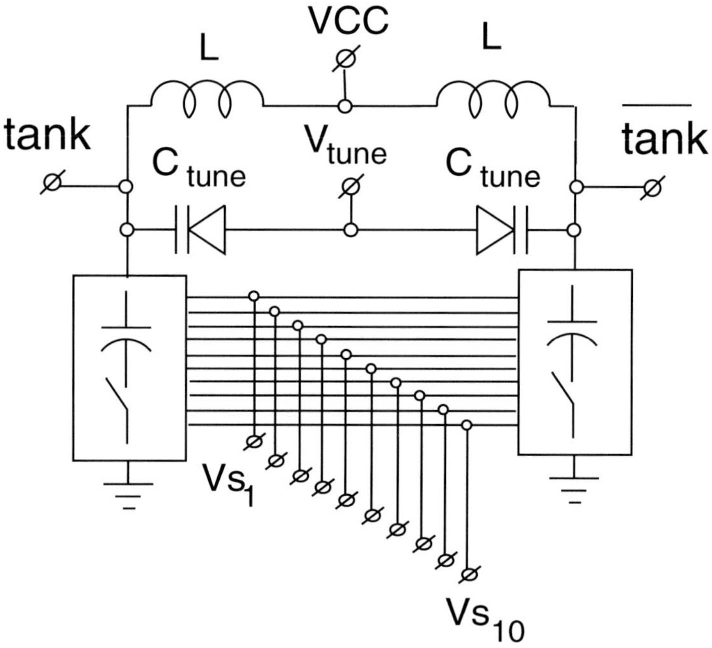

Figure 6.27 Resonator with switched capacitors. This design is used for the LO of a Low-IF FM radio receiver.

Resistance $R_{Q_{on}}$ needs to be minimized for a high Q_b, and this can be achieved by applying maximum V_{switch} and maximizing W. However, since the parasitic capacitances of transistor Q_{switch} are proportional to W, $C_{bandmin}$ places an upper bound on W. A design procedure for a switched capacitor array, optimizing for quality factor and tuning range, while taking into account the discussed constraints, is described in [121].

> Like capacitors, inductors can be switched too. However, since the chip area of inductors usually is large and on-chip inductors are more difficult to design than capacitors, switching between frequency bands with capacitors is preferred in practice to band-switching with inductors.

Example of resonator with switched capacitors

An example of an application where band-switching can be used to cover a large tuning range is FM radio. As shown in Table 4.1 on p. 69, the consumer FM radio band ranges from 87.5 MHz to 108 MHz, which is more than 20% relative tuning range. In Figure 6.27, the resonator design of a LO for a low-IF FM radio receiver

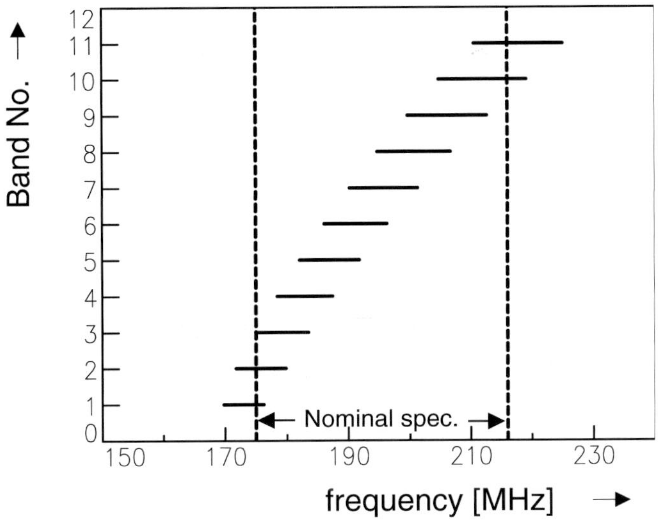

Figure 6.28 Measured frequency bands of the FM-LO

is shown [120]. The FM-LO is designed for a frequency twice the FM radio band, and a divider is cascaded that generates quadrature signals at $f_{osc}/2$. Since in the implementation technology, an 11 GHz, 0.6 μm BiCMOS process, no varactor was available with sufficient $\alpha_{varactor}$, band-switching was used. Varactor C_{tune} in Figure 6.27 was constructed with an NPN-transistor as explained the previous section, and its $C_{min} = 4.9$pF at $V_{tune} = 0.2V$ (with $VCC = 2.7V$) and $C_{max} = 8.5$pF at $V_{tune} = 2.5V$. Ten band-switches were used to divide the tuning range into 11 bands. The band-switch is constructed using a capacitor C_{switch} of 1.5 pF and an NMOS with a W/L of 300/0.6 and 12 gate fingers. Simulated C_Q was around 370 fF and the band-switch quality factor $Q_b > 23$. Inductance L_p was chosen 40 nH and constructed with an external PCB-coil ($\approx$32 nH), some package inductance ($\approx$5 nH) and bond-wire inductance ($\approx$3 nH). Note that the external PCB coil makes the application of the FM-LO flexible. For example, the center frequency of the oscillator can easily be adapted to the Japanese FM radio band, which starts at 76 MHz and ends at 90 MHz. The measured tuning range of the FM-LO is shown in Figure 6.28, revealing sufficient overlap between the frequency bands.

A SEM photo of the FM-LO is shown in Fig. 6.29 and its size is approximately 0.7 mm^2. The active oscillator part is a cross-coupled pair with a tail current of 2.25 mA. Therefore, the dissipation of the FM-LO was 6.1 mW, with a supply voltage of 2.7 V. The worst case measured $\mathcal{L}(10\,\text{kHz})$ was -95 dBc/Hz (at the maximum frequency). This figure improves 6 dB, since the LO-signal is divided by 2 in the low-IF architecture. This shows that band-switching can be used to achieve a large tuning range combined with excellent phase noise performance, at the cost of a higher complexity. It is

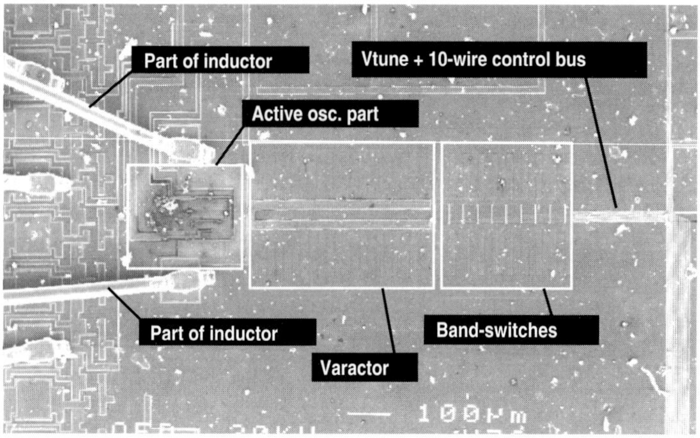

Figure 6.29 SEM photo of the FM oscillator.

interesting to mention that in the case of this particular design, the band-switches and varactor (via a D/A converter) were controlled by a microprocessor of the application, in which the FM receiver was embedded (e.g. a mobile phone or MP3 player) [120].

Continuous active capacitive tuning

Circuit topologies can be used to emulate a varactor function. Large tuning ranges can be achieved at high frequencies using variable impedance converters [188, 189].

> Since active devices are part of active capacitive tuning solutions, the additional noise of the active devices can degrade the spectral purity of an oscillator severely, compared to passive tuning solutions.

To investigate the properties of a variable capacitance solution, an architecture based on current multiplication is explored. The concept and transistor implementation are shown in Figure 6.30(a) and (b), respectively. From the behavioral model in Figure 6.30(a), we can derive the value of the active varactor

$$\frac{1}{j\omega C_{active}} = \frac{v_x}{i_x} = \frac{1}{j\omega(\frac{C}{2}(1+\alpha_{cur}))}, \tag{6.39}$$

where C is the fixed capacitance and α_{cur} the variable current multiplication factor. The concept can be realized with a Gilbert cell, which implements the current multiplier (Figure 6.30(b)).

The simulated capacitance curve of the variable capacitance topology is shown in Figure 6.31 (solid curve). The simulated $\alpha_{varactor}$ is equal to 7.6 (5.3 pF/0.7 pF). Note that the tuning curve can be linearized by driving the current multiplier tuning voltage

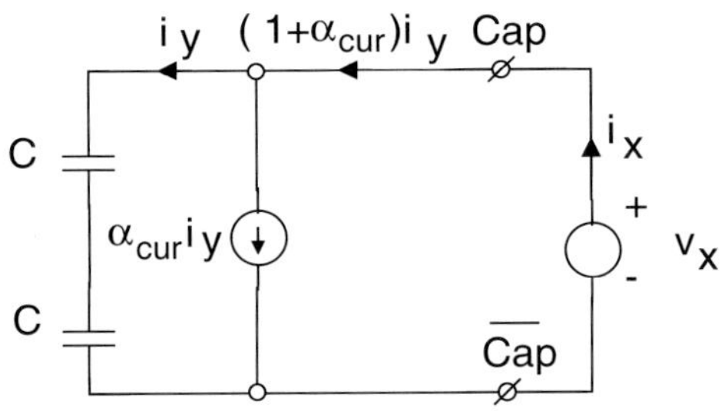

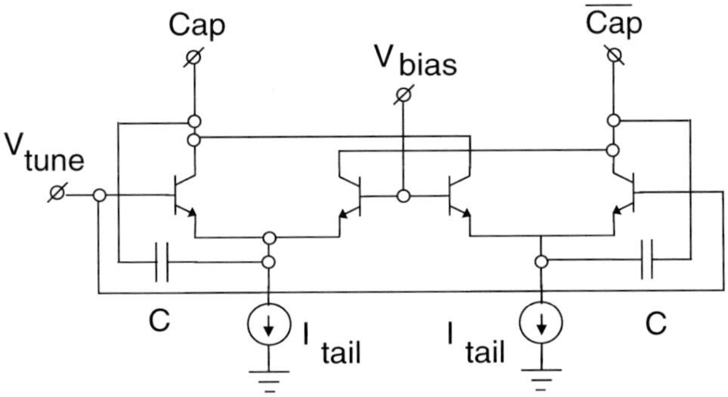

Figure 6.30 An active variable capacitance topology. The behavioral model (a) and its implementation with a Gilbert cell (b).

via two diode-connected transistors [87]. Figure 6.31 (dashed curve) shows the equivalent input noise current density $(\overline{i^2_{n_{tot}}})^{-\frac{1}{2}}$ of the active varactor versus tuning voltage V_{tune}. When the Gilbert cell is in balance (α_{cur}=0), $(\overline{i^2_{n_{tot}}})^{-\frac{1}{2}}$ reaches its maximum of 24 pA/$\sqrt{Hz}$. The varactor capacitance is 3 pF at this point. In order to be able to compare this varactor with passive varactor quality factors , it is useful to define an "effective quality factor" $Q_{C_{active}}$. This quality factor definition also takes active device noise into account. If we model the active varactor as a parallel circuit of C_{active}, a resistance R and a current noise source with current density $\overline{i^2_{n_{tot}}}$, the quality factor of an active varactor can be expressed as follows.

The effective quality factor of an active varactor is defined as

$$Q_{C_{active}} = Q_{varactor}\sqrt{\frac{\overline{i^2_{n_R}}}{\overline{i^2_{n_{tot}}}}}, \tag{6.40}$$

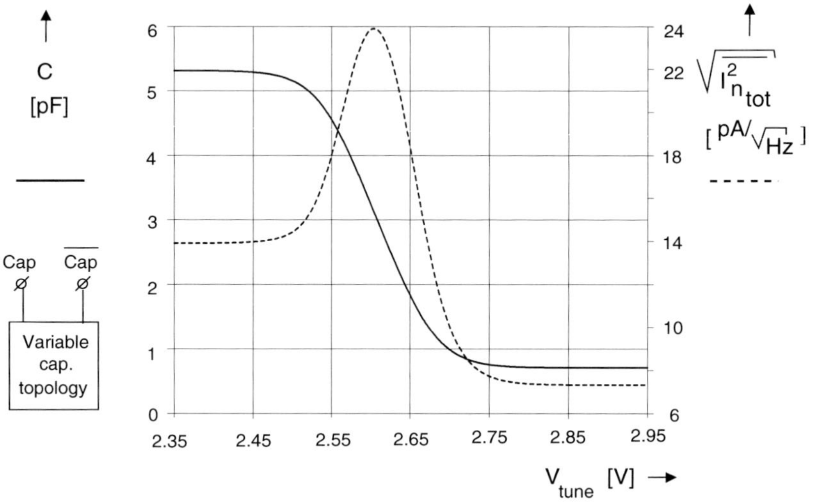

Figure 6.31 Simulated capacitance characteristic of the active varactor and its equivalent input noise current density versus V_{tune}.

with $Q_{varactor} = \omega C_{active} R$, and $\overline{i^2_{nR}}$ the noise current density of R: $4kT/R$. The formula for $Q_{varactor}$ is identical to the quality factor definition of a passive varactor with equivalent parallel resistance R. Since for a passive varactor $\overline{i^2_{nR}} = \overline{i^2_{ntot}}$, (6.40) reduces to $Q_{varactor}$. An active varactor will have $\overline{i^2_{ntot}} > \overline{i^2_{nR}}$, and this can result in a significant decrease of $Q_{C_{active}}$. For example, when $(\overline{i^2_{ntot}})^{-\frac{1}{2}}$ reaches its maximum of 24 pA/$\sqrt{Hz}$ (see Figure 6.31), the simulated R is 1kΩ and C_{var} is 3 pF (simulated). $Q_{varactor}$ is equal to 5.8, whereas $Q_{C_{active}}$ drops to 1.

> Especially in the middle of the tuning range, poor phase noise performance can be expected of active capacitive tuned LC oscillators.

In Section 9.3 the design of a reference oscillator for a double-loop tuning system is described. This tuning system can be used in digital satellite TV receivers. The tuning range specification of this oscillator is from 237 MHz to approximately 310 MHz. Several varactor options to realize this tuning range are investigated in Section 9.3. One of them was the active variable capacitance topology from 6.30(b). An LC oscillator with this varactor was realized in an 11 GHz f_T bipolar technology, to verify expectations concerning the poor phase noise of active varactors. The active part of this oscillator is described in Section 9.3 and shown in Figure 9.22 on p. 238. With a supply voltage of 3.5 V, it consumes about 4 mA. On top of that, the active varactor draws also 4 mA. Hence, compared to passive solutions, additional power also is a drawback of active varactor solutions. The frequency and $\mathcal{L}$(10kHz) of the LC

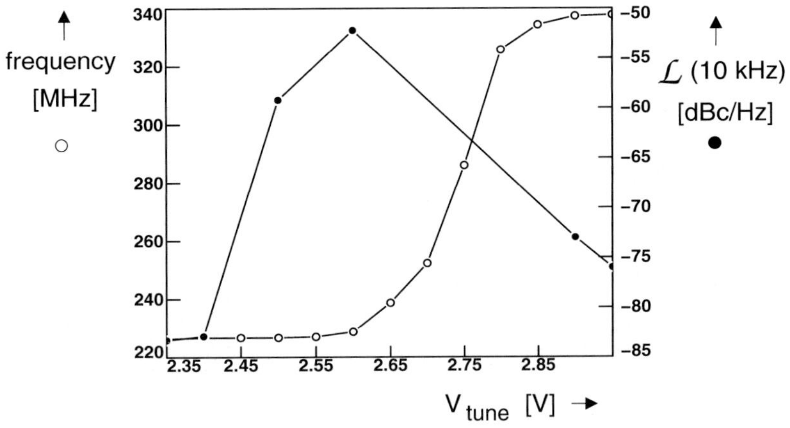

Figure 6.32 Measured frequency and $\mathcal{L}(10\text{kHz})$ of a 300 MHz LC oscillator with air coils and the integrated active varactor from Figure 6.30.

oscillator with integrated active varactor and two 32 nH external air coils are shown in Figure 6.32. As expected, $\mathcal{L}(10\text{kHz})$ varies significantly across the whole tuning range and drops to below -55 dBc/Hz, around $V_{tune} = 2.6V$. At minimum and maximum frequency $\mathcal{L}(10\text{kHz})$ is more than 20 dB lower. At these extreme frequencies, phase noise in oscillators with active tuning solutions is often measured [188], but as Figure 6.32 illustrates, worst case $\mathcal{L}(f_m)$ is measured somewhere in between. Therefore, especially for active varactor solutions, $\mathcal{L}(f_m)$ should be evaluated across the whole tuning range. The die-photo of the discussed oscillator is shown in Figure 6.33, and has an active die-area of 0.26 mm^2.

> As (6.35) suggests, it is important to note that the quality factor of active varactors (and likewise any varactor) can be improved by placing high-quality capacitors in series or in parallel. Therefore, when only a very small tuning range is required, active capacitive tuning may be a viable solution.

A good example of a design achieving excellent phase noise performance with active varactors, uses an active varactor topology to tune a crystal oscillator for an FM Stereo Decoder [122].

Continuous passive inductive tuning

Figure 6.34 shows an implementation of a differential resonator, which uses continuous passive inductive tuning [123]. Inductors L_2 have a different value than L_1, and in series with L_2, NMOS control transistors are connected. For a high control voltage $V_{control}$, the control transistors are on most of the time, and the total single-ended

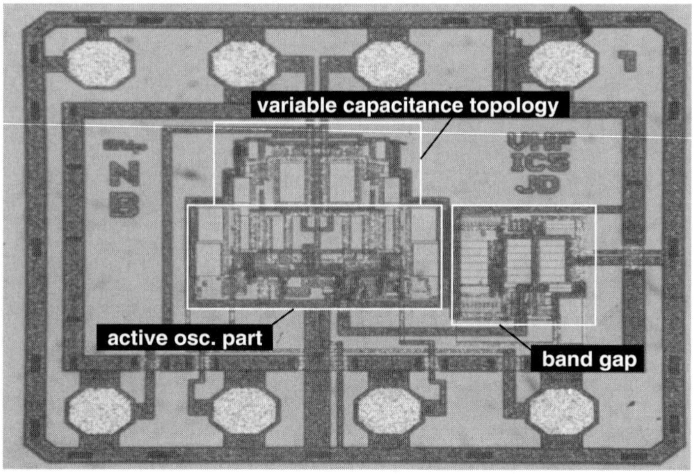

Figure 6.33 Micrograph of the LC oscillator with active capacitive tuning.

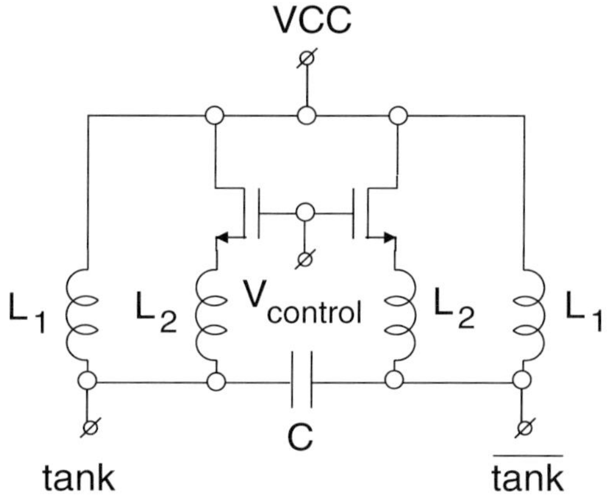

Figure 6.34 Continuous passive inductive tuning.

inductance approximately consists of L_1 and L_2 in parallel. As a result the resonance frequency is high. For a low $V_{control}$ value, L_2 is not active and the resonance frequency is low. In other words, $V_{control}$ determines the fraction of inductance L_2 that is switched in parallel with L_1 and can be used for continuous tuning.

Using continuous passive inductive tuning, a tuning range of 1.34 GHz to 2.14 GHz can been realized [123]. This constitutes 46 % of the relative tuning range, significantly more than normally is achieved in LC oscillators.

A major drawback of passive inductive tuning is its inferior phase noise performance compared to passive capacitive tuning methods.

The resistance of the NMOS control transistors (this resistance is approximately equal to $1/(\mu_n C_{ox}(W/L)(V_{control} - V_s - V_{TH})$ in the triode region) degrades the quality factor of L_2, and therefore the overall resonator quality factor. Reported $\mathcal{L}(1\text{MHz})$ varies from -117 dBc/Hz at 1.34 GHz to -102 dBc/Hz at 2.14 GHz [123].

Continuous active inductive tuning

Inductance L_p in Figure 6.1, or part of it, can be implemented with a tunable active inductor. For high frequencies, a bipolar transistor with shorted base and collector becomes an active inductor [124–126]. Likewise, a MOS-transistor with an intrinsic or externally added gate resistance and shorted gate/drain can emulate an inductance at high frequencies. Also, more complex active inductances can be constructed, for example gyrator-based[21] active inductances [128, 129].

First order calculation of the admittance Y_{ind} considering the combined base-collector terminal of a bipolar transistor shows that its value is approximately

$$Y_{ind} \approx \frac{1}{R_{base}} + \frac{g_{m_{bipo}}}{j\omega C_\pi R_{base}} + j\omega C_{js}. \tag{6.41}$$

Equation (6.41) is derived assuming $\omega\, r_\pi\, C_\pi\, R_{base} \gg R_{base} + r_\pi$. The inductive part of Y_{ind}, $(C_\pi\, R_{base})/g_{m_{bipo}}$, can be tuned by varying the total base resistance (i.e. by adding an external variable resistance) or transconductance $g_{m_{bipo}}$. Thus, by making deliberate use of the parasitic phase shift of a transistor at high frequencies, a tunable active inductor is obtained.

The simulated impedance of an active inductor (Q_L) in parallel with a parasitic capacitance C_{par} is shown in Figure 6.35, to illustrate active inductive tuning. The transistor size used for simulation was a bipolar transistor with emitter area 4 × 0.9 μm × 6.2 μm (four emitter contacts), with a base resistance of 35 Ω, and a peak f_T of 30 GHz around a current level of 9 mA. Figure 6.35 shows that transistor Q_{L1} forms a resonator with C_{par} (chosen 100 fF) with a resonance frequency that can be varied over a large range by varying the external base resistance (R_{tune1}). As mentioned, alternatively current I_{bias} can be varied to tune the resonance frequency [229].

Similar to the quality factor definition of an active capacitance, an effective quality factor can be defined for active inductances. If we model the active inductance as a parallel circuit of L_{active}, a resistance R and a current noise source with current density $\overline{i^2_{n_{tot}}}$, the quality factor of an active inductance can be defined as

[21] The impedance seen at the input of a gyrator with gyrator-constant G, terminated at the output with a capacitor C, is an inductance equal to C/G^2 [127].

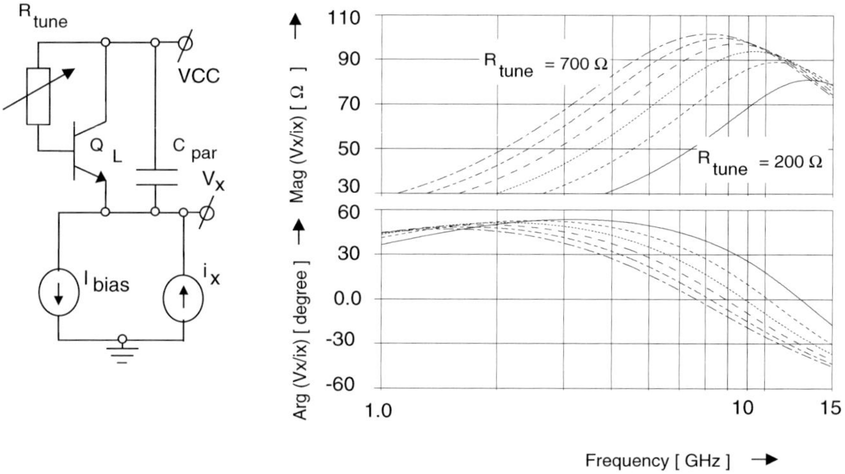

Figure 6.35 The magnitude and phase of a tunable active inductor in parallel with 100 fF parasitic capacitance with R_{tune1} varying from 200 Ω to 700 Ω.

$$Q_{L_{active}} = Q_{inductance}\sqrt{\frac{\overline{i_{n_R}^2}}{\overline{i_{n_{tot}}^2}}}, \tag{6.42}$$

with $Q_{inductance} = R/(\omega L_{active})$, and $\overline{i_{n_R}^2}$ the noise current density of R: $4kT/R$. As the reader will intuitively realize, using active tunable inductance is a bad idea for low-phase-noise oscillator design. For example, an LC oscillator with gyrator-based active differential inductance achieves a simulated $\mathcal{L}(100\text{kHz})$ of -65 dBc/Hz with 8.6 mW dissipation at 3 V supply voltage [128]. This is more than 40 dB below the GSM specification. On a positive note, the oscillator does have a large tuning range: from 400 MHz to 1.16 GHz. In Section 9.4, the simulated tunable active inductance in Figure 6.35, will be used in a 10 GHz ring oscillator. As will be shown, ring oscillators have poorer phase noise performance than LC oscillators (given the same amount of power), which means that the contribution of the active inductor to the total $\mathcal{L}(f_m)$ is smaller (it is not necessarily the dominant cause of phase noise, whereas this most likely would be the case in an LC oscillator design).

Phase shift tuning

As is evident from (5.11) on p. 95, and Figure 5.9 on p. 100, an N-stage LC oscillator can be tuned by changing resonator phase shift ϕ_{res}. In principle, this tuning method can also be employed in single-phase LC oscillators. In Section 6.3 it will be shown

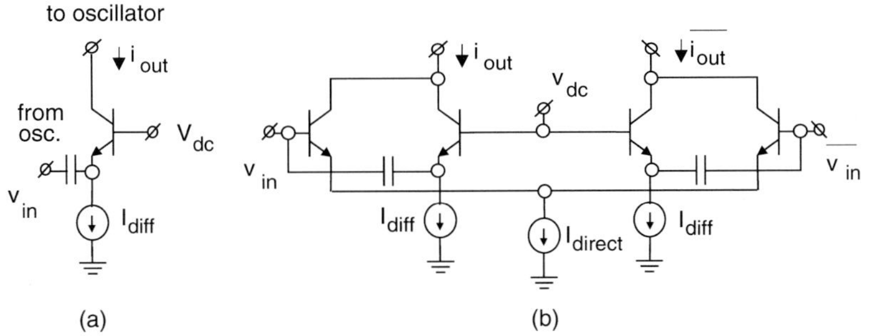

Figure 6.36 A single-ended phase shifter with fixed phase shift (a), and a differential tunable phase shifter (b).

that $\mathcal{L}(f_m)$ is at its optimum, if θ in Figure 6.5 is equal to $\pm 180/N$. Therefore, an intentional phase-shifter can be present between the oscillator stages. In any case, coupling circuits are already present in N-stage LC oscillators, and phase shift tuning is more easily implemented, compared to single-phase LC oscillators.

In Section 6.3, analysis of the phase noise in two-stage I/Q LC oscillator leads to the insight that minimum $\mathcal{L}(f_m)$ is obtained with a phase shift θ of $\pm 90°$ (See Figure 6.5 on p. 119) . As a $\pm 90°$phase shift for sinusoidal signals is similar to an integrating or differentiating action, it is possible to implement the phase shifters based on an integrator or a differentiator. Figure 6.36(a) shows the circuit implementation of a differentiator. The circuit not only provides the required phase shift, but also provides isolation between the oscillator stages. The current i_{out} flowing into the collector shows a -90° phase shift compared to the input voltage v_{in} up to a certain frequency. Above this frequency, the phase will deviate from -90° due to additional poles in the transistor. This effect can be compensated for by adding a direct path between v_{in} and i_{out}, as is shown in Figure 6.36(b). In principle, any phase shift can be made by proper dimensioning of the currents I_{direct} and I_{diff} of the direct path and the differentiating path, respectively. At high frequencies, however, parasitic effects limit the range of the phase variation, and hence the tuning range.

At this point it is instructive to write down the quality factor of N-stage LC oscillators (modeled by the linear model on p. 119). This quality factor will be discussed on p. 161 and can be written as

$$Q_{NLC} \approx N \cdot Q_p \cdot \cos(\phi_{res}). \tag{6.43}$$

Quality factor Q_{NLC} is maximal, if $\phi_{res} = 0$. Phase shift tuning can be performed, but at the price of a reduced minimum quality factor, resulting in an increase in phase noise.

Oscillator switching

Band-switching using capacitors or inductors has already been discussed. Instead of switching resonator elements on and off, oscillators can be switched on and off [130, 131]. Each oscillator covers one of the overlapping frequency bands, and the varactor or varactors in each oscillator cover the frequencies within one band.

> A disadvantages of oscillator switching as tuning method are that much more chip area is needed and circuitry is needed to select and switch unused oscillators off, as well as circuitry that connects the selected oscillator to the blocks where its signal is used (a "multiplexer"). An advantage of oscillator switching is that the overall phase noise performance will be superior compared to single oscillator designs.

Each oscillator can be optimized for maximum $\mathcal{L}(f_m)$ independent of the other oscillators. Activating and deactivating an oscillator can be as simple as switching a tail current on and off, and this can be easier to realize than making high-quality switched capacitors at high frequencies.

6.2.2 Ring oscillators

Tuning of a practical two-integrator oscillator (modeled by Figure 6.10) can be realized by varying the transconductance g_{m_t}, g_{m_a}, capacitor C or resistor R, as indicated by (6.18). Transconductance g_{m_t} is normally used for tuning, similar to the ideal two-integrator oscillator model discussed in Chapter 5. If the transconductances are implemented by bipolar differential pairs, the frequency will vary linearly with the tail current. If MOS differential pairs are used in saturation, the frequency will be proportional to the square root of the tail current. Figure 6.37 shows the tuning behavior of a bipolar implementation (see Figure 6.12) and the effect of transistor parasitics. A large part of the nonlinearity is determined by the diffusion capacitance part of C_π, which is current dependent. Since g_{m_t} can be varied over a large range, the tuning range can easily be made more than an octave [52, 132], and can even span four decades [133].

If two-integrator oscillators operate in strongly nonlinear mode, linear equations such as (6.18) and (6.19) become invalid, and the frequency of the oscillator should be expressed in terms of stage delay (see 5.14 on p. 97). In that case, similar to N-stage oscillators with $N \geq 3$, two-stage oscillators can be tuned by any means, which changes the stage delay τ_{delay}, and several methods are discussed below.

Ring oscillators (see Figure 6.15 on p. 128 for a behavioral model for $N \geq 3$) can be tuned in at least four different ways. The first two were already discussed in Chapter 5. By changing resistance R or capacitance C in Figure 6.15, the oscillation frequency changes. The third method is to change the delay in a practical stage by changing parasitic time-constants of the circuit implementation of transconductance g_m in Figure 6.15. The fourth possibility is to use two signal paths per stage with different delays, and summing the outputs of the separate paths with a controllable weighting factor. This latter technique is called delay interpolation.

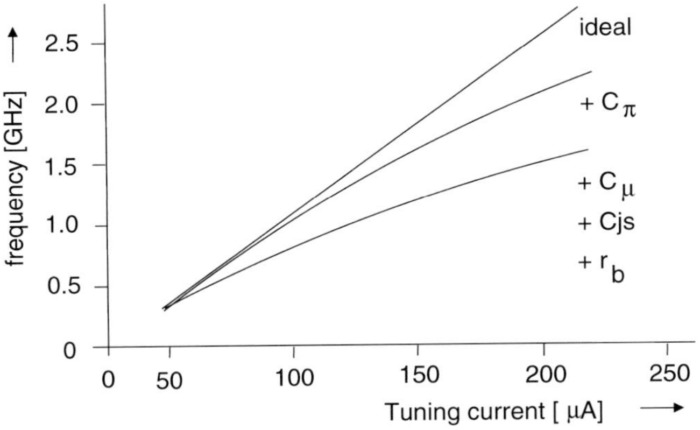

Figure 6.37 Typical tuning characteristic of a two-integrator oscillator.

Capacitive tuning can be realized using the techniques described in Section 6.2.1. The tuning range then is rather limited and nonlinear compared to the other available techniques for tuning ring oscillators. Capacitive tuning also limits the maximum oscillation frequency, since a varactor with a certain minimum capacitance must be added. For these reasons, capacitive tuning is rarely used in combination with ring oscillators.

If the transconductances in Figure 6.15 are implemented by bipolar differential pairs, for example, the parasitic delay (controlled by the tail current) can be used for tuning (see (6.28) on p. 131). However, the tuning range is very small, especially when t_{delay} is dominated by the stage output-time-constant determined by R and C.

> Linear tuning ranges of several octaves are possible when employing resistive tuning.

Two examples are given in Figure 6.38. The capacitors should be omitted in these circuit diagrams for a maximum oscillation frequency. In Figure 6.38(a), the cross-coupled pair implements a negative resistance, which is in parallel with the collector resistors and input resistance of the next stage. Therefore, by varying I_{tune} the total resistance and thus the frequency is varied. A second possibility is shown in Figure 6.38(b). The PMOS transistors operate as variable resistances controlled by V_{tune} in the triode region. Since it is not only the frequency that varies with both methods but also the gain in each stage, the amplitude level varies significantly over the tuning range. This means that $\mathcal{L}(f_m)$ will also vary. By increasing I_{level} in Figure 6.38(b) when V_{tune} decreases, the swing variation across the tuning range can be decreased [134].

Figure 6.39 shows an example of ring oscillator tuning applying delay interpolation [39]. The differential pair controlled by V_{tune} determines the ratio of the output

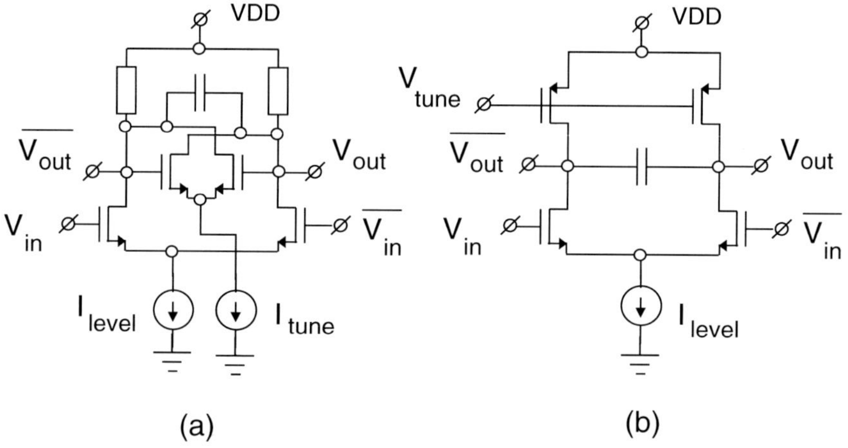

Figure 6.38 Examples of resistive tuning in a stage using a negative controllable resistance (a) or MOS transistors in the triode region (b).

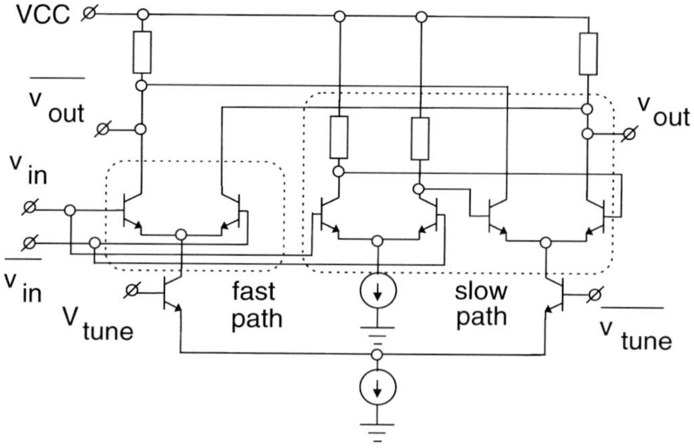

Figure 6.39 Bipolar implementation of a delay interpolation stage.

current of the fast and the slow path. The slow path has one additional delay compared to the fast path, implemented by a differential pair.

> Ring oscillators, just like LC oscillators, can have large tuning constants (i.e a large K_{vco} or K_{cco}). A large gain constant means severe requirements for the output noise of a V/I-converter in front of a CCO, or for the maximum noise voltage on the input of a voltage controlled ring oscillator.

Band-switching can reduce the tuning constant of ring oscillators, in the same way it can reduce the tuning constant of LC oscillators. Instead of switching capacitors with a MOS-transistor, switchable current sources are often used [52]. For example, the output current of a V/I converter can be summed with the output of current DAC (iDAC) and this current controls the frequency of a ring oscillator [135]. Obviously, the price for a lower tuning constant is an increase in complexity; the current switches or iDAC must be controlled by some sort of tuning mechanism.

In Appendix H, a number of ring oscillators reported in literature are summarized. Their relative tuning range are listed. Several ring oscillators achieve a relative tuning range of 80% or more, and most designs have a relative tuning range above 30%.

6.3 $\mathcal{L}(f_m)$: linear time-invariant modeling

One of the first phase noise models for oscillators was proposed by Leeson in 1966 [14]. Leeson's formula includes many characteristics of real oscillators, such as the -6 dB per octave decay of the phase noise sidebands close to the carrier. Since no formal proof was given by Leeson, and his formula includes a noise figure as a "fit" factor, this formula is generally considered to be heuristic.

Since Leeson (1966), many publications and phase noise theories have raised the understanding of phase noise generation mechanisms to a mature level.

It is important to stress that in general, a time-varying theory is needed to predict oscillator phase noise accurately. Bias conditions vary significantly in an oscillator working in strongly nonlinear mode and many nonlinear phase noise generation mechanisms will then add to the total oscillator phase noise. Nevertheless, many design insights provided by recent phase noise theories [29,30,33,34,104], which include nonlinearity and time-varying aspects, can be estimated and explained using linear time-invariant (LTI) oscillator modeling. This notion is underlined by several publications reporting a reasonable match (less than 1–4 dB difference) between $\mathcal{L}(f_m)$ expressions obtained with linear modeling and $\mathcal{L}(f_m)$ measurements [26, 136, 137].

The phase noise to carrier ratio of LC and ring oscillators will be analyzed using linear time invariant modelling in this section. This analysis will be followed by a discussion of neglected nonlinear and time-variant phase noise aspects in Section 6.4.

6.3.1 LC oscillators

The following $\mathcal{L}(f_m)$ calculations are based on the noise shaping theory for single-stage and multi-stage oscillators described in Section 3.4.1 on p. 58. Since this theory is derived at a system level, it allows a unified approach to $\mathcal{L}(f_m)$ calculations of LC and ring oscillators (Section 6.3.2). The phase noise generated by noise shaping gives rise to a -6 dB/octave slope of the oscillator spectrum, which usually is the biggest part of the oscillator sideband (see Figure 4.3 on p. 73).

Single-phase LC oscillators

Consider the oscillator model in Figure 6.1 on p. 113. As shown in Appendix I, $\mathcal{L}(f_m)$ is equal to

$$\mathcal{L}_{LC}(f_m) = \frac{1}{2} \cdot \frac{1}{4Q_p^2} \left(\frac{f_{osc}}{f_m}\right)^2 \frac{\overline{i_n^2}}{i_{carrier}^2}, \qquad (6.44)$$

for this model in which $i_{carrier}^2$ the squared rms carrier current. It is instructive to see whether we can rewrite (6.44) into Leeson's equation for the phase noise of an oscillator [14]. Indeed if the noise factor F is defined as $\overline{i_n^2}/(4kT/R_p)$ and P_{RF} as $i_{carrier}^2 \cdot R_p$, Leeson's equation appears:

$$\mathcal{L}(f_m) = \frac{1}{2} \cdot \frac{1}{Q_p^2} \left(\frac{f_{osc}}{f_m}\right)^2 \frac{FkT}{P_{RF}}. \qquad (6.45)$$

The terms $\overline{i_n^2}$ and $i_{carrier}^2$ in (6.44) can be derived from an implementation of the transconductance g_m (the active oscillator part) in Figure 6.1.

Consider for example the bipolar and MOS LC oscillator shown in Figure 6.40. Both cross-coupled differential pairs have a tail current I_{tail} and a transconductance of $-g_{m_{bipo}}/2$ and $-g_{m_{MOS}}/2$ (see Section 3.4.2 on p. 62), respectively. The total noise $\overline{i_n^2}$ current of the bipolar LC oscillator in Figure 6.40(a) is

$$\overline{i_n^2} \approx \frac{\overline{i_{n_{bipo}}^2}}{2} + \frac{4kT}{R_p} = \frac{4kT}{R_p}\left(\frac{1}{2}\alpha_{ol} + 1\right), \qquad (6.46)$$

with $\overline{i_{n_{bipo}}^2} = 2kT/g_{m_{bipo}}$, if we neglect the base resistance noise and assume that current gain $\beta_O \gg 1$. In order to assure start-up of the oscillator, the transconductance $-g_{m_{bipo}}/2$ always is larger than the value required to compensate all losses $(1/R_p)$. This is taken into account in (6.46) by the open loop gain α_{ol}. The factor α_{ol} is one or

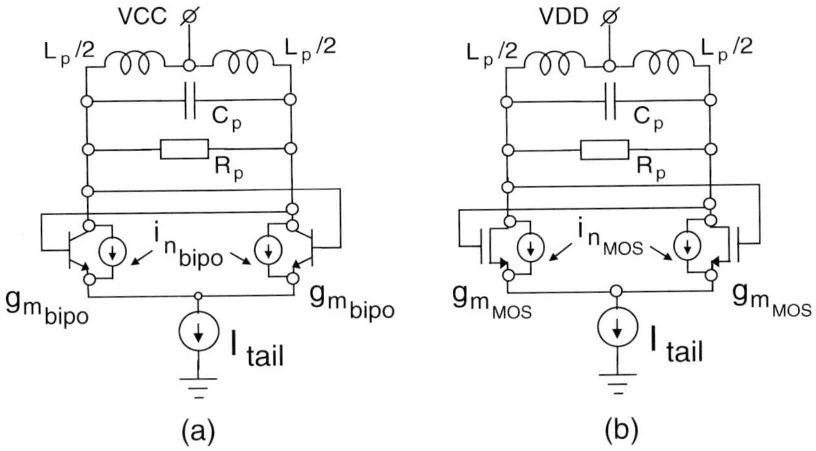

Figure 6.40 Bipolar (a) and NMOS cross-coupled LC oscillator (b).

larger and, in practice, in the range of two to five for self-limiting oscillators. Alternatively, α_{ol} can be close to unity when an AGC-loop is used to stabilize the oscillator amplitude.

The total noise current $\overline{i_n^2}$ of the MOS LC oscillator in Figure 6.40(b) is

$$\overline{i_n^2} \approx \frac{\overline{i_{n_{MOS}}^2}}{2} + \frac{4kT}{R_p} = \frac{4kT}{R_p}\left(\frac{2}{3}\alpha_{ol} + 1\right), \tag{6.47}$$

in which the factor $2/3$ is only accurate for long-channel devices and can be significantly higher in short-channel devices [138]. Furthermore, (6.47) only holds when the differential pair is operating in the strong inversion region [139]. This equation also is an approximation since noise sources such as gate and bulk resistive noise are neglected. As can be seen from (6.46) and (6.47), the noise sources of a bipolar and MOS cross-coupled LC oscillator only differ by a small factor when using first order modeling. For completeness, it is interesting to note that the total noise current $\overline{i_n^2}$ of the complementary CMOS LC oscillator in Figure 6.4(a) on p. 117 is also given by (6.47), if we assume that $g_{m_{Qn}} = g_{m_{Qp}}$ in this circuit.

In the following formulas, where (6.44) will be further expanded, only $\mathcal{L}(f_m)$ of the bipolar cross-coupled pair of Figure 6.40(a) will be considered, but adaptation to a MOS implementation is straightforward.

> The maximum peak carrier current of the bipolar differential pair is reached when it is fully switching, and is equal to $2/\pi \cdot I_{tail}$. Note that this is $2I_{tail}$ for a Colpitts oscillator [34], and $4/\pi \cdot I_{tail}$ for the complementary LC oscillator in Figure 6.4(a).

Depending on the value for α_{ol}, the oscillator may no longer switch at high frequencies, where transistor parasitics reduce the internal base voltages. To some extent this can be compensated by increasing α_{ol}. We will use the maximum carrier current as a first order estimate. With $i_{carrier} = (\sqrt{2}/\pi)I_{tail}$ and $\overline{i_n^2}$ given by (6.46), $\mathcal{L}(f_m)$ of a bipolar LC oscillator becomes

$$\mathcal{L}(f_m)_{bipo} \approx \frac{1}{2} \cdot \frac{1}{Q_p^2} \left(\frac{f_{osc}}{f_m}\right)^2 \frac{kT(\frac{1}{2}\alpha_{ol}+1)}{R_p \frac{2}{\pi^2} I_{tail}^2}. \tag{6.48}$$

Alternatively, η_{LC}, the efficiency of converting DC power to RF power for an LC oscillator can be introduced

$$\eta_{LC} = \frac{P_{RF}}{P_{DC}} = \frac{R_p\, i_{carrier}^2}{I_{tail}\, VCC}, \tag{6.49}$$

which means that $\mathcal{L}(f_m)_{bipo}$ also can be denoted as

$$\mathcal{L}(f_m)_{bipo} \approx \frac{1}{2} \cdot \frac{1}{Q_p^2} \left(\frac{f_{osc}}{f_m}\right)^2 \frac{kT(\frac{1}{2}\alpha_{ol}+1)}{\eta_{LC} P_{DC}}. \tag{6.50}$$

To determine how well (6.48) estimates the phase noise of a bipolar cross-coupled oscillator, SpectreRF simulations were performed on an implementation of the oscillator in a 30 GHz f_T bipolar technology, and with an ideal (noiseless) tail current source. SpectreRF is a circuit simulator capable of phase noise simulations, which include all nonlinear effects in oscillators that will be discussed in Section 6.4. In practice, phase noise simulation results produced by SpectreRF show a good matching with measurements [124].

First, transient simulations needed to be performed to determine the exact rms carrier current $i_{carrier}$. In Figure 6.41 the simulation results are shown together with the approximation $i_{carrier} = \sqrt{2}/\pi \cdot I_{tail}$. The first simulation result is for $\alpha_{ol} = 1.3$ (i.e $(gm_{bipo}/2)R_p = 1.3$) and I_{tail} is equal to 24.9 μA. The differential pair is not yet switching, and the calculated $i_{carrier}$ of 45 μA therefore overestimates the actual carrier current. For larger α_{ol}, the differential pair is practically switching, and the calculated $i_{carrier}$ matches with simulations well[22].

Calculation and simulation results of $\mathcal{L}(100\text{kHz})$, for the oscillator under discussion, are shown in Table 6.3 on p. 160. Column $\mathcal{L}_{calc1}$ lists the results calculated with (6.48). Column $\mathcal{L}_{calc2}$ lists the results calculated with (6.48), with $i_{carrier} = \sqrt{2}/\pi \cdot I_{tail}$ replaced by the simulated $i_{carrier}$ from Figure 6.41. The last column of

[22] As long as saturation mechanisms or voltage limiting by VCC do not limit the oscillation amplitude, the oscillator is in the current limited region, as is the case for the simulation results discussed. A further increase of α_{ol} eventually leads to operation in the voltage limited region (see Section 6.6) and the calculation $i_{carrier} = \sqrt{2}/\pi \cdot I_{tail}$ then overestimates the actual carrier current , because $i_{carrier}$ no longer is a function of I_{tail}.

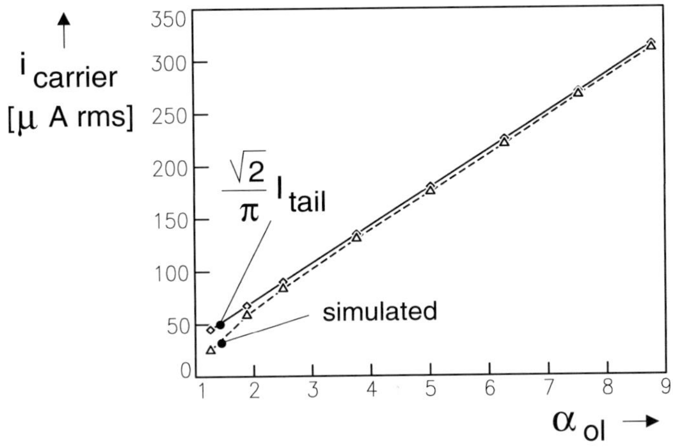

Figure 6.41 Calculated and simulated $i_{carrier}$ of the oscillator in Figure 6.40(a), versus α_{ol}. The oscillator's frequency is 1 GHz and its Q_p is 10 ($L_p = 20$nH and $R_p = 1.26$kΩ). It is implemented in a 30 GHz f_T bipolar technology.

Table 6.3 displays the simulation results of SpectreRF. For $\alpha_{ol} = 1.3$, $\mathcal{L}_{calc1}(100\text{kHz})$ is -89dBC/Hz, and is too optimistic simply because of the error in $i_{carrier}$, shown in Figure 6.41. The value for $\mathcal{L}_{calc2}(100\text{kHz})$, using the simulated $i_{carrier}$, is within 1 dB accurate. When I_{tail} is increased the $\mathcal{L}_{calc1}(f_m)$ becomes approximately identical to $\mathcal{L}_{calc2}(f_m)$, because $\sqrt{2}/\pi \cdot I_{tail}$ approximates actual $i_{carrier}$ accurately.

> For high α_{ol} the differential pair is fully switching and the shot noise of the transistors reduces as the transistors switch on and off. In addition, noise around harmonics folds back, but the shot noise reduction due to transistor switching is dominant for the discussed design example (see Table 6.3).

Both effects are covered by Section 6.4. Equation (6.48) estimates the phase noise over the full range of α_{ol} from 1.3 up to 8.8 with an error less than 5.5 dB. For oscillators working in weakly nonlinear mode (small α_{ol}) it estimates the phase noise accurately within one dB, provided that the actual carrier current (e.g. obtained by simulation) is used in (6.48), instead of the approximation $\sqrt{2}/\pi \cdot I_{tail}$.

A number of observations and remarks can be made regarding the preceding phase noise equations obtained with linear modeling:

- Equations (6.44), (6.45), (6.48), and (6.50) all model the -6 dB per octave slope (proportional to $1/f^2$) of an oscillator. This slope continues for large offset frequencies until it hits the white noise floor of the physical resistors R_{ls} and

α_{ol}	I_{tail} μA	$\mathcal{L}_{calc1}(100\text{kHz})$ dBc/Hz	$\mathcal{L}_{calc2}(100\text{kHz})$ dBc/Hz	$\mathcal{L}_{SpectreRF}(100\text{kHz})$ dBc/Hz
1.3	100	-89	-83.9	-83.7
1.9	150	-91.8	-90.6	-91.2
2.5	200	-93.7	-93.1	-94.3
3.8	300	-96.2	-96	-98
5.0	400	-97.8	-97.6	-100.5
6.0	500	-98.9	-98.8	-102.4
7.5	600	-100	-100	-104
8.8	700	-100.8	-100.8	-105.2

Table 6.3 Calculated and simulated (with SpectreRF) $\mathcal{L}(100\text{kHz})$ of the oscillator in Figure 6.40(a) oscillating at 1 GHz and with $Q_p = 10$, implemented in a 30 GHz f_T bipolar technology.

R_{cs} in Figure 6.2. In practice, the white noise floor is determined by buffers cascaded after the oscillator. Hence, the point where the oscillator sideband hits the white noise floor depends on the buffer design. The measurement equipment used to characterize an oscillator can also introduce a dominant white noise floor.

- From (6.48) it is clear that a doubling of the Q_p in principle gives 9 dB improvement in $\mathcal{L}(f_m)$. The term Q_p^2 contributes 6 dB, but R_p is also doubled, which accounts for 3 dB improvement. In practice, the improvement can also be 6 dB as indicated by (6.44). If $\overline{i_n^2}$ is dominated by device noise, the reduction by a factor two of the tank noise source $4kT/Rp$ caused by the doubling of R_p is negligible, and $\mathcal{L}(f_m)$ improves by 6 dB if the quality factor is doubled.

- Fundamentally, $\mathcal{L}(f_m)$ is 3 dB lower if the power is doubled (see (6.50)). Conceptually, this can be understood by putting two identical oscillators in parallel. In such a case, properties like tuning range and center frequency remain unchanged, but the power is doubled. Since the signal will add voltage-wise and the noise power-wise, a 3 dB improvement in signal-to-noise is obtained. However, as Table 6.3 makes clear, doubling of the power without changing the oscillator circuit (putting two in parallel) can improve $\mathcal{L}(f_m)$ more than 3 dB in a practical oscillator circuit. For example, $\mathcal{L}_{SpectreRF}(100\text{kHz})$ becomes as much as 6 dB better (see Table 6.3) when I_{tail} is increased from 300 μA to 600 μA (see also the final bullet of this list).

- From the latter two conclusions and (6.44) it is clear that the quality factor and the signal power should be maximized for a minimum $\mathcal{L}(f_m)$. This means that every component in the resonator (inductors, varactors, etcetera) should be optimized for minimal losses.

- To make the phase noise estimate calculated by (6.50) more accurate, the loaded quality factor can be substituted in this equation. In practice, the loading by the active circuit may be negligible, and this can be checked with (6.13) on p. 117. For example, the loaded quality factor of the oscillator in the discussed phase noise simulation example is approximately 9.9 instead of 10, at 1 GHz.

- Table 6.3 makes clear that linear modeling with (6.48) provides a reasonable estimate of $\mathcal{L}(f_m)$, but is too pessimistic for large open loop gains. SpectreRF simulations reveal that the relative contribution of the collector shot noise decreases for increasing nonlinear operation of LC oscillators. This effect (modulation of noise sources) will be discussed qualitatively and confirmed quantitatively in Section 6.4. In Section 6.4 noise folding will be discussed as well, which increases with increased nonlinearity [31] and adds to $\mathcal{L}(f_m)$.

> Table 6.3 shows that this effect is not dominant and that the total improvement of increasing α_{ol} can clearly outweigh the negative effect of noise folding.

If one only takes into account the effect of noise folding, one can come to the erroneous conclusion that α_{ol} should be kept small (e.g. close to 2) [100]. In practice, when generation of some harmonics is allowed, a significantly higher α_{ol} should be chosen for optimum $\mathcal{L}(f_m)$ (as long as no other phase noise mechanism starts becoming dominant: eventually when I_{tail} and thus α_{ol} is increased, the LC oscillator will enter the voltage limited region; see Section 6.6).

Multi-phase LC oscillators

Consider the multi-phase oscillator model in Figure 6.5 on p. 119. As shown in Appendix I, $\mathcal{L}(f_m)$ for this model is equal to

$$\mathcal{L}_{NLC}(f_m) = \frac{1}{2} \cdot \frac{1}{4Q_{NLC}^2} \left(\frac{f_{osc}}{f_m}\right)^2 \frac{N \cdot \overline{i_n^2}}{i_{carrier}^2}, \tag{6.51}$$

where $i_{carrier}^2$ is the squared rms carrier current and Q_{NLC} is approximately equal to

$$Q_{NLC} \approx N \cdot Q_p \cdot \cos(\phi_{res}), \tag{6.52}$$

with ϕ_{res} the LC resonator phase shift [106]. Equation (6.52) is a good approximation of a more complex expression given in Appendix I. It has an error less than 1% for ϕ ranging from 0 to 70° [106]. At 80° and 89° the error is 3.8% and 70%, respectively. Equation (6.52) reaches its maximum when the phase shift ϕ_{res} of the resonator in each stage is zero. In that case, the maximum quality factor of an N-stage LC oscillator is

$$Q_{NLC}|_{\phi_{res}=0} = N \cdot Q_p. \tag{6.53}$$

Intuitively this result could be expected, since an LC resonator has its maximum phase slope $d\phi/d\omega$ at zero phase shift. Now the importance of the phase shifter with phase shift θ in Figure 6.5 becomes clear.

> The quality factor of an N-stage multi-phase LC oscillator can be maximized by incorporating a phase shift θ in each stage, with θ equal to $\pm 180°/N$.

The value $\pm 180°/N$ follows from the phase condition for oscillation, and for this value $\phi_{res} = 0$. The phase shift θ can be deliberately implemented by a phase shifter, but can also be partly or completely implemented by parasitic phase shift of active devices, especially at high frequencies.

Substitution of (6.52) in (6.51) yields

$$\mathcal{L}_{NLC}(f_m) \approx \frac{1}{2} \cdot \frac{1}{4N \cdot Q_p^2 \cos^2(\phi_{res})} \left(\frac{f_{osc}}{f_m}\right)^2 \frac{\overline{i_n^2}}{i_{carrier}^2}. \tag{6.54}$$

In a similar way as was done for the single-phase LC oscillator, $\overline{i_n^2}$ and $i_{carrier}^2$ can be expanded for a given N-stage LC oscillator topology. Since an N-stage LC oscillator consists of LC oscillator stages, the preceding conclusions regarding phase noise in single-phase LC oscillators are also of importance for N-stage LC oscillators: obviously, carrier power and quality factor should be maximized in a N-stage LC oscillator too for minimum $\mathcal{L}(f_m)$. In addition to the remarks regarding phase noise in a single-phase LC oscillator, the following remarks and conclusions can be made for a multi-phase LC oscillator:

- From (6.54) it follows that the resonator phase shift must be minimized for a minimum $\mathcal{L}(f_m)$. This can be achieved by making the phase shift θ equal to $\pm 180°/N$ in an N-stage LC oscillator. Note that (6.54) is also valid for $N = 1$. Equation (6.54) with $N = 1$ is identical to (6.44), if the resonator phase shift ϕ_{res} is zero.

> For a single phase oscillator, the phase shift of the active part should be minimized for a minimal $\mathcal{L}(f_m)$.

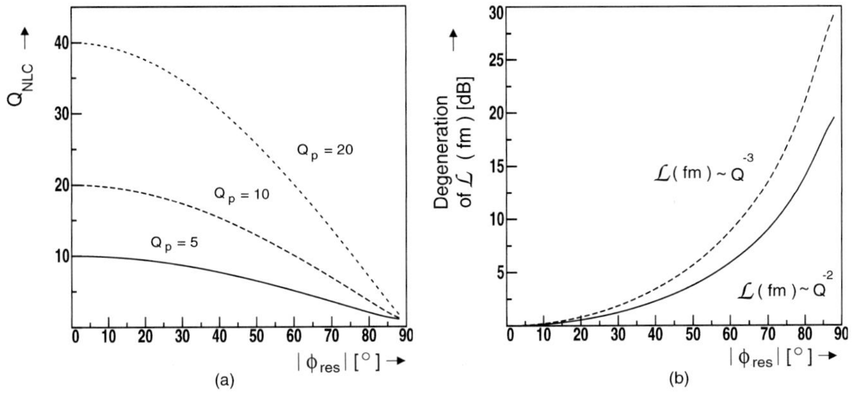

Figure 6.42 Quality factor Q_{NLC} versus the resonator phase shift, for Q_p is 5,10 and 20 and $N = 2$ (a). On the right, the degeneration of the $\mathcal{L}(f_m)$ versus the resonator phase shift, for $N = 2$, $Q_p = 5$, and for $\mathcal{L}(f_m)$ proportional to Q^{-3} and Q^{-2} (b).

○ Figure 6.42(a) shows the Q_{NLC} as a function of the resonator phase shift ϕ_{res} and the related degradation in $\mathcal{L}(f_m)$ for a quadrature LC oscillator ($N = 2$). As mentioned during the discussion of the phase noise of a single phase oscillator, $\mathcal{L}(f_m)$ can be proportional to $1/Q_p^2$ or $1/Q_p^3$. Both cases are plotted in Figure 6.42(b). At around 45° and less, the degradation is smaller than 5 dB. For large ϕ_{res}, the degradation can amount to more than 25 dB. Hence, the phase shift θ in Figure 6.5 on p. 119, which takes care of the required phase shift per section ($\pm 180°/N$) to obey the phase condition for oscillation at $\phi_{res} = 0$, is needed for optimal performance. At high frequencies, where the parasitic phase shift of the active devices is relatively large, part of the required phase shift per section can be provided by these devices.

○ Equation (6.54) indicates that $\mathcal{L}_{NLC}(f_m)$ improves by 3 dB, each time the number of stages is doubled. However, this improvement comes at the expense of an N-times increase in power. Therefore, given the same power budget, the achievable $\mathcal{L}_{LC}(f_m)$ is in principle identical to $\mathcal{L}_{NLC}(f_m)$.

○ The phase shifters in Figure 6.5 on p. 119 when given a phase shift $\theta = \pm 180°/N$, lower the risk of spurious oscillation in a multi-phase LC oscillator. If the LC resonators in each oscillator stage are not at zero phase shift, more than one oscillation mode will exist as explained on p. 121, which can cause multi-oscillation and can hamper proper operation of the circuit. If and when this phenomenon occurs, depends on the value of the coupling current and the degree of nonlinearity in the multi-phase LC oscillator [185].

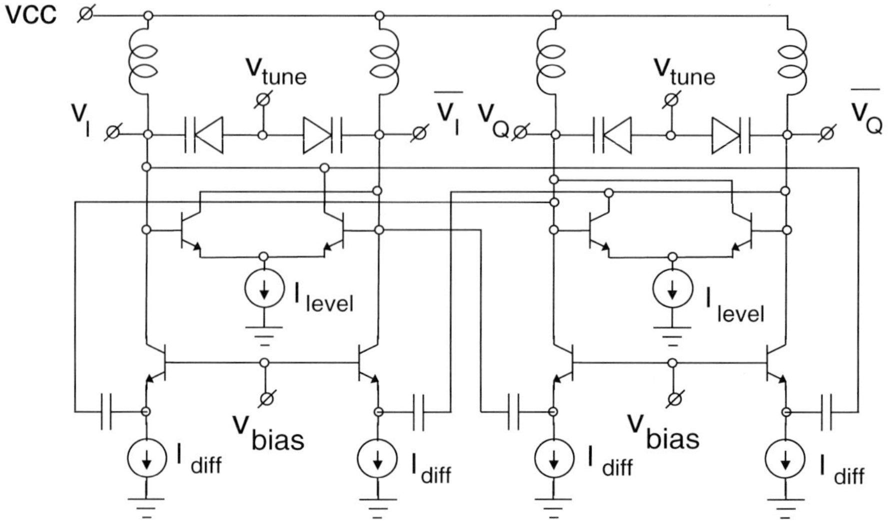

Figure 6.43 Simplified circuit diagram of realized I/Q LC oscillator with phase shifters.

In Appendix G the measured $\mathcal{L}(f_m)$ of a large number of LC oscillators is listed. This should give the reader a good impression about what can been realized for various power budgets, resonator quality factors and IC technologies.

Quadrature LC oscillator implementation example

Two-stage I/Q LC oscillators are optimally coupled if each stage is connected with a phase shift of $\pm 90°$. An example of a I/Q oscillator with phase shifters is shown in Figure 6.43 [140]. The fixed phase shifters previously discussed on p. 151 are used to couple the two single-phase LC stages, and thus to realize zero resonator phase shift operation for each resonator.

The I/Q LC oscillator is implemented in a BiCMOS process with a 30 GHz cut-off frequency (f_T) [141]. Supply voltage VCC was set to 2.7 V and I_{level} was set to 1.44 mA. The two inductors in each stage in Figure 6.43 are implemented as one balanced coil with a center tap, which is connected to VCC. Constraints in design time led to the reuse of a well-characterized resonator. At 5 GHz the quality factor (Q_p) of this resonator is only 4 because the resonator was optimized for lower frequencies. Furthermore, its large parasitic capacitance reduces the effective tuning range of the PN-type varactors. Obviously, a resonator with higher Q_p and lower parasitics will improve $\mathcal{L}(f_m)$ as well as the tuning range. For example, an optimally coupled 1.57 GHz I/Q

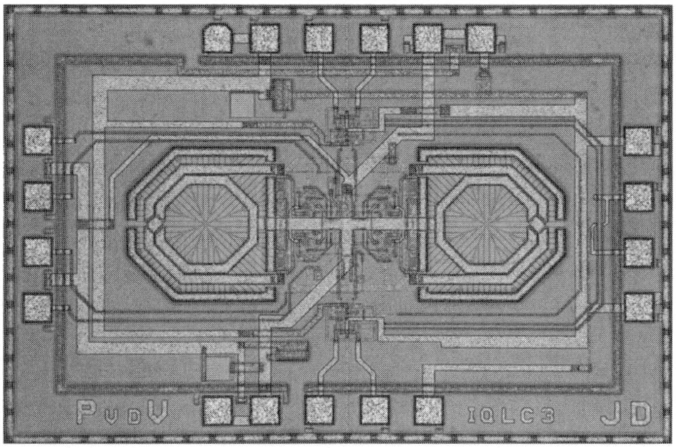

Figure 6.44 Die photo of the 5 GHz I/Q LC oscillator with phase shifters.

oscillator has been reported with a Q_p of 20 achieving $\mathcal{L}(600\,\text{kHz}) = -130.5\,\text{dBc/Hz}$ and a tuning range of 24 % [190].

To obtain a value for the $\mathcal{L}(f_m)$ of the I/Q LC oscillator, the circuit simulator SpectreRF was used. At 5 GHz the simulated $\mathcal{L}(2\,\text{MHz})$ is about -114 dBc/Hz. This figure was compared with $\mathcal{L}(f_m)$ of the same bipolar design, but coupled without phase shifters (i.e, coupling as is realized in the MOS implementation of Figure 6.9). Phase noise simulations showed a degradation of 4.3 dB at 5 GHz compared to the implementation of Figure 6.43. Even at 5 GHz and with a moderate resonator quality factor Q_p, the proposed architecture with phase shifters has improved performance in comparison to conventionally coupled I/Q LC oscillators.

Figure 6.44 shows a micrograph of the I/Q LC oscillator with phase shifters. The active chip area is 1450 μm$\times$ 2280 μm and the VCO core dissipation is 21.2 mW with a supply voltage of 2.7 V.

Measurement results of the oscillation frequency and $\mathcal{L}(2\,\text{MHz})$ versus varactor tuning voltage V_{tune} are shown in Figure 6.45. The $\mathcal{L}(2\,\text{MHz})$ is better than -113 dBc/Hz over the complete tuning range, and the phase noise simulation result of -114 dBc/Hz matches well with the measurements. Measurements of $\mathcal{L}(2\,\text{MHz})$ and the tuning range were performed using a spectrum analyzer.

The $\mathcal{L}(f_m)$ measurements performed with a spectrum analyzer were verified with a phase noise sideband measurement set-up based on the HP 3048. The measurement accuracy of this set-up is $\pm$2dB. Figure 6.46 shows the measured phase noise sideband of the I/Q LC oscillator at an oscillation frequency of 4.9 GHz.

6.3.2 Ring oscillators

For the linear time invariant analysis of phase noise in ring oscillators the same approach will be followed as with LC oscillators.

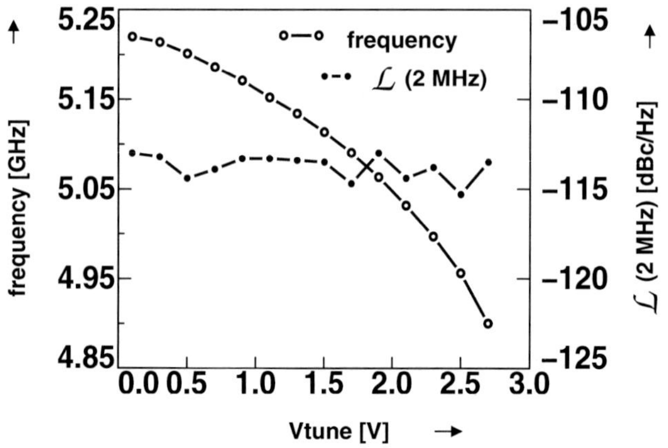

Figure 6.45 Frequency and $\mathcal{L}(2\,\text{MHz})$ versus the tuning voltage V_{tune}.

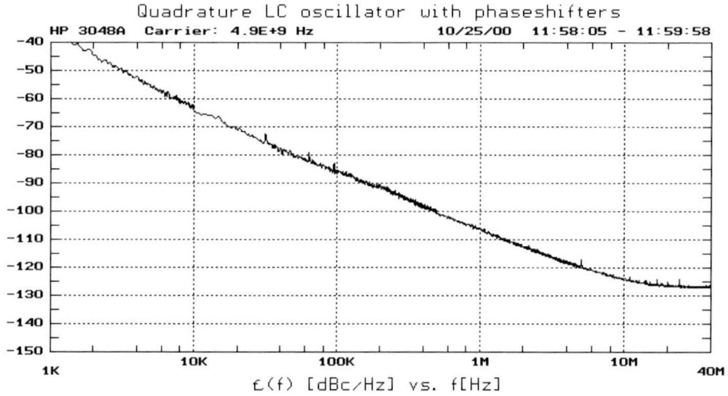

Figure 6.46 $\mathcal{L}(f_m)$ versus the offset frequency f_m. The carrier frequency is 4.9 GHz.

Based on the noise shaping theory for single-stage and multi-stage oscillators described in Section 3.4.1 on p. 58, $\mathcal{L}(f_m)$ of the two-integrator oscillator and N-stage ring oscillators, with $N \geq 3$, are derived in Appendix J.

For the two-integrator oscillator model in Figure 6.10 on p. 123, $\mathcal{L}(f_m)$ is equal to

$$\mathcal{L}_{two-int}(f_m) = \frac{1}{2} \cdot \frac{1}{4} \left(\frac{f_{osc}}{f_m} \right)^2 \frac{2 \cdot \overline{i_n^2}}{i_{carrier}^2}, \tag{6.55}$$

with quality factor $Q_{two-int} = 1$. Note that factor two in front of $\overline{i_n^2}$ takes care of modeling the noise of both stages of the oscillator.

As shown in Appendix J, $\mathcal{L}(f_m)$ of the N-stage ring oscillator model in Figure 6.15 on p. 128 can be written as

$$\mathcal{L}_{Nring}(f_m) = \frac{1}{2} \cdot \frac{1}{4Q_{Nring}^2} \left(\frac{f_{osc}}{f_m}\right)^2 \frac{N \cdot \overline{i_n^2}}{i_{carrier}^2}, \tag{6.56}$$

in which Q_{Nring} becomes [125],

$$Q_{Nring} = \frac{1}{2} N \sin\left(\frac{\pi}{N}\right). \tag{6.57}$$

When N goes to infinity, Q_{Nring} reaches its maximum[23] $\pi/2$. Since $Q_{Nring} = 1$ for $N = 2$ and (6.57) is for $N = 2$ identical to (6.56), (6.56) describes the phase noise to carrier ratio due to noise shaping in ring oscillators for $N \geq 2$, and after substitution of (6.57) is becomes,

$$\mathcal{L}_{ring}(f_m) = \frac{1}{2} \cdot \frac{1}{N \sin^2\left(\frac{\pi}{N}\right)} \left(\frac{f_{osc}}{f_m}\right)^2 \frac{\overline{i_n^2}}{i_{carrier}^2}. \tag{6.58}$$

When a specific ring oscillator topology is chosen, $\overline{i_n^2}$ in (6.58) can be expanded. For example, for the CML stage from Figure 6.16(a), the differential[24] output noise current $\overline{i_n^2}$ is

$$\overline{i_n^2} \approx \frac{1}{2}(2qI_{tail} + \frac{4kT}{R_c}), \tag{6.59}$$

if we take only transistor shot noise into account. For a source coupled differential NMOS stage with PMOS loads working in the triode region (see Figure 6.16(b)), $\overline{i_n^2}$ is equal to

$$\overline{i_n^2} \approx \frac{1}{2}(\frac{2}{3}4kTg_{m_{Qn}} + 4kTg_{m_{Qp}}). \tag{6.60}$$

Note that the factor $2/3$ in (6.60) is only valid for long-channel MOS devices and can be significantly larger for short channel devices (for example, this factor is about 2.5 for some 0.25 μm MOS devices [89]).

Equation (6.59) –and (6.60) in a similar way– can be expressed in terms of open loop gain $\alpha_{ol_{ring}}$, R_c and N as

$$\overline{i_n^2} \approx \frac{2kT}{R_c}(\frac{1}{2}\alpha_{ol_{ring}}\sqrt{1 + \tan^2(\pi/N)} + 1), \tag{6.61}$$

[23]Note that this value is obtained by linear modeling. An effective Q can be defined, which includes the phase noise reducing effect of fast switching and voltage clipping in ring oscillators: $Q_{eff} = 9/8\sqrt{(\pi V_{pp})/(2VDD)}$, with V_{pp} the peak to peak voltage swing [142, 143]. For a modern 0.18 μm CMOS process an effective Q of about 3.6 is predicted for a fully switching ring oscillator at 900 MHz [142].

[24]We define the rms current $i_{carrier}$ differentially. For example, if the rms output voltage of the CML stage is $v_{carrier}$, then $i_{carrier} = v_{carrier}/(2 \cdot (R_c \,||\, (j\omega C_L)^{-1}))$ if we neglect transistor parasitics. This is consistent with the definition of the rms carrier current in LC oscillators.

with $\alpha_{ol_{ring}}$ in an N-stage CML ring oscillator equal to

$$\alpha_{ol_{ring}} = \frac{g_{m_{bipo}} R_c}{\sqrt{1 + \tan^2(\pi/N)}}. \tag{6.62}$$

Analogous to the open loop gain α_{ol} in LC oscillators, $\alpha_{ol_{ring}}$ is unity when the gain condition for steady-state oscillation is met. If we introduce the efficiency for the CML ring oscillator as

$$\eta_{ring} = \frac{R_c i_{carrier}^2}{I_{tail} VCC} = \frac{R_c i_{carrier}^2}{P_{DC}/N}, \tag{6.63}$$

and substitute this definition and (6.61) in (6.58), we arrive at the following approximation for $\mathcal{L}(f_m)$ of a CML ring oscillator:

$$\mathcal{L}_{CML}(f_m) \approx \frac{1}{2} \cdot \frac{1}{\sin^2\left(\frac{\pi}{N}\right)} \left(\frac{f_{osc}}{f_m}\right)^2 \frac{2kT\left(\frac{1}{2}\alpha_{ol_{ring}}\sqrt{1+\tan^2(\pi/N)} + 1\right)}{\eta_{ring} P_{DC}}. \tag{6.64}$$

Equation (6.64) has been evaluated for a 3-stage CML ring oscillator implemented in 30 GHz f_T bipolar process for $\alpha_{ol_{ring}}$ ranging from 1.2 to 8 and compared with SpectreRF simulations. For all loop gain values, the difference between predictions made by (6.64) and the simulation results were found to be smaller than 5 dB.

A number of conclusions and remarks can be made about the derived equations for ring oscillators:

- Equation (6.58) models the -6 dB per octave slope (proportional to $1/f^2$) of a ring oscillator. This slope continues for large offset frequencies until it hits the white noise floor. In practice, this white noise floor is determined by buffers cascaded after the oscillator.

- Similar to LC oscillators, the phase noise becomes 3 dB lower if the power is doubled, which is expressed by (6.64). If two identical ring oscillators are combined, the phase noise lowers 3 dB, since the carrier adds correlatedly and the noise adds uncorrelatedly.

- The quality factor can be slightly improved by increasing the number of stages N. However, when N is increased, the number of noise sources and the power dissipation are increased as well.

- Since the quality factor of ring oscillators varies between unity (two-integrator oscillator) and $\pi/2$ (N-stage ring oscillator with an infinite number of stages), it is clear that LC oscillators have superior phase noise for a given power budget. Compare for example (6.64) with (6.50) on p. 158. In practice, Q_p at 1 or 2 GHz of integrated resonators is in the range of 4 to 20 (see for example Table G.1 on p. 277), and depends on the resonator technology, whereas Q_{ring} is technology independent. In other words, given a certain $\mathcal{L}(f_m)$ specification, the power dissipation in a ring oscillator is much higher than in an LC oscillator.

○ Equation (6.64) suggests that a minimum number of stages is optimal for minimum $\mathcal{L}(f_m)$.

> It is better to spend power to improve $\mathcal{L}(f_m)$ in a short ring oscillator, than to add more stages.

For instance, for a large open loop gain, (6.64) predicts that a 3-stage CML ring oscillator can have a 2.3 dB better phase noise performance than a 6-stage CML ring oscillator, for the same amount of power. Note that since we use linear modeling a number of effects, such as down-conversion of noise, are not included and that this figure only is a first order approximation. However, as we will note in the following section, nonlinear phase noise modeling shows that a minimum number of stages is indeed optimal for optimum $\mathcal{L}(f_m)$ in differential ring oscillators. The two-integrator and other two-stage ring oscillators represent this optimum case (and give the highest oscillation frequency as well), although the difference in $\mathcal{L}(f_m)$ in dBs compared to a 3-stage ring oscillator is small.

○ One linear phase noise mechanism (which can also be nonlinear and is therefore also discussed in Section 6.4), other than the modeled noise shaping, has not been mentioned yet: AM-to-PM conversion. Consider, for example, the two-integrator oscillator in Figure 6.12 on p. 126. If current source I_{tune} has amplitude noise this will result in a modulation of the frequency. The frequency of the two-integrator oscillator is determined by the 0 dB point of the gain characteristic, the gain varies due to a noisy I_{tune} and thus phase noise is generated. By proper design (e.g. using low-noise current sources with sufficient degeneration), AM-to-PM can be reduced such that noise shaping is the dominant phase noise generation mechanism.

> In Appendix H the measured $\mathcal{L}(f_m)$ of a number of ring oscillators is listed. This should give the reader a good impression of what can been realized for various numbers of ring oscillator stages, power budgets and IC technologies.

6.4 $\mathcal{L}(f_m)$: linear time-variant and nonlinear modeling

The majority of oscillators in integrated transceivers use self-limiting as an amplitude stabilizing mechanism and the active devices of these oscillators work in strongly nonlinear mode. With an AGC the output voltage swing oscillators can be limited to the linear operating region of the active circuits in the oscillator.

> Linear time invariant modeling gives a good first approximation for $\mathcal{L}(f_m)$. However, practical oscillators in transceivers often operate in the strongly nonlinear region.

Nonlinear operation can lead to phase noise improvement. For example, on p. 160 in Table 6.3, we saw a reduction of active device noise in a bipolar LC oscillator due the dependence of the transistor shot noise on the switching collector current.

Intuitively one can understand that design insights concerning the minimization of $\mathcal{L}(f_m)$ obtained with linear modeling, such as maximization of the signal power and quality factor (and thus operation at zero resonator phase shift), and minimization of noise sources, are still valid for practical nonlinear oscillators. However, due to the presence of additional phase noise reduction and generation mechanisms, and the time-varying character of all bias points in the oscillator, linear modeling is not suitable to explain all phenomena occurring in oscillators that influence $\mathcal{L}(f_m)$. Most important nonlinear and time-variant $\mathcal{L}(f_m)$ considerations will first be discussed qualitatively. Next, several phase noise theories, which provide accurate quantitative predictions of $\mathcal{L}(f_m)$ in oscillators working in strongly nonlinear mode, will be briefly discussed. These theories will extend the design insights concerning phase noise minimization obtained with linear modeling.

6.4.1 Qualitative analysis

The most important time-variant and nonlinear mechanisms influencing achievable $\mathcal{L}(f_m)$ are the following:

- **Up-conversion of noise**

  The $1/f$-noise of the devices in the oscillator core contribute to $\mathcal{L}(f_m)$ via up-conversion. This is illustrated in Figure 6.47. For example, in the case of an LC oscillator with a cross-coupled pair, the transistors may be fully switching on and off and will mix the low frequency noise to ω_{osc} and higher harmonics.

  > The $1/f$-noise corner of the devices in the oscillator does not have to coincide with the $1/f$-noise corner $f_{1/f}$ (see Figure 4.3 on p. 73) of the oscillator spectrum. For details see Section 6.4.2.

- **AM-PM conversion**

  This phase noise generation mechanism can be explained by considering an LC oscillator (e.g. Figure 6.3) with tail current I_{tail}. The transistor that implements the tail current source has $1/f$-noise. In the first instance, this $1/f$-noise only causes AM-modulation. However, nonlinear junction capacitances and especially varactors with high tuning slopes convert the AM noise into PM noise.

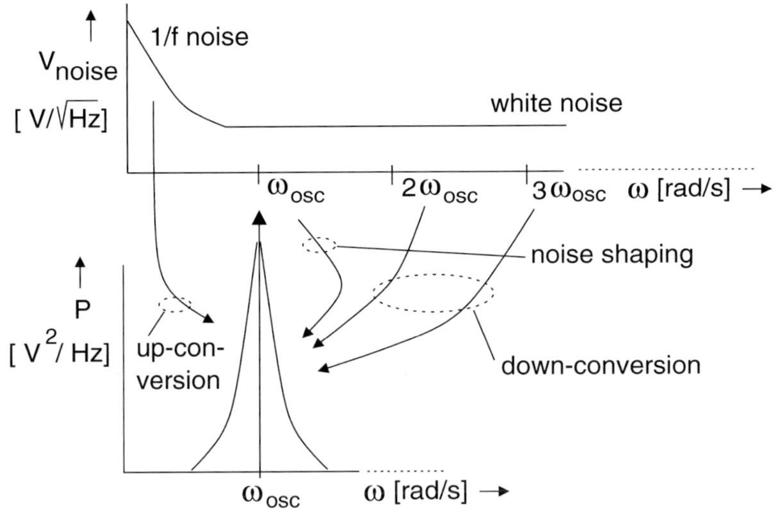

Figure 6.47 Up and down-converted noise add to the total $\mathcal{L}(f_m)$.

The $\mathcal{L}(f_m)$ resulting from AM-PM conversion due to the varactor slope can be expressed using modulation theory. Substitution of $\delta f^2 = K_{I_{tail}}^2 \overline{i_{n_{tail}}^2}$ in (4.9) on p. 76, and taking $NBW = 1\,\text{Hz}$ yields

$$\mathcal{L}(f_m) = \frac{K_{I_{tail}}^2 \overline{i_{n_{tail}}^2}}{2 f_m^2}, \tag{6.65}$$

in which $K_{I_{tail}}$ is the sensitivity of the oscillation frequency for tail current variations in Hz/A, and $\overline{i_{n_{tail}}^2}$ is the noise of the tail current source in $A_{\text{rms}}^2/\text{Hz}$. If AM to PM conversion is dominant, the contribution of this effect can be lowered by adding fixed linear capacitance (and lowering the tank inductance or the varactor capacitance to keep the same oscillation frequency). Lowering the varactor capacitance also lowers the available tuning range. Alternatively, the tuning range can be divided into multiple bands, which lowers the varactor slope within one band, as discussed on p. 140. Flicker noise originating from NMOS tail current sources can be reduced for a given W/L ratio by implementing the current source with a PMOS device: since they have an inherently lower flicker noise compared to NMOS devices [191].

○ **Down-conversion of noise**

Similar to up-conversion, the noise at higher harmonics of ω_{osc} contributes to the total phase noise at ω_{osc} due to down-conversion or inter-modulation between the carrier and the wide-band noise at multiples of the oscillation frequency (Figure 6.47). This mechanism is also referred to as noise folding [31].

The number of harmonics that should be taken into account for an accurate estimation of this mechanism to the total $\mathcal{L}(f_m)$ depends on the bandwidth of the active devices and the frequency of operation and the degree of non-linear behavior (e.g. sinusoidal waveform or square-wave waveform). In practice, if noise contributions up to the fifth harmonic are included, results are sufficiently accurate.

Tail current noise in LC oscillators at $2\omega_{osc}$ is an example of noise that contributes to the phase noise at ω_{osc} [187].

> A useful technique to reduce the tail current noise is application of a series inductor to the VCO core, and a capacitor to ground [187].

This way, the noise has a low-ohmic path to ground and a high-ohmic path to the oscillator core without disturbing the DC conditions. Alternatively, inductive degeneration can be an excellent method to lower phase noise generated by the tail current source [144].

○ **Modulation of noise sources**

In contrast to noise generated by resistors, which is stationary, the channel noise of a MOS transistor and the shot noise (base and collector) of a bipolar transistor are cyclo-stationary. The amplitude of the noise of the latter two sources varies, since the current through these devices varies significantly every oscillation period. This must be taken into account for accurate $\mathcal{L}(f_m)$ predictions.

> Modulation of the drain current in MOS transistors by switching them on and off, can reduce the $1/f$ noise generated by MOS transistors considerably [91].

○ **Saturation effects**

The oscillator designer usually tries to maximize the carrier level to maximize $\mathcal{L}(f_m)$. This results in a large voltage swing across the varactors and active devices in the oscillator. Phase noise performance can degrade by many dBs if the oscillator enters the voltage limited region (see Section 6.6) during the oscillation period, or, in other words, if the oscillation amplitude is no longer tail current limited. In addition, if tuning diodes are used, these should be used in the reverse region only, to minimize their leakage currents and noise contribution. On the other hand, too high a reverse voltage can put the tuning diode in the avalanche region, which is a very noisy and bad region for operation as well.

6.4.2 Quantitative analysis

There are several phase noise theories that take into account the time-varying and nonlinear character of practical oscillators. First, two theories will be briefly discussed that were developed for a cross-coupled LC oscillator and for LC oscillators in general (but demonstrated on a Colpitts LC oscillator), respectively. Next, a generalized phase noise theory is discussed together with some important design implications that follow from this theory.

Phase noise in cross-coupled LC oscillators

In [33], the $\mathcal{L}(f_m)$ of a cross-coupled MOS LC oscillator is investigated taking into account resonator noise, tail current noise, and the noise of the differential pair, assuming the oscillator is fully switching. The switching action of the differential pair causes down-conversion of noise from harmonics, contributing to the phase noise at the fundamental. It also causes periodic modulation of the noise sources of the transistors; these will only contribute to the total noise when they are on. The time window when the transistors are on is set by the tail current I_{tail} and the slope of the output waveform at the zero crossings. This consideration and the calculations of the noise influence of the tail current transistor lead to the following noise figure for a switching LC oscillator [187]

$$F = 1 + \frac{4\gamma R_p I_{tail}}{\pi V_o} + \gamma \frac{8}{9} g_{m_{tail}} R_p, \tag{6.66}$$

with V_o the peak voltage across the resonator, $g_{m_{tail}}$ the transconductance of the tail current source, and γ the noise factor of a MOS transistor in saturation, as mentioned, typically $2/3$ for a long-channel MOS. This noise figure can be used for $\mathcal{L}(f_m)$ predictions in Leeson's equation, (6.45) on p. 156.

Equation (6.66) allows us to explain the difference between $\mathcal{L}(f_m)$ predictions based on (6.48), and the simulations results in Table 6.3 on p. 160, for high open loop gains (thus when the oscillator is switching). Although (6.66) was derived for a MOS oscillator, its underlying theory is technology independent, and we can also apply it to bipolar cross-coupled LC oscillators. Multiplication of (6.66) with $4kT/R_p$ and taking $\gamma = 1/2$ (which follows from comparing (6.46) with (6.47)), yields the noise current of a switching bipolar cross-coupled LC oscillator

$$\overline{i_n^2} \approx \frac{4kT}{R_p}\left(\frac{2R_p I_{tail}}{\pi V_o} + 1\right), \tag{6.67}$$

assuming an ideal tail current source ($g_{m_{tail}} = 0$). As we saw in Table 6.3, for $\alpha_{ol} = 8.8$ the $\mathcal{L}(f_m)$ prediction by (6.48) of -100.8 dBc/Hz was about 5 dB too pessimistic. If we use (6.67) to model $\overline{i_n^2}$ in (6.48), i.e. replacing $(1/2 \cdot \alpha_{ol} + 1)$ by $(2R_p I_{tail}/(\pi V_o) + 1)$ in (6.48), $\mathcal{L}(100\,\text{kHz})_{bipo}$ approximately becomes -106.3 dBc/Hz for $\alpha_{ol} = 8.8$. This matches closely with the simulated value of -105.2 dBc/Hz, in the last row of Table 6.3.

Phase noise in a single-ended Colpitts oscillator

By thorough analysis involving Fourier theory and Taylor expansion, the $\mathcal{L}(f_m)$ of LC oscillators can be exactly calculated as shown in [35]. The analysis takes the time-varying character of all bias currents into account and does not assume the presence of some form of instantaneous AGC. As a result, this phase noise modeling approach also takes into account the fact that transistor noise sources are affected by the duty cycle in which the transistor conducts current (noise modulation). A MOS Colpitts oscillator (see Figure 2.10(a) on p. 27 for the circuit diagram of a bipolar Colpitts oscillator) is used as an example to illustrate how additive noise converts into phase noise.

> The derivation of $\mathcal{L}(f_m)$ of the Colpitts oscillator is cumbersome, and is therefore omitted.

In principle this calculation method of $\mathcal{L}(f_m)$ for LC oscillators, with active devices working in the strongly nonlinear region, is also applicable to other types of LC oscillators. For more details and the closed form expression of $\mathcal{L}(f_m)$ for a Colpitts oscillator the reader is referred to [34], as the phase noise theory is extensively explained in this paper.

A generalized phase noise theory

Until now, we have not discussed a phase noise theory yet that also is capable of accurate $\mathcal{L}(f_m)$ predictions in ring oscillators working in strongly nonlinear mode.

> In principle, the generalized phase noise theory proposed in [29,30], can model the phase noise of any electrical oscillator.

The major difference between the linear $\mathcal{L}(f_m)$ modeling performed in Section 6.3 and this model, is its time-varying nature. It also assumes linearity, not for the total oscillator, but only as far as the noise-to-phase transfer function is concerned [145]. The linear relation between these two variables, the noise (input variable) and the phase (output), can be easily verified by simulation [29]. In other words, the oscillator is modeled as a linear, but periodically time-varying (LTV) system.

The heart of the generalized phase noise theory is the impulse sensitivity function (ISF) $\Gamma(\omega_{osc}t)$, which models the phase shift $\Delta\phi$ occurring as a result of a voltage variation ΔV across a node in the oscillator with a node capacitance C_{node} [230],

$$\Delta\phi = \Gamma(\omega_{osc}t)\frac{\Delta V}{V_{swing}} = \Gamma(\omega_{osc}t)\frac{\Delta q}{q_{max}} \tag{6.68}$$

with V_{swing} the peak voltage swing across the capacitor and $q_{max} = C_{node}V_{swing}$. The ISF function $\Gamma(\omega_{osc}t)$ therefore is time-varying and is periodic in 2π. The shape of a

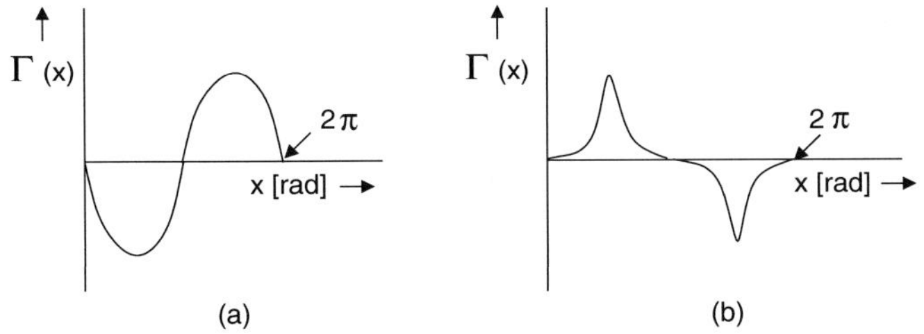

Figure 6.48 Examples of the shape of the ISF of an LC oscillator (a) and a ring oscillator (b). The ISF of the ring oscillator reaches its extremes during transitions.

typical ISF of LC and ring oscillators is shown in Figure 6.48(a) and (b), respectively. In order to include the effect of modulation for noise sources, the cyclo-stationary effect of these noise sources is included in the LTV model by multiplication of $\Gamma(\omega_{osc}t)$ with $\alpha(\omega_{osc}t)$,

$$\Gamma_{\text{eff}}(\omega_{osc}t) = \Gamma(\omega_{osc}t)\alpha(\omega_{osc}t), \tag{6.69}$$

where $\alpha(\omega_{osc}t)$ is a periodic unitless function with a peak value of unity [145]. For stationary noise sources $\alpha(\omega_{osc}t)$ simply is unity throughout the whole oscillation period. The rms value of Γ_{eff}, Γ_{rms} can be found by evaluation of

$$\Gamma_{\text{rms}}^2 = \frac{1}{2\pi}\int_0^{2\pi} \Gamma_{\text{eff}}^2(x)dx. \tag{6.70}$$

It can be shown that $\mathcal{L}(f_m)$ can be written in terms of Γ_{rms}^2 as [230],

$$\mathcal{L}(f_m) = \frac{1}{8\pi^2 f_m^2} \cdot \frac{1}{q_{max}^2} \cdot \sum_n \left(\overline{i_n^2} \cdot \Gamma_{\text{rms,n}}^2\right). \tag{6.71}$$

Obviously, accurate determination of Γ_{rms}^2 is vital for the accuracy of $\mathcal{L}(f_m)$ predictions with this equation. Determination of the ISF is most straightforwardly performed by replacing the noise sources in an oscillator with impulsive sources of small width (charge) and measuring the resultant phase shift[25]. Repeating this process by injecting the impulse at various times during a period and measuring the resultant phase shift allows the ISF to be calculated [104]. This method is the most accurate one and can be performed using circuit simulators. Two other methods that yield an approximation of the ISF are described in [29].

Linear modeling is incapable of predicting the $1/f$-noise corner $f_{1/f}$ of phase noise sidebands (see Figure 4.3 on p. 73), and therefore we ignored this part of the oscillator spectrum so far. One major contribution of the phase noise model proposed

[25] Injecting a very large charge would violate the linearity assumption between the injected charge and the phase. If the injected charge is doubled, the observed phase deviation should also double.

in [29] is its ability to predict $f_{1/f}$. If we define $f_{1/f-device}$ as the $1/f$ noise corner of a device in the oscillator, then $f_{1/f}$ as a result of the noise generated by this device can be written as

$$f_{1/f} = f_{1/f-device} \cdot \left(\frac{\Gamma_{\mathrm{DC}}}{\Gamma_{\mathrm{rms}}}\right)^2, \tag{6.72}$$

in which Γ_{DC} is equal to

$$\Gamma_{\mathrm{DC}} = \frac{1}{2\pi}\int_0^{2\pi} \Gamma_{\mathrm{eff}}(x)dx. \tag{6.73}$$

As explained in [29], the value of Γ_{DC} is closely related to the symmetry properties of the waveform of the oscillator. Especially the rise and fall times are important. If they differ substantially this results in a large Γ_{DC} value and a large $f_{1/f}$. In other words, symmetry in rise and fall times can reduce $f_{1/f}$ significantly. The push-pull configuration of the complementary MOS LC oscillator in Figure 6.4(a) on p. 117 has received renewed interest because of this design insight. As symmetry is important for each noise source, every half-circuit of the oscillator should be as symmetrical as possible. Through proper sizing of the PMOS and NMOS transistor, the complementary oscillator can have a very low value of Γ_{DC} value and thus a very low $f_{1/f}$ corner.

For a detailed discussion on the general phase noise theory and analytical expressions of the phase noise in specific LC and ring oscillator topologies, the reader is referred to [104]. Here we close the discussion on this specific phase noise theory with two important remarks. The insights regarding $\mathcal{L}(f_m)$ optimization, obtained with the linear behavioral models (LTI models) used in Section 6.3, are confirmed by the LTV model proposed in [29]. The carrier in an oscillator should be maximized as well as the resonator quality factor, resonator phase shift should be close to zero or, better, zero, and the number of stages in differential ring oscillators should be kept to a minimum[26].

The linear time invariant model shows that energy sustaining the oscillation in an oscillator should be restored impulsively during a cycle when the ISF is at its minimum. It is at this point that the oscillator is the least sensitive to the impulse of noisy energy.

In a Colpitts oscillator the ISF can be minimized at the point where the core transistor delivers energy to the resonator, whereas a normal cross-coupled pair has the disadvantage that it delivers energy in a range where the ISF varies between zero and maximum. Exploiting the "noise-shifting" behavior of the Colpitts oscillator[27], a 2.1 GHz differential Colpitts oscillator has been demonstrated with a core dissipation of 10 mW, an inductor quality factor of 6, and with a $\mathcal{L}(3\,\mathrm{MHz})$ of $-138.2\,\mathrm{dBc/Hz}$ [192].

[26] For single-ended ring oscillator topologies the LTV model predicts no dependence of $\mathcal{L}(f_m)$ on N [230].
[27] And the fact that its maximum tank amplitude is $2I_{bias}R_p$.

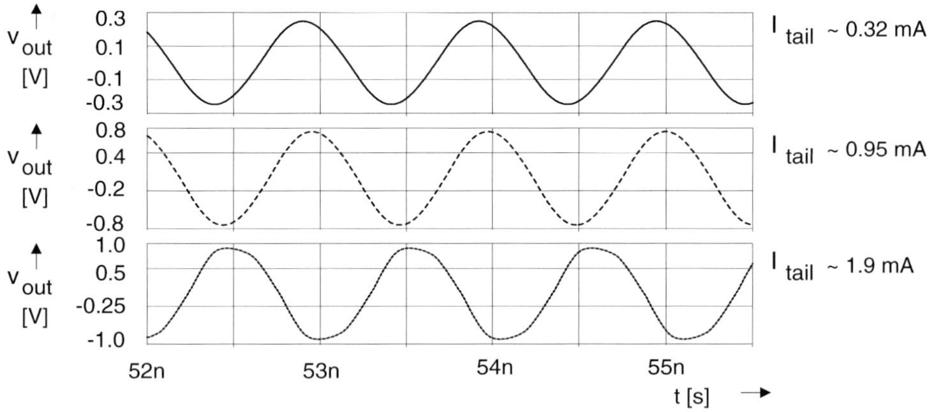

Figure 6.49 Simulated output voltage of bipolar cross-coupled pair LC oscillator for three different tail currents and with $VCC = 3$ V. The oscillator runs at 1 GHz. $Q_p = 10$ and $R_p = 1.25\,\text{k}\Omega$.

6.5 Waveform

Given modern technologies where f_T's of 30 GHz up to 100 GHz in standard Bipolar, BiCMOS or sub-micron CMOS can be achieved, and the frequency range of many transceivers (see for example Table 4.1 on p. 69), the waveform generated by oscillators under normal operating conditions approximate the waveforms discussed in Section 5.3 quite well. However, for frequencies say beyond 1/5 of the f_T, it is much harder to make the transistors in an oscillator core switch. The oscillation voltage swing on a gate or base may be much larger than the voltage required at low frequencies for switching, but internal voltage division may prevent a device from fully reaching the on or off state. Furthermore, the bandwidth limiting effect of parasitics of the active and passive device plus the interconnect reduce the harmonic content of the current and voltages in an oscillator.

Practical issues such as the limiting effect of the supply voltage, PN-junctions entering the forward region or MOS transistors, which should stay in saturation, entering the linear region, can have a huge impact on the oscillator waveform. Figure 6.49 shows the output voltage of the bipolar LC oscillator implementation of Figure 6.3 on p. 115, for increasing tail current. The output voltage of the oscillator is approximately $2/\pi \cdot I_{tail}R_p$ for $I_{tail} = 0.32\,\text{mA}$. For a tail current of 1.9 mA, the collector-base junction starts limiting the output voltage, as is evident from the flat tops of v_{out} in Figure 6.49. Since this junction starts conducting current, one can imagine that this not a good operating point for optimum $\mathcal{L}(f_m)$ performance.

In Figure 6.50 the simulated drain current of one of the two transistors in the NMOS LC oscillator from Figure 6.4 on p. 117 is plotted for three tail current values.

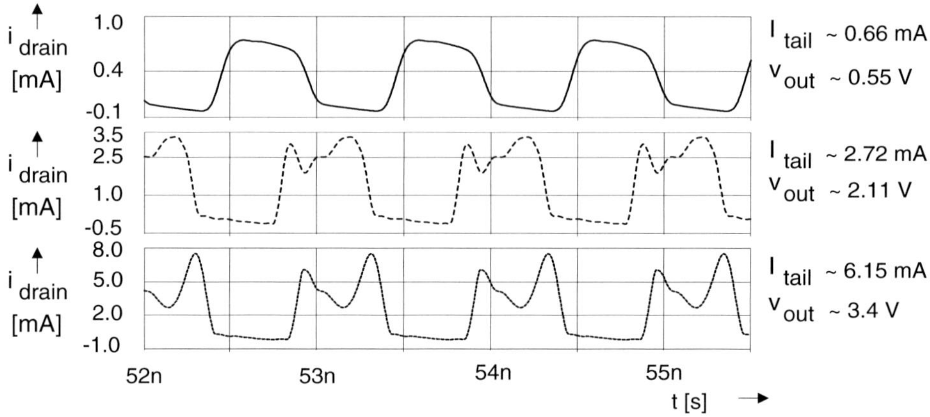

Figure 6.50 Simulated drain current of one transistor in a NMOS cross-coupled pair LC oscillator for three different tail currents and with $VDD = 1.8\,\text{V}$. The oscillator runs at 1 GHz. $Q_p = 10$ and $R_p = 1.25\,\text{k}\Omega$.

For a tail current of 0.66 mA the drain current approximates a square wave quite well. When the tail current is increased and v_{out} starts approaching $2\,VDD$, the NMOS pair enters the linear region for a certain period when v_{out} peaks, which results in the dip in the drain current. As will be explained in Section 6.6, the oscillator now operates in the voltage limited region. The simulated oscillator has a Q_p of 10. Because the transconductance is reduced for high tail currents (and not the voltage is limited as was the case in Figure 6.49), the output voltage waveforms, corresponding to the drain currents of Figure 6.50, approximate ideal sine waves. Hence, looking only at the voltage waveform in a high Q oscillator may not directly reveal whether the oscillator has entered the voltage limited region.

Circuit topologies, which allow the designer to improve the symmetry of the waveform generated by the oscillator, should be used if a low $f_{1/f}$ is important for the application. Symmetric waveforms yield a low Γ_{DC} and therefore a low $f_{1/f}$ (see (6.72)).

6.6 Carrier amplitude and power

The carrier amplitude and carrier power in LC and ring oscillators depend on many factors and can only be obtained accurately with simulation. Among other things, these factors in a given topology include the region of operation (linear[28], weakly nonlinear, or strongly nonlinear), oscillation frequency, implementation technology,

[28]Employing an AGC.

transistor size and parasitics. Provided that the oscillator is switching and is in normal operation mode (e.g. no voltage limiting), the carrier current can be approximated with a square wave and first order approximations, similar to the ones given for single-ended behavioral models in Table 5.1 on p. 109 can be used. In Section 6.3 we approximated the output voltage v_{out} of a ("non-complementary") cross-coupled LC oscillator as $2/\pi \cdot I_{tail}R_p$, whereas the maximum output voltage of the complementary LC oscillator on p. 117 is $4/\pi \cdot I_{tail}R_p$. At high frequencies, the carrier current is better approximated by a sinusoidal waveform, due to the limited bandwidth of the transistors in an oscillator. For the above-mentioned oscillators this means that v_{out} is approximately $1/2 \cdot I_{tail}R_p$ and $I_{tail}\ R_p$, respectively. Similarly, in ring oscillators v_{out} can be approximated by $I_{tail} \cdot (R \,||\, (j\omega C)^{-1})$, with R and C the resistance and capacitance seen at one single-ended node of a ring oscillator stage (e.g. R_c and C_L in Figure 6.16(a) on p. 130).

In the above-mentioned oscillators, the carrier current and therefore the carrier amplitude is a function of one tail current. In multi-phase LC oscillators, two-integrator oscillators and many ring oscillator topologies this is not the case. Here the total carrier current consists of a sum of various currents that must be summed to find the total current, taking into account the relative phase of each current. Consider, for example, the I/Q LC oscillator in Figure 6.9 on p. 122. If we assume that transistors Q_{nc} and Q_{nl} are identical and have no parasitic phase shift, the differential carrier current contributed by the cross-coupled pair ($2/\pi \cdot I_L$, assuming the stage is switching) is 90° out of phase with carrier current provided by the coupling transistors. The total peak carrier current therefore is $2/\pi \cdot \sqrt{I_L^2 + I_C^2}$, assuming all transistors are switching. For an arbitrary angle ϕ between the two currents, the maximum peak carrier current I_{peak} can be written as [108]

$$I_{peak} = 2/\pi \cdot \sqrt{I_L^2 + I_C^2 + 2I_L I_C \cos\phi}. \tag{6.74}$$

For an optimally coupled[29] I/Q LC oscillators, ϕ will be zero and I_{peak} becomes $2/\pi \cdot 2I_L = 2/\pi \cdot 2I_C$, if $I_L = I_C$. We can conclude that not only the quality factor of an optimally coupled multi-phase LC oscillator is at its maximum, but also its carrier current. Under optimal coupling conditions the current contributed by the cross-coupled pair in a stage adds in phase with the current contributed by the coupling transistors.

An oscillator works in the current limited region if the amplitude grows more or less linearly with the tail current. If the output voltage of the oscillator starts to saturate, the oscillator is said to work in the voltage limited region.

> Operation in the current limited region results in better $\mathcal{L}(f_m)$ than operation in the voltage limited region.

In the voltage limited region, tail current is wasted since it does not contribute to a higher voltage swing. As a matter of fact, $\mathcal{L}(f_m)$ usually degrades in the voltage

[29] Each LC stage is coupled with phase shifters having a phase shift $\theta = \pm 180/N$.

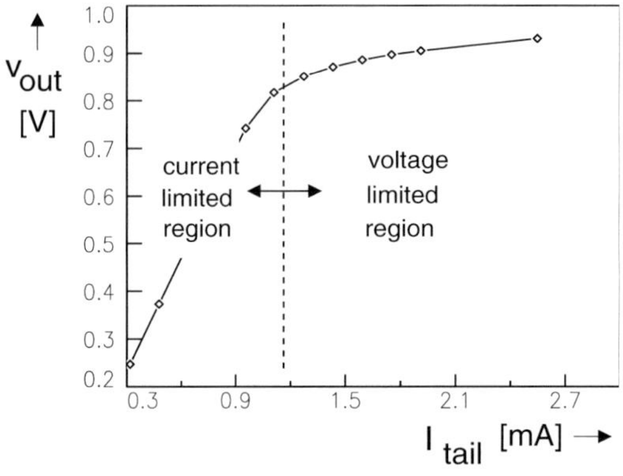

Figure 6.51 Simulated differential peak output voltage of a bipolar cross-coupled pair LC oscillator versus tail current I_{tail}, with $VCC = 3\,V$. At an output voltage of $\approx$0.85 V the collector-base junction starts limiting the output voltage.

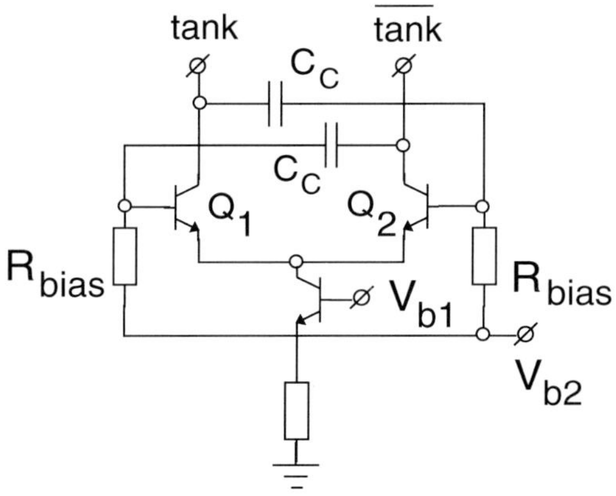

Figure 6.52 AC-coupled bipolar cross-coupled pair to avoid forward biasing of the collector-base junctions.

limited region, due to –among other causes– increased device noise. In general, the optimum bias tail current results in a voltage swing, which is as large as possible without entering the voltage limited region.

In Section 6.5, the voltage limiting effect of the collector-base junction in a 1 GHz bipolar LC oscillator (Figure 6.3 on p. 115) was discussed. The output voltage versus I_{tail} characteristic of this oscillator is shown in Figure 6.51. Indeed, the normal cross-coupled bipolar LC oscillator enters the voltage limited region around a 0.85 V voltage swing. If a larger swing is required, MOS transistors can be used as they do not clamp

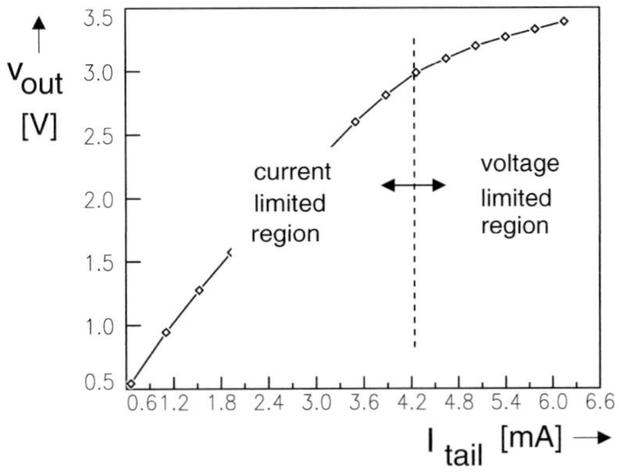

Figure 6.53 Simulated differential peak output voltage of an NMOS cross-coupled pair LC oscillator versus tail current I_{tail}, with $VDD = 1.8\,V$. The maximum output voltage is approximately 2 times VDD.

the voltage swing. Alternatively, the oscillator topology in Figure 6.52 can be used. Here, the tank circuit is AC-coupled to the active oscillator part, and the collector-base junctions can be reverse-biased to allow a higher amplitude.

The output voltage versus tail current characteristic of the MOS oscillator example discussed in Section 6.5 is shown in Figure 6.53. When the output voltage starts approaching 2 times VDD, the transistors in the oscillator start entering the triode region. Increasing the tail current from 4.8 mA to 6.0 mA only marginally increases V_{out} and it is clear that the oscillator, when biased in this region, is voltage limited. So the maximum voltage swing in a MOS cross-coupled LC oscillator should stay somewhat below $2VDD$. For the complementary MOS LC oscillator this value is VDD.

6.7 Power dissipation and supply voltage

The power dissipation in most differential oscillators can easily be derived by counting tail currents. In an N-stage LC or ring oscillator the power dissipation P_{DC} is equal to

$$P_{DC} = N \sum_i I_i V_{supply}, \tag{6.75}$$

with i the index of the various current sources in one stage. For a simple single-phase LC oscillator P_{DC} simply is $I_{tail} V_{supply}$. Lowering V_{supply} lowers the total power dissipation. A minimum number of stages is optimal if we consider the power dissipation only. We already encountered various drawbacks of a low V_{supply}. Examples are a lower maximum output voltage, a smaller varactor tuning range, and higher tuning constant. Furthermore, there is less voltage room for circuit techniques (e.g. stacking of transistors), which might improve PSRR and other performance aspects. Generally speaking, a large V_{supply} makes the life of an oscillator designer easier.

6.8 Summary

The practical properties of LC and ring oscillators have been described in this chapter. The main focus of this chapter was on oscillation frequency, tuning range and the phase noise to carrier ratio, in the context of parasitics, noise and other non-ideal aspects, which are present in integrated circuits. In addition, waveforms, carrier amplitude and power dissipation of oscillators have been addressed.

The oscillation frequency of LC and ring oscillators are influenced by parasitic elements present in IC technology. Parasitic capacitances contribute to the total tank capacitance in an LC oscillator. Losses in resonator components can cause a frequency shift compared to the nominal frequency expected when taking only the reactive tank elements into account. Furthermore, loading effects, by for example buffers, must be considered. Two circuit examples of single-phase LC oscillator have been presented and analyzed. A number of methods for quadrature signal generation were discussed and one of them is utilization of a multi-phase LC oscillator. The same non-idealities that influence the frequency of single-phase oscillators also play a role in multi-phase oscillators. In addition, the phase and relative amplitude of the currents provided by the negative transconductance in each stage, and the coupling transconductances are of importance. A MOS I/Q LC oscillator implementation was presented.

A first order estimation of the frequency of two-integrator oscillators and N-stage ring oscillators can be made on the basis of f_{MAX} and a single pole model of a ring oscillator stage, respectively. These frequency estimators were verified and several ring oscillator circuit examples were discussed.

Parasitic capacitance reduces the effective tuning range that can be obtained with a varactor in an LC oscillator. The quality factor of the tuning element should be maximized, just like the quality factor of every component in the LC resonator, for low $\mathcal{L}(f_m)$. Practical aspects of the following tuning methods have been highlighted: continuous passive capacitive tuning, discrete passive capacitive (or inductive) tuning (band-switching), continuous active capacitive tuning, continuous passive inductive tuning, continuous active inductive tuning, phase shift tuning and oscillator switching. An effective quality factor has been introduced for the active tuning solutions, and generally in the worst case this factor is very low, resulting in a poor $\mathcal{L}(f_m)$. Band-switching and oscillator switching can be used to realize a large tuning range and also lower the tuning constant of an oscillator. Band-switching and active capacitive tuning have been demonstrated with an integrated circuit design.

In practice, the two-integrator oscillator has losses and a negative transconductance can be introduced to compensate these losses. This transconductance or the transconductance of the integrator can be used for tuning, but normally the integrator transconductance is used for this purpose. The negative transconductance should be kept as small as possible, because it introduces extra noise. In N-stage ring oscillators resistive tuning is most used. This tuning method results in large, often quite linear, tuning ranges. It does require circuitry to keep the amplitude constant, which can be of importance for optimum $\mathcal{L}(f_m)$.

Expressions for $\mathcal{L}(f_m)$ for LTI LC and ring oscillators models have been derived. LTI modeling provides a first order estimate of $\mathcal{L}(f_m)$ and provides a number of important design insights, which are confirmed by LTV models that also capture non-linearity and the effect of varying bias currents in practical oscillators. Despite noise folding in oscillators, strong nonlinear operation (by setting the open loop gain much larger than 2) significantly improves the $\mathcal{L}(f_m)$ of cross-coupled LC oscillators. This improvement stems from modulation of the noise sources of the active devices in the oscillator.

The quality factor, and therefore $\mathcal{L}(f_m)$, in N-stage LC oscillators is at a maximum when each stage is coupled with a phase shift of $\pm 180°/N$. A 5 GHz I/Q oscillator test-IC has been realized and presented to demonstrate optimal coupling of LC oscillators. Ring oscillators have a quality factor between 1 and $\pi/2$, which makes clear why LC oscillators that normally have much higher quality factors, have superior $\mathcal{L}(f_m)$. Fast switching in ring oscillator stages increases the quality factor of ring oscillators beyond the maximum of $\pi/2$ that is obtained with linear modeling. In LC oscillators the quality factor and the carrier amplitude should be maximized for a low $\mathcal{L}(f_m)$. Furthermore, the resonator phase shift should stay close to or be made equal to zero degrees. A minimum number of ring oscillator stages is optimal for minimum $\mathcal{L}(f_m)$ given a certain power budget.

Many nonlinear phase noise mechanisms, among other things the up and down-conversion of noise, add to the phase noise predicted by LTI modeling. The cyclo-stationary nature of some noise sources in oscillators can cause a decrease of the contribution of these sources to $\mathcal{L}(f_m)$. Several LTV phase noise models have been discussed and especially the generalized phase noise theory built around the definition of an impulsive noise function ISF is of high interest. This model is capable of accurately modeling all oscillators within the scope of this book in all regions of operation (weakly nonlinear as well as strongly nonlinear). In addition to insights obtained with LTI modeling, two important design guidelines are obtained by this theory. First, energy that is lost each cycle should be restored impulsively when the ISF corresponding to that noise source is at a minimum. Second, the waveform of an oscillator should be made as symmetrical as possible for a low $f_{1/f}$ corner of the oscillator spectrum.

In the current limited region of operation, the voltage swing of an oscillator grows linearly with the tail current. In the voltage limited region a saturation mechanism prevents a further increase of the amplitude. In the latter region current is wasted and noise increases. For the lowest phase-noise, an oscillator should operate in the current limited region. The total power dissipation of an oscillator increases with the number of tail current sources, the number of stages and the supply voltage. A high supply voltage has a number of advantages. It allows a high voltage swing, lower K_{vco} compared to low supply voltages, and a larger tuning range for varactor-tuned LC oscillators.

7

Figures of merit

DESIGN FOMS and benchmark FOMs were introduced in Chapter 3. As mentioned in Section 3.3.2, design FOMs compare an estimation of a performance aspect of an oscillator with its functional specification. The design margin is indicated by the value of the FOM. At the cost of reduced accuracy, relatively simple first-order calculations can be used to formulate design FOMs. In other words, the complexity of the design FOM is reduced at the cost of some accuracy. Once all important design FOMs are defined and evaluated for several oscillator topologies, for example the design FOMs concerning frequency, tuning range and $\mathcal{L}(f_m)$, these topologies can be ranked on the basis of the design FOM values. With 100% accurate modeling design FOMs would lead to hard design decisions: a negative design margin would mean that the oscillator topology under consideration does not meet the functional specifications and that it can be discarded. In practice, there is modeling uncertainty and design FOMs thus have a value with a small enough error-band to be useful. The topology with highest design FOM values is the best option for further optimization given the set of functional specifications, as illustrated in Figure 7.1.

> Design FOMs of oscillator topologies provide qualitative insight concerning the relations between the design parameters, and quantitative insight when evaluated.

In addition, design FOMs can be used to find the minimum value of one design parameter given all other design parameters, by solving FOM=0 (in dBs) for the design

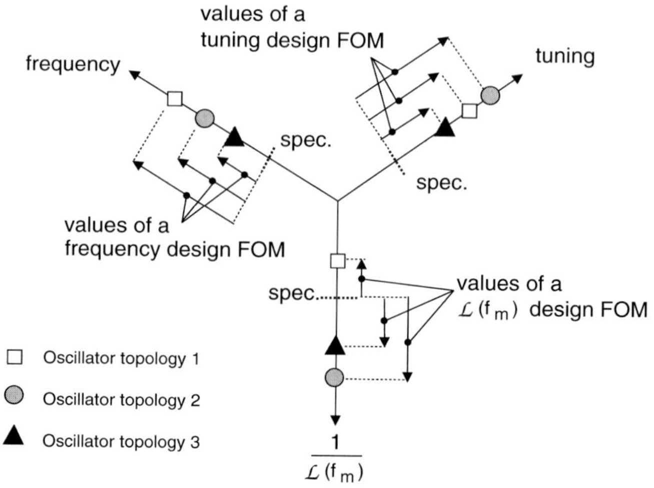

Figure 7.1 Topology selection with design FOMs. Topology 1 has a negative design margin for $\mathcal{L}(f_m)$. For oscillator topology 2, all design margins are positive and higher than topology 3, which makes it the best candidate for further optimization.

parameter of interest. For example, with a design FOM for the $\mathcal{L}(f_m)$ of an LC oscillator, the minimum quality factor for the LC resonator can be calculated. Using parts of the theory described in Chapter 5 and Chapter 6, several design FOM examples are given in Section 7.1.

In Section 7.2 four benchmark FOMs are discussed.

> Benchmark FOMs normalize performance aspects of an oscillator to allow fair comparison with other oscillator designs, or to compare them against a theoretical performance limit.

Two benchmark FOMs, Oscillator Number (Osc-No) and Normalized Phase-Noise (Nor-PN) are known from literature and are often used to compare the measured $\mathcal{L}(f_m)$ of an oscillator design with previously published designs. This comparison can also be performed with the novel benchmark FOMs: LC oscillator design efficiency (LC-ODE) and ring oscillator design efficiency (ring-ODE). However, ODE compares the achieved $\mathcal{L}(f_m)$ with an estimation of a theoretical $\mathcal{L}(f_m)$ limit, thus providing much more information for the designer, as will become clear.

7.1 Design FOMs

In this section several frequency, tuning and $\mathcal{L}(f_m)$ design FOMs are presented, making use of the analysis of the various LC and ring oscillator properties and topologies presented in the Chapters 5 and 6. A number of these design FOMs is used in Chapter 9, highlighting four oscillator designs.

7.1.1 Frequency design FOMs

The maximum frequency of an LC oscillator is achieved, assuming a fixed resonator inductance L_p, when the varactor has its minimum capacitance value. This maximum should be higher than the specified maximum oscillation frequency. Making use of (6.4) on p. 114, with $C_{fixed} = 0\,\text{fF}$, we can express this in a frequency design FOM as

$$FOM_{freq-LC} = \frac{\left(2\pi\sqrt{L_p\left(C_{tune}|_{minimum} + C_{par}\right)}\right)^{-1}}{f_{max,spec}}. \tag{7.1}$$

Capacitance C_{par} can be further expanded for a specific LC oscillator topology. For example, (6.11) on p. 116 and (6.15) on p. 118 can be used for a first order estimation of C_{par} for a bipolar LC oscillator and a complementary MOS LC oscillator, respectively. The choice of L_p has a big influence on the maximum oscillation frequency. Often, a smaller L_p has a higher quality factor and a higher resonance frequency (thus a smaller contribution to C_{par})[1]. Furthermore, a smaller L_p allows a higher minimum varactor capacitance for a certain maximum oscillation frequency, and therefore a higher tuning range (see Section 7.1.2). However, the choice of the inductor value also influences $\mathcal{L}(f_m)$. A guideline for choosing L_p is given in Section 7.1.3.

On p. 126 we saw that the oscillation frequency of the two-integrator oscillator from Figure 6.12 can be estimated using the technology FOM $f_{MAX}/4$. Therefore, we can define a frequency design FOM to estimate whether we can meet the maximum frequency specification given a certain technology as

$$FOM_{freq-two-int} = \frac{\frac{f_{MAX}|_{optimum}}{4}}{f_{max,spec}}. \tag{7.2}$$

For bipolar implementations of the two-integrator f_{MAX} is optimal (i.e. at its maximum) at a bias current equal to the peak f_T current of the transistors. In the case of a MOS implementation f_{MAX} is maximized by maximizing over-drive voltage $V_{gs} - V_{TH}$ and by using minimum length transistors[2].

For N-stage ring oscillators that obey dominant pole behavior, we found that (6.27) on p. 130 accurately estimates the oscillation frequency. Similar to the definition of the frequency design FOM for the two-integrator oscillator we define

[1] Smaller inductors have a smaller number of turns. A single-turn coil needs no bridge to connect the most inner turn. Normally this bridge represents a relatively large DC resistance. See Figure 9.3 on p. 219 for quality factor measurements of three coils (in SOA technology) with 1, 2 and 7 turns.

[2] This is true for MOS transistors obeying long-channel approximations. The f_{MAX} of short-channel devices in saturation does not depend on W and bias conditions [146].

$$FOM_{freq-Nring} = \frac{\left(2N0.8\tau|_{minimum}\right)^{-1}}{f_{max,spec}}. \quad (7.3)$$

The time constant of the dominant pole τ can be expanded for a specific ring oscillator stage topology. For example, we found that the time constant for a CML stage is approximated by (6.28) on p. 131, and for a MOS stage with PMOS loads in the triode region by (6.31) on p. 132. For bipolar implementations τ is a function of bias conditions, as is evident from (6.28), and minimum τ is obtained around the peak f_T current of the transistors in the oscillator (see for example Figure 6.18 on p. 132).

7.1.2 Tuning design FOMs

In Chapter 6, a number of tuning possibilities for LC oscillators were discussed. If continuous passive capacitive tuning is investigated as a first way to realize a specified tuning range, its required capacitance ratio can be assessed with the following tuning FOM

$$FOM_{tuning-LC} = \frac{\frac{C_{max}+C_{par}}{C_{min}+C_{par}}}{\left(\frac{f_{max,spec}}{f_{min,spec}}\right)^2}, \quad (7.4)$$

in which C_{max} and C_{min} are the maximum and minimum capacitance of a simple varactor, or a compound varactor with fixed series and/or parallel capacitance (see Figure 6.20 on p. 134). Similar to C_{par} in $FOM_{freq-LC}$, this parameter can be expanded for a given LC oscillator topology.

7.1.3 $\mathcal{L}(f_m)$ design FOMs

In Section 6.3, linear modeling of phase noise was discussed, whereas in Section 6.4 several phase noise theories were discussed that take into account the nonlinear time-varying nature of oscillators. We saw that linear modeling yields a reasonable estimation of $\mathcal{L}(f_m)$. Utilizing the linear phase noise models for LC and ring oscillators (equations (6.44), (6.51), (6.55) and (6.56)) obtained in the previous chapter, we define a design FOM to assess $\mathcal{L}(f_m)$ of an oscillator as

$$FOM_{\mathcal{L}(f_m)} = \mathcal{L}(f_m)_{spec} \Bigg/ \left(\frac{1}{2}\cdot\frac{1}{4Q^2}\left(\frac{f_{osc}}{f_m}\right)^2 \frac{N\cdot\overline{i_n^2}}{i_{carrier}^2}\right), \quad (7.5)$$

with quality factor Q depending on the oscillator type and topology under consideration.

Examples of quality factors as function of the oscillator type are

$$N Q_p \cos(\phi_{res}) \quad \text{for LC oscillators with passive LC resonator,} \tag{7.6}$$

$$Q_{C_{active}} \quad \text{for an active capacitance (see (6.40)),} \tag{7.7}$$

$$Q_{L_{active}} \quad \text{for an active inductance (see (6.42)),} \tag{7.8}$$

$$\frac{1}{2} N \sin\left(\frac{\pi}{N}\right) \quad \text{for ring oscillators.} \tag{7.9}$$

A first order estimation of parameters $\overline{i_n^2}$ and $i_{carrier}^2$ in $FOM_{\mathcal{L}(f_m)}$ can easily be calculated for a given oscillator topology. For example, based on (6.48) on p. 158, we can write[3] the following for a bipolar LC oscillator (e.g. the oscillator in Figure 6.3 on p. 115):

$$FOM_{\mathcal{L}(f_m),LC-bipo} = \mathcal{L}(f_m)_{spec} \Big/ \left(\frac{1}{2} \cdot \frac{1}{Q_p^2} \left(\frac{f_{osc}}{f_m} \right)^2 \frac{kT(\frac{1}{2}\alpha_{ol} + 1)}{R_p \frac{2}{\pi^2} I_{tail}^2} \right). \tag{7.10}$$

As with any design FOM, (7.10) can be used to assess whether the $\mathcal{L}(f_m)$ is met with enough of a design margin, but also to derive requirements of design parameters. For example, if we set $FOM_{\mathcal{L}(f_m),LC-bipo} = 1$ (zero design-margin) and solve Q_p we obtain an estimation of the minimum quality factor of the resonator, given all other design parameters in (7.10), such as frequency, power dissipation, etcetera. To solve Q_p all design parameters, except Q_p, have to be substituted in (7.10), including R_p. Effective parallel resistor R_p can be replaced by $Q_p \omega_{LC} L_p$, which indicates an important degree of freedom in $\mathcal{L}(f_m)$ optimization. The inductor value is a design freedom. At first it may seem there are rather conflicting reports on whether L_p should be maximized or minimized for optimum $\mathcal{L}(f_m)$. References [34] and [193] advise maximization L_p for minimum phase noise given a certain power budget[4]. References [31] and [183] suggest minimization of L_p for optimum $\mathcal{L}(f_m)$ for a certain power budget[5].

Evaluation of the variation in $FOM_{\mathcal{L}(f_m),LC-bipo}$ as a result of an increased L_p in Table 7.1 reveals that both suggestions, maximization as well as minimization of L_p, can be correct. Whether $FOM_{\mathcal{L}(f_m),LC-bipo}$ improves or not when L_p is increased depends on the nature of the inductor design and on the dominant noise source. It is important to note that Table 7.1 only covers the current-limited region. In the voltage-limited region either the tail current of the LC oscillator should be reduced or, for minimum $\mathcal{L}(f_m)$, the value of L_p in the design should be lowered to the point that the oscillator operates in the current-limited region. Table 7.1 shows that there indeed is a

---

[3] Assuming zero resonator phase-shift.

[4] In this case, the maximum L_p is limited by the tuning range as $FOM_{tuning-LC}$ decreases for increasing L_p.

[5] In this case, the minimum L_p is determined by the oscillator start-up condition α_{ol} or a minimum output voltage ($\propto I_{tail} R_p$) condition.

	Q_p increases for decreasing L_p	Q_p stays constant for decreasing L_p
$\mathcal{L}(f_m)$ is dominated by resonator noise	$L_p \uparrow \Rightarrow FOM_{\mathcal{L}(f_m),LC-bipo} \uparrow$	$L_p \uparrow \Rightarrow FOM_{\mathcal{L}(f_m),LC-bipo} \uparrow$
$\mathcal{L}(f_m)$ is dominated by device noise	$L_p \uparrow \Rightarrow FOM_{\mathcal{L}(f_m),LC-bipo} \downarrow$	$L_p \uparrow \Rightarrow FOM_{\mathcal{L}(f_m),LC-bipo}$ –

Table 7.1 In the current-limited region, $FOM_{\mathcal{L}(f_m),LC-bipo}$ (for a MOS implementation the table is identical), can decrease (↓), stay constant (–) or increase (↑), depending on the dominant phase noise source and the relation between the inductance value and its quality factor.

situation that L_p should be minimized for optimum $\mathcal{L}(f_m)$, as suggested in [183]. If the quality factor of inductor-design is higher for lower inductance values[6] and the active device noise is dominant in a bipolar LC oscillator[7], $FOM_{\mathcal{L}(f_m),LC-bipo}$ improves for decreasing L_p. However, in all other situations mentioned in Table 7.1 maximization of L_p is best or at least not harmful as long as the oscillator operates in the current limited region. In all cases the inductance value should be chosen large enough to ensure that the cross-coupled pair is switching, since we saw in Table 6.3 on p. 160 that this reduces the active device noise significantly (and this reduction is predicted by (6.67) on p. 173).

7.2 Benchmark FOMs

The benchmark FOMs "Oscillator Number" (Osc-No), "Normalized Phase Noise" (Nor-PN), "LC oscillator design efficiency" (LC-ODE) and "ring oscillator design efficiency" (ring-ODE) are discussed in this section.

[6]In [183], EM simulations of planar on-chip inductors indeed show some improvement in the quality factor for decreasing L_p values.

[7]And also in a MOS LC oscillator. Replacement of $1/2\alpha_{ol}$ by $2/3\alpha_{ol}$ in (7.10) yields a similar $\mathcal{L}(f_m)$ design FOM definition for a MOS implementation.

The first two benchmark FOMs provide a first order normalization of $\mathcal{L}(f_m)$ for some parameters that influence $\mathcal{L}(f_m)$, and can be used to compare oscillator designs: relative benchmarking. Oscillator design efficiency provides absolute benchmarking, because its compares the achieved $\mathcal{L}(f_m)$ with an estimation of a theoretical limit of $\mathcal{L}(f_m)$. To illustrate the four benchmark FOMs, the performance data on LC and ring oscillator designs from Appendix G and Appendix H is used. The numbers in the benchmark plots that follow below refer to the references on these designs in the back of this work.

7.2.1 Oscillator number

Based on linear noise modeling, for example see (3.6) on p. 59, we know that $\mathcal{L}(f_m)$ is proportional to $(f_{osc}/f_m)^2$. The benchmark FOM "Oscillator Number" (Osc-No) normalizes for f_{osc} and f_m to allow comparison between oscillators running at different frequencies and evaluated at different offset frequencies [50, 147]. The assumption here is that $\mathcal{L}(f_m)$ is measured where $\mathcal{L}(f_m)$ of the oscillator has a -6 dB per octave slope. The benchmark FOM Osc-No is defined as

$$FOM_{Osc-No} = 10\log(\mathcal{L}(f_m)) + 20\log\left(\frac{f_m}{f_{osc}}\right). \tag{7.11}$$

One could say that FOM_{Osc-No} is the extrapolated value of $\mathcal{L}(f_m)$ at an offset frequency equal to the carrier frequency.

In other words, the oscillator sideband with a -6 dB slope per octave is extrapolated, to end up with a figure that is usually below the white-noise floor of the oscillator. A lower FOM_{Osc-No} (i.e. a more negative value in dBs) indicates a better oscillator design concerning $\mathcal{L}(f_m)$ performance given a certain power budget.

As the dependence $\mathcal{L}(f_m)$ on f_{osc} and f_m is the same for LC and ring, FOM_{Osc-No} can be used for both types of oscillators. Figure 7.2(a) and Figure 7.2(b) show the FOM_{Osc-No} values for the LC and ring oscillators listed in Appendix G and Appendix H, respectively. Designs [187, 204] and [191] achieve the best value, around -200 dB.

As ring oscillators are much noisier, the best encountered FOM_{Osc-No} for ring oscillators is worse compared to LC oscillators

Design [237] achieves a value of about -180 dB. Notice the large number of LC oscillator designs in Figure 7.2(a) around 1.8 GHz. Many mobile phone standards operate near this frequency or near half of this frequency.

7.2.2 Normalized phase noise

The benchmark that we will refer to as "Normalized Phase Noise" (Nor-PN) is much more often used in the literature for oscillator comparison than FOM_{Osc-No}. This

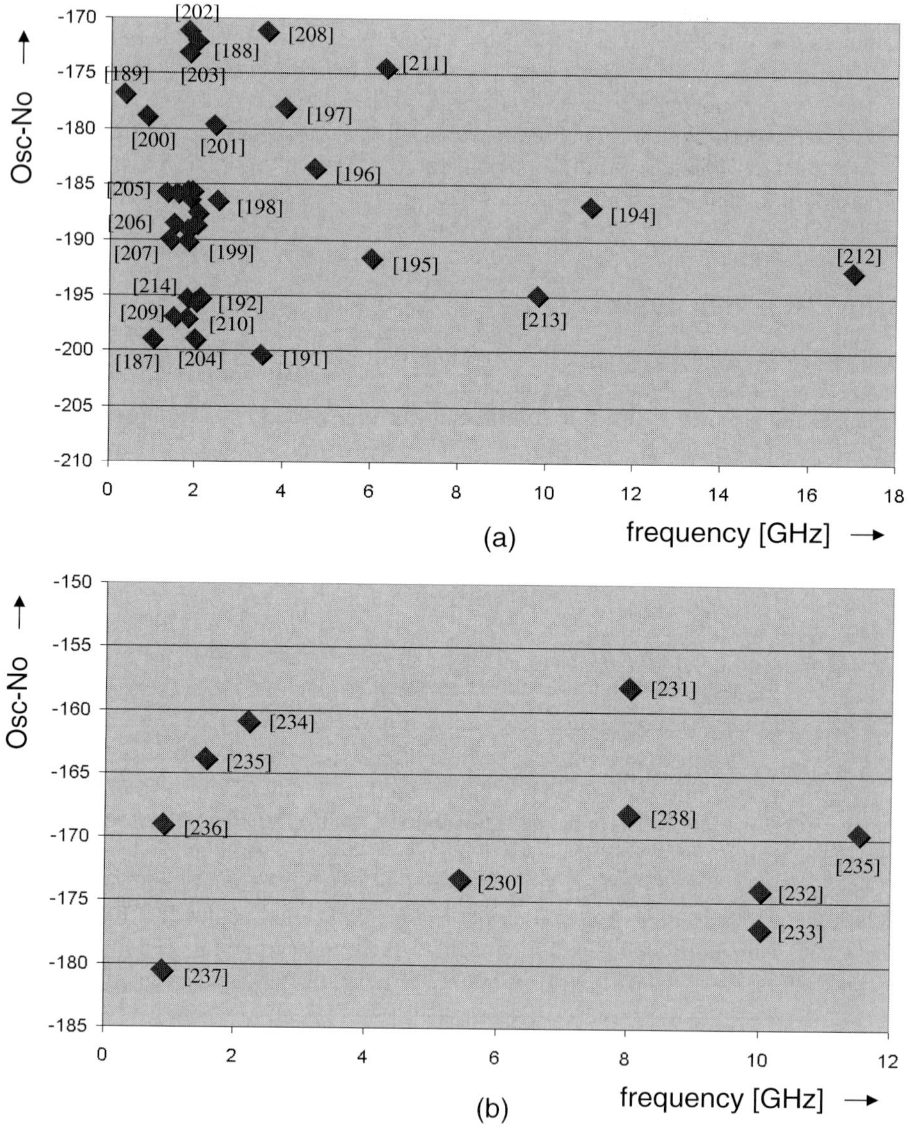

Figure 7.2 Benchmark FOM $Osc - No$ versus f_{osc} of a number of LC oscillator designs (a) and for several ring oscillator designs (b).

is because it takes into account the power dissipation P_{DC} of the oscillator core too, in addition to f_{osc} and f_m. From (6.50) on p. 158 and (6.64) on p. 168 it is clear that $\mathcal{L}(f_m) \propto 1/P_{DC}$. To take this dependence into account, FOM_{Nor-PN} normalizes the oscillator power dissipation to a reference power dissipation of 1 mW. Hence FOM_{Nor-PN} is defined as [195, 208],

$$FOM_{Nor-PN} = 10\log(\mathcal{L}(f_m)) + 10\log\left(\frac{f_m^2}{f_{osc}^2}\frac{P_{DC}}{1\,\text{mW}}\right). \tag{7.12}$$

A lower FOM_{Nor-PN} (i.e. a more negative value in dBs) indicates better phase noise performance given a certain power budget.

Figure 7.3(a) shows FOM_{Nor-PN} values of referenced LC oscillators. In Figure 7.3(b) the values of this benchmark FOM for referenced ring oscillators are presented. Compared to Figure 7.2 the values shift up or down somewhat depending on the power dissipation of the oscillator core. For example design [234] in Figure 7.3(b) moved up in ranking quite a bit compared to Figure 7.2(b), because of its relatively modest power dissipation of 11.8 mW (see Table H.1 on p. 280). Another major difference is that the encountered FOM_{Nor-PN} values are about 10 to 15 dB lower than FOM_{Osc-No}. This is simply due to the fact that most oscillator designs use more than 1 mW power. Since with more power $\mathcal{L}(f_m)$ can always be improved, FOM_{Nor-PN} should always be preferred to FOM_{Osc-No} for fair oscillator comparison.

7.2.3 Oscillator design efficiency

The benchmark FOM FOM_{Osc-No} normalizes for f_{osc} and f_m, and FOM_{Nor-PN} also normalizes for P_{DC}. Both FOMs can be used to rank LC and ring oscillators.

Ranking with FOM_{Osc-No} and FOM_{Nor-PN} allows the oscillator designer to assess whether his design belongs to the top of the class or hangs at the bottom. However, it leaves the designer with a number of interesting and valuable unanswered questions.

Questions such as: "Is it feasible to achieve the $\mathcal{L}(f_m)$ specification with this technology (quality factor) and power budget?", and "How difficult is it to meet this $\mathcal{L}(f_m)$ specification, i.e. is it close to the maximum achievable performance?". In this section we propose two figures of merit, LC-ODE and ring-ODE, which provide an answer to these questions.

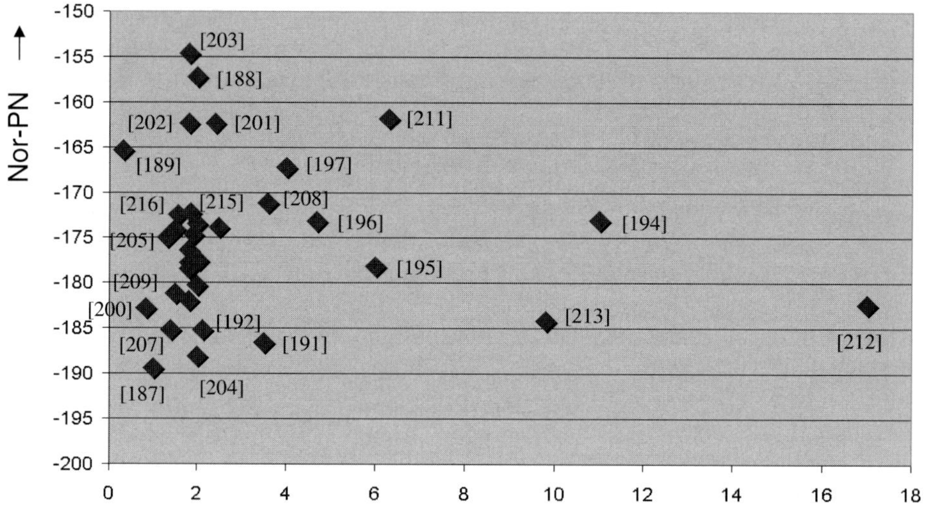

(a)

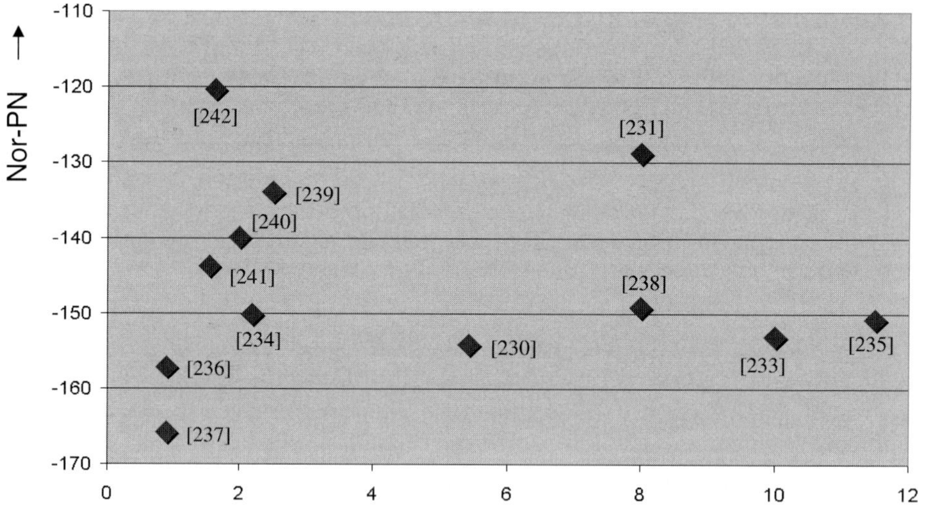

(b)

Figure 7.3 Benchmark FOM $Nor-PN$ versus f_{osc} of a number of LC oscillator designs (a) and benchmark FOM $Nor-PN$ versus f_{osc} for several ring oscillator designs (b).

LC oscillator design efficiency

To answer the questions posed in the introduction of this section, we need absolute benchmarking of $\mathcal{L}(f_m)$ of an oscillator design: the achieved $\mathcal{L}(f_m)$ needs to be compared with an estimation of the minimum $\mathcal{L}(f_m)$ that can be achieved. Linear phase noise modeling provides a first order estimate of this limit. Consider (6.50) on p. 158 that models $\mathcal{L}(f_m)$ due to noise shaping for LC oscillators. To be on the conservative side we can neglect active device noise by setting $\alpha_{ol} = 0$ in this equation, taking the efficiency $\eta_{LC} = 100\%$ and taking the unloaded resonator quality for Q_p factor (at zero resonator phase shift). With these settings and considerations in mind, we define the benchmark FOM LC-ODE as [148]

$$FOM_{LC-ODE} = -10\log(\mathcal{L}(f_m)) + 10\log\left(\frac{kT}{2P_{DC}}\frac{1}{Q_p^2}\frac{f_{osc}^2}{f_m^2}\right). \quad (7.13)$$

Since linear phase noise modeling neglects all additional phase noise generation mechanisms discussed in Section 6.4, and we set all cyclo-stationary active noise sources to zero, FOM_{LC-ODE} should always be negative. If the FOM_{LC-ODE} value of a certain oscillator design is larger than 0 dB, the theoretical limit that was defined using best case design parameters is crossed and this is extremely unlikely, if not impossible. A FOM_{LC-ODE} value of 0 dB implies 100% design efficiency. As we will see, an ODE of 10% (or -10 dB) already is an excellent oscillator compared to the majority of LC oscillator designs. On p. 163 we discussed that, given the same power budget, achievable $\mathcal{L}(f_m)$ is identical for single-phase and multi-phase LC oscillators. Hence FOM_{LC-ODE} can also be used for multi-phase LC oscillators.

In Figure 7.4(a) FOM_{LC-ODE} versus f_{osc} is plotted for most[8] of the single-phase and multi-phase LC oscillator designs in Appendix G. The majority of oscillators in Figure 7.4(a) has a FOM_{LC-ODE} value between -10dB (10% design efficiency) and -20 dB (1% design efficiency). If a set of specifications and a given technology[9] deliver a FOM_{LC-ODE} better (closer to 0 dB) than -10 dB, the specification is really challenging and may result in a significantly longer design trajectory compared to more moderate FOM_{LC-ODE} values. A few LC oscillator designs achieve an ODE value better than -10 dB. For example, designs [187], [192] and [191] achieve values of -7.4, -7.2 dB and -5.8 dB, respectively. Design [192] is the noise-shifting differential Colpitts oscillator that was briefly discussed on p. 176. Design [213] is a 10 GHz CMOS oscillator; using the design parameters reported in [213] its FOM_{LC-ODE} value comes to -2.2. This is of course highly unlikely, since an oscillator at 10 GHz in a 0.35 μm process will have sinusoidal waveforms and therefore significantly more active noise contributions to $\mathcal{L}(f_m)$ compared to the noise-shifting Colpitts oscillator

[8] For calculation of FOM_{LC-ODE} the quality factor of the resonator is needed in addition to parameters $\mathcal{L}(f_m)$, f_m, and P_{DC}. Therefore, only for the designs in Appendix G with these parameters available (i.e. reported in the referenced paper) FOM_{LC-ODE} is plotted.

[9] Which determines available quality factor values.

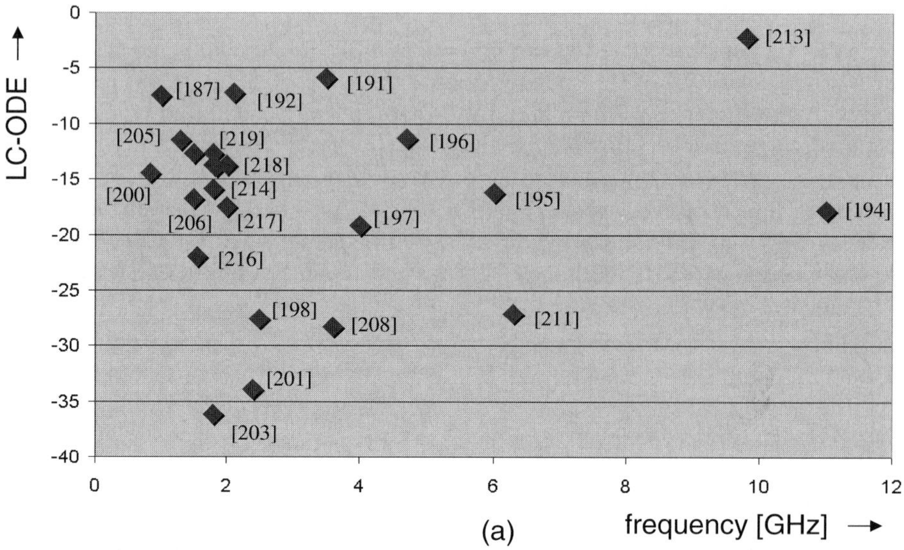

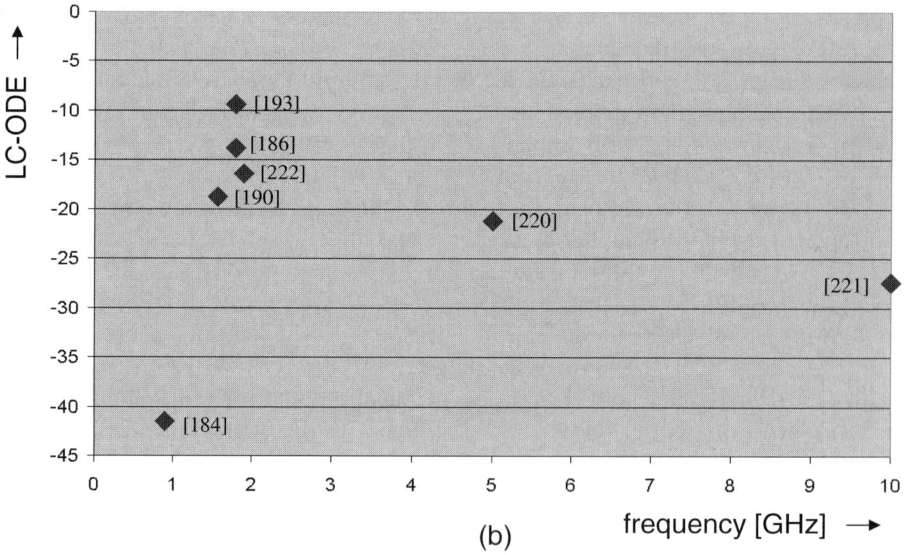

Figure 7.4 Benchmark FOM $LC-ODE$ versus f_{osc} for a number of single-phase LC (a) and multi-phase LC oscillator designs (b).

for example. Several reasons can lead to this inconsistent FOM value. For example, the actual quality factor of the used resonator can be significantly higher than the one reported.

> Whenever FOM_{LC-ODE} reaches a positive or unrealistic negative value, it is worthwhile to reassess oscillator key parameters and the simulation or measurement results.

The preceding example demonstrates how absolute benchmarking with FOM_{LC-ODE} provides the oscillator designer with a consistency check.

The lowest FOM_{LC-ODE} value (-36 dB) is linked to design [203] (also described in [58]). This oscillator employs an interesting implementation of phase-shift tuning using two resonator circuits. In the middle of its tuning range, the resonator phase shift for both resonator circuits in the oscillator is large. As we saw in Figure 6.42 on p. 163 this results in a grave quality factor reduction.

> A low FOM_{LC-ODE} value for a design can be the result of not taking full advantage of the available maximum resonator quality factor. In addition, other phase noise mechanisms than the unavoidable ones can be dominant, and possibly available power is not used efficiently.

The preceding examples demonstrate the added value of FOM_{LC-ODE} compared to the benchmark FOMs FOM_{Osc-No} and FOM_{Nor-PN}. Only FOM_{LC-ODE} provides the designer with a consistency check tool, a good indication of how difficult it will be to achieve the $\mathcal{L}(f_m)$ functional specification and alerts him when other phase noise mechanisms than the unavoidable ones are dominant or when available resonator quality factor is not fully employed.

Ring oscillator design efficiency

Analogous to the definition of FOM_{LC-ODE}, we can base an oscillator design efficiency FOM for ring oscillators on (6.64) on p. 168. By setting α_{ol} to zero in this equation we only include thermal noise of the collector or drain resistors in a ring oscillator stage in our absolute limit of $\mathcal{L}(f_m)$ for ring oscillators[10]. If we take a maximum ring oscillator quality factor $\pi/2$ and 100 % efficiency in P_{DC} to RF carrier power conversion, $FOM_{ring-ODE}$ can be defined as

$$FOM_{ring-ODE} = -10\log\left(\mathcal{L}(f_m)\right) + 10\log\left(\frac{N^2kT}{4P_{DC}}\frac{1}{(\frac{\pi}{2})^2}\frac{f_{osc}^2}{f_m^2}\right) \qquad (7.14)$$

[10]Note that single-ended and differential inverter-based oscillators have quite different topologies from ring oscillators based on differential pairs (like a CML ring oscillator). We should keep this in mind when we use a FOM based on (6.64) for inverter-type ring oscillators (for example design [237]).

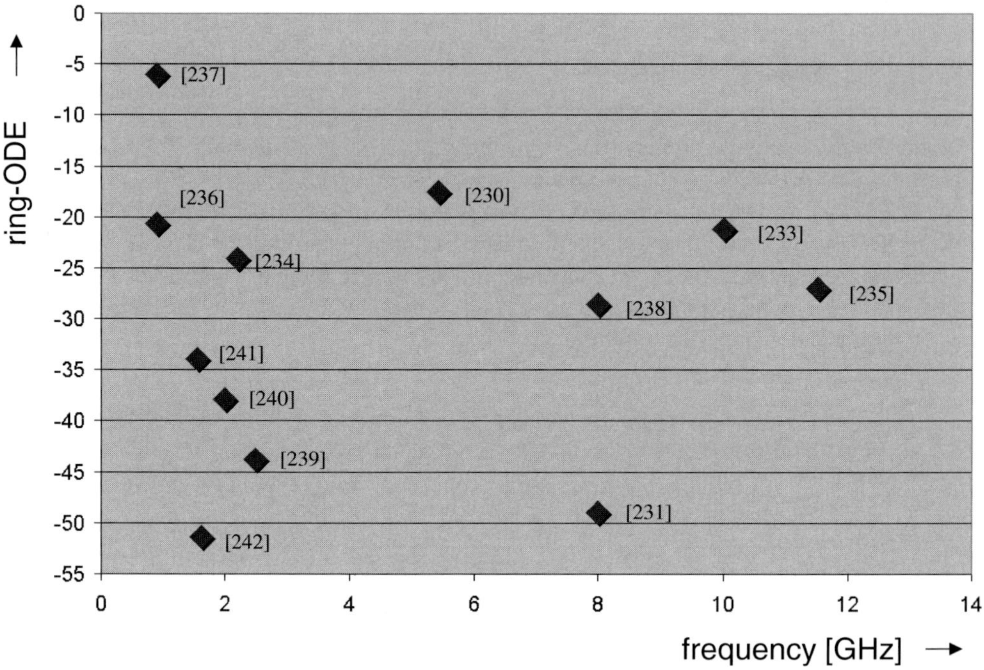

Figure 7.5 Benchmark FOM $ring - ODE$ versus f_{osc} for several ring oscillator designs.

A $FOM_{ring-ODE}$ value of 0 dB implies 100% design efficiency. $FOM_{ring-ODE}$ is plotted versus the oscillation frequency for various ring oscillator designs in Figure 7.5. It should be noted that compared to FOM_{LC-ODE}, $FOM_{ring-ODE}$ is less powerful since the first order estimation we use for the theoretical $\mathcal{L}(f_m)$ limit does not include technology parameters[11], whereas FOM_{LC-ODE} includes resonator technology. Therefore the additional information provided by $FOM_{ring-ODE}$ compared to FOM_{Osc-No} and FOM_{Nor-PN} is less, which is demonstrated by the fact that the ranking of oscillators in Figure 7.5 is almost identical to Figure 7.3(b). Nevertheless, $FOM_{ring-ODE}$ does provide a consistency check. A simulated or measured

[11] However, by using the quality factor defined in [142] for switching ring oscillators instead of $\pi/2$, the technology (characterized by f_T, V_{TH} and VDD for MOS designs) can be included. For an inverter-type based ring oscillator (that is not loaded by a buffer) this quality factor is given by

$$Q_{eff} = \frac{9}{8}\sqrt{\frac{\pi f_T}{2f_{osc}(1+\alpha_p)}\left(1 - \frac{V_{TH}}{VDD}\right)}, \tag{7.15}$$

with f_T and V_{TH} the transition frequency and threshold voltage of the NMOS devices in the ring oscillator and with α_p the ratio between the W/L of the PMOS transistors in the ring oscillator and the W/L of the NMOS transistors (typically 2.5 to 3).

$FOM_{ring-ODE}$ value greater than 0 dB is highly unlikely. Design [237] is extremely good and has a $FOM_{ring-ODE}$ value of -6 dB. It utilizes a new type of inverter-based differential ring oscillator stage. According to [237] the fast-slewing delay cell that performs full switching shows this good $\mathcal{L}(f_m)$ performance because the transistors are off (and thus generate no noise) during a large part of one oscillation period. It should be noted that if we take the reported $\mathcal{L}(f_m)$ reading of the free-running spectrum (instead of the $\mathcal{L}(f_m)$ reported measurement value that was performed with a spectrum analyzer with additional HP 85671A phase noise measurement software), the $FOM_{ring-ODE}$ value for this oscillator comes to -12.4 dB. This is closer to what one would expect for a ring oscillator and to the other values in Figure 7.5.

> Figure 7.5 indicates that a $FOM_{ring-ODE}$ value much higher than -15 dB will be hard to achieve in practice.

Design [242] has a $FOM_{ring-ODE}$ value of -51 dB. This design is a four-stage ring oscillator with multiple feedback paths. It features a high tuning range of 88%, but at the cost of a complex structure and many transistors per stage, which obviously results in a poor $FOM_{ring-ODE}$ value.

7.3 Summary

Design FOMs provide a powerful means for first order topology ranking and selection. Once all relevant design FOMs are defined for a set of topologies, the most promising oscillator topology given a set of functional specifications can be determined with design FOMs. In addition, design FOMs can be used to find the minimum value of one design parameter given all other design parameters, by solving design FOM=0 dB for the design parameter of interest. Design FOMs of an oscillator topology provide qualitative insight into the relations between the design parameters, quantitative insight when the FOM is evaluated, and FOMs document design knowledge. In this chapter, a number of frequency, tuning and $\mathcal{L}(f_m)$ design FOM examples have been described, taking advantage of the behavioral models and theory introduced and developed in Chapter 5 and Chapter 6.

Four benchmark FOMs have been described in this chapter. The benchmark FOM "Oscillator Number" normalizes the $\mathcal{L}(f_m)$ of an oscillator for oscillation frequency f_{osc} and offset frequency f_m. The FOM "Normalized Phase Noise" also includes a normalization for power dissipation and this benchmark FOM is used most often in the literature. Absolute benchmarking of $\mathcal{L}(f_m)$ of LC and ring oscillators can be performed with "LC-ODE" and "ring-ODE", respectively. These benchmark FOMS compare $\mathcal{L}(f_m)$ with a first order estimation of the best case achievable $\mathcal{L}(f_m)$. If an ODE value is above 0 dB or unrealistically close to 0 dB, most likely something went wrong during the simulation or measurement of $\mathcal{L}(f_m)$. ODE can also be used to check whether a set of functional specifications are consistent in the sense that it is possible to achieve the $\mathcal{L}(f_m)$ specification. All four benchmark FOMs were demonstrated using the performance data on LC and ring oscillator in Appendices G and H, respectively.

8

AC phase noise simulation tool

DIMENSIONING and optimization of an oscillator constitutes an important part of the total design time. CAD-tools play a crucial role in this process. Built-in advanced transistors models (see for example Table 3.1 on p. 64) of a circuit simulator nowadays allow for an excellent prediction of all important oscillator specifications, when combined with accurate passive device modeling and accurate technology parameters. Many oscillator aspects can be simulated with traditional circuit-analysis types such as DC, AC and transient analysis. For example, start-up conditions can be conveniently verified in AC analysis, and the actual start-up and steady-state waveforms can be verified in transient analysis[1].

Fast analysis of oscillator phase noise was developed in the second half of the 90's and now is part of commercial tools such as SpectreRF[2] [77] and ADS[3]. The costs of these tools are relatively high, but their performance in terms of speed and accuracy is very good.

[1]Since circuit noise is not modeled in transient analysis, a small transient has to be given in transient analysis for start-up, e.g. in the form of a short pulse to the supply voltage or a current pulse in one of the nodes of the oscillator.

[2]Phase noise is simulated using the Periodic Steady State (PSS) analysis type followed by a periodic noise (pnoise) analysis. These specialized analysis types capture nonlinear and time-varying phase noise aspects. Note that the $\mathcal{L}(f_m)$ given by SpectreRF also includes AM components if these are not completely removed by an internal limiting mechanism in the oscillator: it plots the total noise divided by the rms carrier.

[3]Advanced Design System, www.agilent.com

In this chapter the phase noise simulation tool ACPN is proposed, which is based on standard AC noise analysis [149].

The foundation of AC phase noise (ACPN) simulation was laid in a period before phase noise analysis was incorporated in commercial circuit simulators [150]. At that time it was one of the few[4] methods to obtain an estimate of $\mathcal{L}(f_m)$. Although its accuracy and scope are lower compared to the mentioned commercial tools that are nowadays available, coexistence of the ACPN tool is justified as will be explained in Section 8.1. In this section, ACPN simulation is introduced and its scope is clearly defined. All required steps in the ACPN simulation flow are described in Section 8.2. Application of the ACPN tool in oscillator modeling and design is illustrated in Section 8.3 and Section 8.4. This chapter concludes with a short summary in Section 8.5.

8.1 AC phase noise simulation

The ACPN tool was primarily developed to obtain a fast estimation of $\mathcal{L}(f_m)$ of an oscillator. Since it can be implemented in any standard circuit simulator with AC noise analysis capability, it has a number of advantages, but at the same time also a limited scope and accuracy, as explained in Section 8.1.1. The principle of ACPN simulation is highlighted in Section 8.1.2.

8.1.1 Introduction

ACPN simulation makes use of AC noise analysis, which is available in standard circuit simulators such as Spice [67]. The following features can be listed for the ACPN tool:

- The cost of the ACPN tool is extremely low, as free-ware circuit simulators can be used like Spice.

- Noise sources are incorporated in the small-signal device models used in standard AC noise analysis (in contrast to transient analysis in a simulator: large-signal models do not have noise sources incorporated). Therefore, apart from some current sources needed for an ACPN-simulation run (see Section 8.2), the oscillator schematic does not need to be altered for a $\mathcal{L}(f_m)$ simulation using the ACPN tool.

[4] Another method can be implemented in transient analysis (see the slide supplement of reference [52] or [31, 151]). This method uses sinusoidal sources to represent (in amplitude scaled) noise sources and, after a long transient simulation run, Fourier analysis is applied to calculate $\mathcal{L}(f_m)$. Compared to the ACPN tool, this method is more accurate (it can take into account noise folding and modulation of noise sources) but it is much slower.

- The contribution of individual noise sources in the small-signal device models used can be queried after an AC noise analysis. Therefore, the oscillator designer is provided with insight into the dominant phase noise contributors.

- Convergence and simulation speed are excellent, since AC analysis is utilized. This is especially of importance when the ACPN tool is used by a circuit optimizer. In an optimization loop, non-convergence often leads to a premature end of the optimization process.

- Because AC noise analysis is a linear analysis type, the designer knows exactly which phase noise generation mechanisms are included in the simulation results. In other words, only phase noise produced by noise shaping, described in Section 3.4.1 on p. 58 is taken into account. Since ACPN-simulation only takes noise shaping into account, it is extremely suitable for numerical verification of any linear phase noise theory. For example, all phase noise formulas in Section 6.3 were verified this way.

The fact that the ACPN tool uses standard AC noise analysis, is the main reason for its limited scope. The tool neglects the time-varying nature of an oscillator and all nonlinear phase noise generation mechanisms that were discussed in Section 6.4. Hence, for practical oscillator circuits working in the strongly nonlinear region, it only provides an estimate of the -6 dB per octave slope ($1/f^2$) of $\mathcal{L}(f_m)$. Nevertheless, this estimate can be quite reasonable even for a switching LC oscillator: less than 5 dB error as we have seen in Section 6.3. It should be clear that the set-up time of an phase noise simulation with the ACPN tool is longer in comparison to built-in phase noise analysis (which is available in SpectreRF, for example).

8.1.2 ACPN simulation principle

ACPN simulation can be used for all LC and ring oscillators within the scope of this work. The primary phase noise generation mechanism (noise shaping) of theses oscillators can be modeled by the theory presented in Section 3.4.1 on p. 58. Without loss of generality, we will explain the principle of ACPN simulation only for a single-phase LC oscillator.

Consider the linear model of an LC oscillator in Figure 8.1. All noise sources are modeled by an equivalent noise voltage $\overline{v_n}$. At start-up $g_{m_{eff}}$ is simply g_m and $|H_{ol}(j\omega_{osc})| = \alpha_{ol} = g_m R_p$ in this model. Open-loop gain $|H_{ol}(j\omega_{osc})|$ needs to be larger than unity for oscillator start-up.

After the oscillator start-up, a practical oscillator reaches steady-state due to an amplitude stabilization mechanism. For example, nonlinear circuit elements (see Section 2.5.1 on p. 31) or AGC (See Section 2.5.2 on p. 34) can stabilize the oscillation amplitude. For a noiseless oscillator (e.g. Figure 8.1 without noise source v_n) steady-state oscillation would be reached if the open-loop gain had been reduced by an amplitude stabilization mechanism to exactly unity: for Figure 8.1 this would mean $g_{m_{eff}} = 1/|H_{ol}(j\omega_{osc})| = 1/(g_m R_p)$. For a practical oscillator output voltage v_{out} and all noise sources (lumped in v_n in Figure 8.1) are finite, which means that

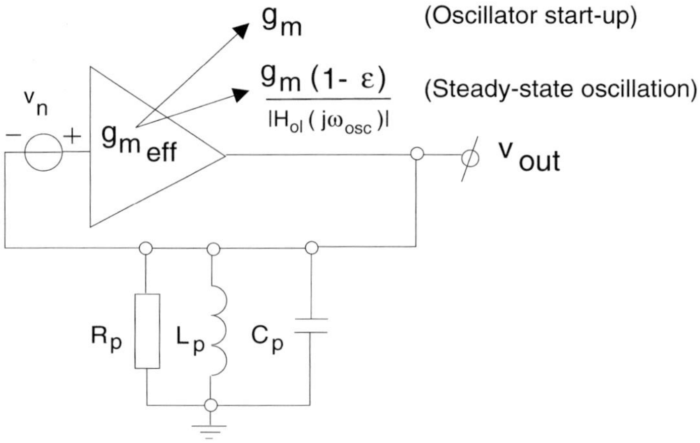

Figure 8.1 Linear behavioral model of an LC-oscillator. The nonlinearity of the transconductance in a practical oscillator modeled is by an effective transconductance that is equal to $g_m \cdot (1 - \varepsilon)/|H_{ol}(j\omega_{osc})|$ for steady-state oscillation.

the closed-loop gain $|H_{cl}(j\omega_{osc})|$ has to be finite. Therefore, $g_{m_{eff}}$ is made equal to $g_m \cdot (1-\varepsilon)/|H_{ol}(j\omega_{osc})|$ in Figure 8.1 to model steady-state oscillation (ε is a constant that is much smaller than unity). For this value of $g_{m_{eff}}$, the open-loop gain for steady-state oscillation is $1-\varepsilon$, that is slightly smaller than unity, resulting in a finite carrier signal.

Given the discussed value for $g_{m_{eff}}$ for steady-state oscillation, the closed-loop gain of the model in Figure 8.1 can be calculated. The squared output noise voltage $\overline{v_{out}^2}$ of the oscillator in Figure 8.1 is equal to the closed-loop gain $|H_{cl}(j\omega_{osc})|^2$ times $\overline{v_n^2}$. Since $|H_{ol}(j\omega_{osc})| = gmR_p$, $Q_p = R_p/(\omega_{osc}L_p)$ and only frequencies close to the carrier are of interest ($|\Delta\omega| \ll \omega_{osc}$), the closed-loop transfer function $|H_{cl}(j\omega)|$ at $\omega = \omega_{osc}$ of the model in Figure 8.1 becomes[5]

$$|H_{cl}(j\omega_{osc})| = \frac{1-\varepsilon}{\sqrt{\varepsilon^2 + (\frac{2\Delta\omega}{\omega_{osc}}Q_p)^2}}. \tag{8.1}$$

Parameter ε can be regarded as a fit parameter that matches the total noise power of the linear model with the actual carrier power, resulting from transient simulations for example (which takes into account all nonlinearities). The output power of an oscillator circuit is the result of frequency selective amplification of $\overline{v_n^2}$. For ε equal to

[5]See Section I.1 on p. 281 for the complete derivation of $|H_{cl}(j\omega_{osc})|$ for practically the same LC oscillator model as the one presented in Figure 8.1.

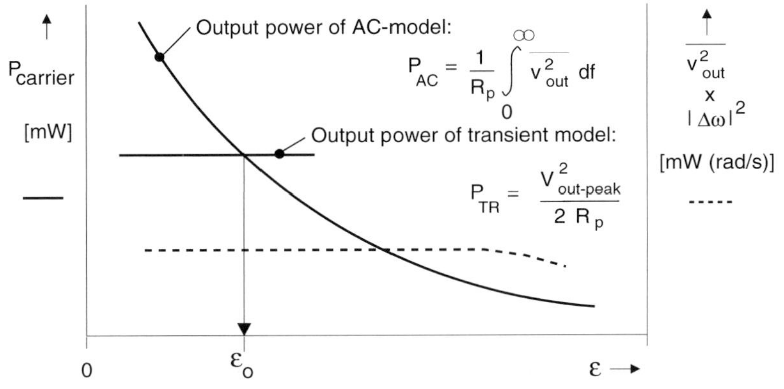

Figure 8.2 Total output power of the behavioral model in Figure 8.1 versus ε. The value ε_o matches the output power of the AC-model to a value obtained with a transient simulation run.

zero, the output power of the linear oscillator model becomes infinite. However, for finite ε values the integrated noise power of the linear oscillator model can be made to match the power predicted by a transient simulation, for a certain value $\varepsilon = \varepsilon_o$, as illustrated in Figure 8.2.

> Equation 8.1 presents the amplification factor of the noise source v_n under closed-loop conditions, resulting in an output $\overline{v_{out}^2}$. This noise shaping has been discussed already in more general terms in Section 3.4.1, but with ACPN a new parameter ε is introduced.

Without parameter ε, closed-loop gain $|H_{cl}(j\omega)|$ would be infinite for the angular offset frequency $\Delta\omega = 0$ and the output spectrum of the oscillator model in Figure 8.1 "blows up". Due to the approximations we made in the linear phase noise modeling approach used in Section 3.4.1 and Section 6.3, $\mathcal{L}(f_m)$ equations in these sections are rather useless when $\Delta\omega$ approaches zero. Fortunately, this does not really limit the usefulness of the $\mathcal{L}(f_m)$-estimates presented in Section 6.3 in practice, but it could potentially give problems in a circuit simulator (i.e. division by zero when $\Delta\omega = 0$ during simulation). A small but non-zero parameter ε prevents divergence of the oscillator output spectrum at the oscillation frequency.

> The oscillator noise, $|H_{cl}(j\omega_{osc})|^2 \cdot \overline{v_n^2}$ represents a Lorentz-function due to the presence of parameter ε in the denominator, which corresponds to an exponential decay of the auto-correlation function in the time domain [152].

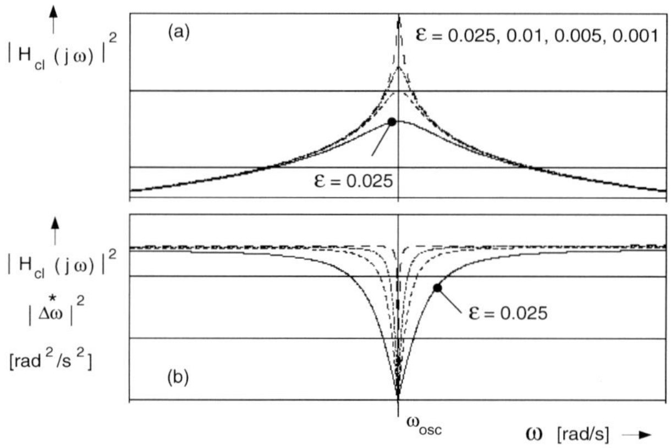

Figure 8.3 $|H_{cl}(j\omega)|^2$ (a) and $|H_{cl}(j\omega)|^2 \cdot |\Delta\omega|^2$ (b) versus ω for four values of parameter ε.

In several ways it can be shown that the power spectrum of a practical oscillator with only white noise sources is a Lorentzian for each harmonic[6] [13, 104, 153]. So, the discussed modeling approach is consistent with the literature.

Figure 8.3(a) shows the function $|H_{cl}(j\omega)|^2$ versus frequency, for ε equal to 0.025, 0.01, 0.005 and 0.001. For $\varepsilon \ll |\Delta\omega|Q_p/\omega_{osc}$, the closed-loop amplification at $\Delta\omega$ is practically the same regardless which value for ε is substituted in (8.1). This is illustrated in Figure 8.3(b): for sufficiently large $\Delta\omega$ the curves $|H_{cl}(j\omega_{osc})|^2 \cdot |\Delta\omega|^2$ converge to the same value. Similar behavior for $\overline{v_{out}^2} \cdot |\Delta\omega|^2$ versus ε was already encountered in Figure 8.2. In sum, the characteristic -6 dB per octave of the oscillator-sideband slopes results in the same constant value for $\overline{v_{out}^2} \cdot |\Delta\omega|^2$ for small arbitrary ε values.

Using the above theory, we can now describe the ACPN simulation principle. If we make (see next section) the open-loop gain equal to $1 - \varepsilon$ somehow for small but arbitrary ε in an oscillator circuit, perform an AC noise analysis and let the simulator calculate the noise at the oscillator output terminals, we obtain the oscillator output spectrum with phase noise sidebands due to noise shaping. If we read the noise power of this spectrum at an offset frequency f_m, while verifying that we obtained this value on the -6 dB per octave slope, divide it by the carrier power (that was for example obtained with a transient analysis) and subtract 3 dB we obtain the -6 dB/octave part of $\mathcal{L}(f_m)$.

[6] A Lorentzian is the shape of the squared magnitude of a one-pole low-pass filter transfer function [153].

> The subtraction of 3 dB results from the fact that a standard AC noise analysis calculates the total noise. In accordance with the convention adopted in this work (see p. 72), $\mathcal{L}(f_m)$ only represents the phase noise of an oscillator and does not include amplitude noise.

8.2 ACPN simulation flow

The following steps can now be identified to simulate the $1/f^2$ slope of $\mathcal{L}(f_m)$ accurately for an oscillator under design, operating in the linear or weakly nonlinear region, or to provide an estimate of $\mathcal{L}(f_m)$ for strongly nonlinear oscillators:

1 Determine the carrier power (*one transient run*).

2 Locate nonlinear element(s) in the oscillator circuit. Insert an attenuator into the output of each element (see Figure 8.1). Usually only one (differential) attenuator is needed per oscillator stage.

3 Determine the open-loop gain[7] $|H_{ol}(\omega_{osc})|$ under closed-loop biasing (with DC conditions as they are for zero oscillation amplitude) and loading conditions. The oscillation frequency ω_{osc} is the frequency where the phase condition for oscillation is met (*one AC run*). Assign a value to the attenuator such that the open-loop gain in the oscillator is (1-ε). See, for example, the attenuator in Figure 8.4 that is implemented as a current source. Its value i_a is equal to the output current I_{cl} of the transconductance g_m, multiplied with $(1-1/|H_{ol}|)\cdot(1-\varepsilon)$.

4 Perform a closed-loop AC noise analysis for a small (e.g. $\varepsilon < 0.01$) and arbitrary ε. The output nodes of the noise analysis are the oscillator output nodes (*one AC run*).

5 Read the noise power values (total noise and/or contributing noise sources) at an offset frequency f_m on the $1/f^2$ slope and divide this by the carrier power. Subtract 3 dB from the result to eliminate AM noise contribution to obtain a value for $\mathcal{L}(f_m)$.

Figure 8.4 illustrates a practical simulation aspect concerning step 4 in the ACPN simulation flow. By giving current attenuator i_a the appropriate value, as shown in Figure 8.4, and step 4 can be performed. Note that ACPN simulation for behavioral models is much faster to set up, compared to an oscillator modeled at circuit level.

---

[7] In [154, 155] the open-loop gain under closed-loop conditions is derived in terms of two-port parameters. Expressed in Z-parameters $H_{ol}(j\omega)$ can be written as $(Z_{21}-Z_{12})/(Z_{11}+Z_{22}-2\cdot Z_{12})$, which is identical to

$$H_{ol}(j\omega) = \frac{S_{21}-S_{12}}{1-S_{11}S_{22}+S_{12}S_{21}-2S_{12}}. \quad (8.2)$$

Note that (8.2) reduces to $S_{21}/(1-S_{11}S_{22})$ for unilateral devices (i.e. $S_{12}=0$).

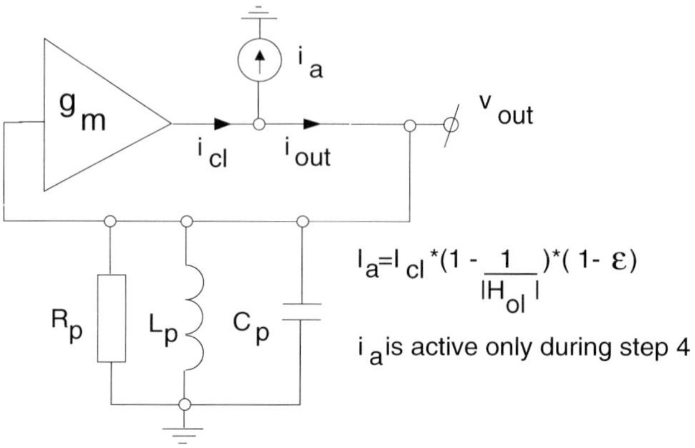

Figure 8.4 Behavioral model of a LC-oscillator with current attenuator i_a that reduces the loop gain to $(1-\varepsilon)$.

Only step 4 and 5 of the ACPN simulation flow need to be performed. We can simply assume a certain carrier current (step 1) and since we are in full control of all parameters of a behavioral model we can set the loop gain of a behavioral model to $1-\varepsilon$ without affecting the amount of device noise generated. In an actual oscillator circuit, step 2 and step 3 are required to fulfill the conditions $|H_{ol}| = 1-\varepsilon$, without affecting the biasing conditions that in turn determine the amount of device noise generated. Two simulation examples are given in the following two sections to illustrate the ACPN tool in practical situations.

8.3 Simulation example I: verification of $\mathcal{L}_{bipo}(f_m)$

In Section 6.3, we made extensive use of linear phase noise modeling to arrive at compact and insightful $\mathcal{L}(f_m)$ expressions for LC and ring oscillators. For example, on p. 158 we derived an estimate of $\mathcal{L}(f_m)$ for a bipolar cross-coupled LC oscillator:

$$\mathcal{L}(f_m)_{bipo} \approx \frac{1}{2} \cdot \frac{1}{Q_P^2} \left(\frac{f_{osc}}{f_m}\right)^2 \frac{kT(\frac{1}{2}\alpha_{ol}+1)}{R_P \frac{2}{\pi^2} I_{tail}^2}. \tag{8.3}$$

Because the ACPN tool uses linear AC noise analysis, it is ideally suited to transparently verify any phase noise formula obtained with linear modeling. In this section, $\mathcal{L}(f_m)_{bipo}$ is compared with the results of the ACPN tool as an example. Other $\mathcal{L}(f_m)$ equations in Section 6.3 can be verified in a similar way.

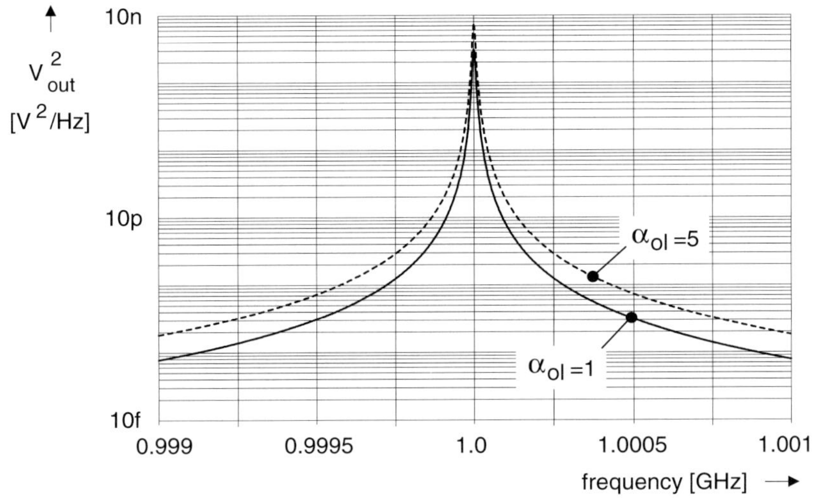

Figure 8.5 Noise voltage density of a differential LC oscillator behavioral model simulated in standard AC noise analysis using the ACPN simulation flow. Both resonator noise and bipolar shot-noise are modeled.

In order to verify $\mathcal{L}(f_m)_{bipo}$ with the ACPN tool, a differential version of the behavioral model from Figure 8.1 was entered in a circuit simulator[8], and in addition to the noise generated by R_p, bipolar shot-noise of a cross-coupled pair was also modeled. This shot-noise was modeled by two current sources (one for every transistor, see Figure 6.40 on p. 157) with a value

$$\overline{i^2_{n_{bipo}}} = \frac{\alpha_{ol}\, 4kT}{R_p}, \tag{8.4}$$

which is equivalent to $2qI_c$ with I_c the collector current of a bipolar transistor. The oscillation frequency was set at 1 GHz, L_p was set to 20 nH, and Q_p was set to 10 ($R_p = 1.26\,k\Omega$). As mentioned in Section 8.2, an ACPN-simulation for a behavioral model is much faster to set up, than for an oscillator modeled at circuit level. For a behavioral model, only step 4 and 5 of the ACPN simulation flow need to be performed.

Figure 8.5 shows the ACPN simulation results for $\varepsilon = 10^{-4}$ of the differential LC oscillator model for α_{ol} in (8.4) 1 and 5. The widening of the spectrum due to increased contribution of the shot noise when α_{ol} is increased from 1 to 5 is evident. The total simulation time needed for analysis was 0.17 s on a HP 9000/785 Unix compute server.

[8]Philips' in-house circuit simulator "Pstar".

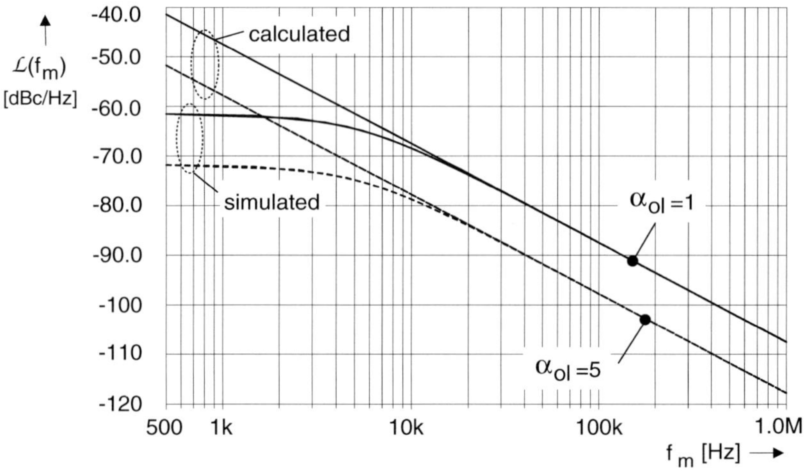

Figure 8.6 $\mathcal{L}(f_m)$ versus f_m of the differential LC oscillator behavioral model simulated with the ACPN tool. The calculated curves are derived using (8.3).

The noise spectrum from Figure 8.5, divided by the estimated carrier power $2/\pi^2 \cdot I_{tail}^2 R_p$ and with 3 dB subtracted (to remove AM noise) is plotted in Figure 8.6 for offset frequency f_m ranging from 500 Hz to 1.0 MHz. Current I_{tail} can be easily calculated for a bipolar cross-coupled LC oscillator once α_{ol} is chosen: $I_{tail} = 4\alpha_{ol}kT/(qR_p)$.

Equation (8.3) is also plotted in Figure 8.6 for α_{ol} equal to 1 and 5. The simulated Lorentzian-shaped $\mathcal{L}(f_m)$ curves starts[9] having a -6 dB slope per octave for offset frequencies above 10 kHz. The values on this slope represent an exact fit with the values calculated with (8.3).

Other $\mathcal{L}(f_m)$ formulas derived in Section 6.3 can be (and have been) verified with the ACPN tool. Two examples of theories on phase noise optimization obtained by using linear phase noise modeling will be mentioned here, because they can be verified and further explored with the ACPN tool. The first theory is described in [156, 157]. Capacitive tapping is used to match the resonator impedance to the impedance of the active part of an oscillator with an series LC resonator to optimize $\mathcal{L}(f_m)$. The second theory is presented in [158, 159]: an optimum ratio between the loaded and unloaded quality factor, namely 2/3, for minimum $\mathcal{L}(f_m)$ is calculated.

[9]In this particular case: $f_{osc} = 1\,\text{GHz}$, $Q_p = 10$ and $\varepsilon = 10^{-4}$.

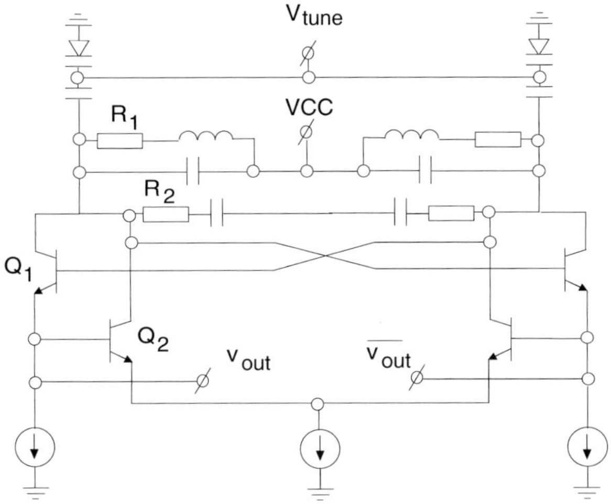

Figure 8.7 Circuit diagram of the balanced 3.6 GHz LC oscillator.

8.4 Simulation example II: $\mathcal{L}(f_m)$ of a SOA LC oscillator

In this example an ultra-low-power oscillator in Silicon On Anything (SOA) technology [160] is simulated with the ACPN tool. SOA is an IC-technology in which the substrate of a complete wafer is removed after processing and replaced by an isolating substrate such as glass [208]. In Section 9.1, an 800 MHz SOA LC oscillator design is described including more details on SOA technology. The SOA LC oscillator investigated here works at 3.6 GHz, dissipates 1 mW and is designed with $|H_{ol}(j\omega_{osc})|$ equal to 2.9. Measured $\mathcal{L}(f_m)$ equals -80 dBc/Hz at 100kHz offset of the 3.6 GHz carrier. Figure 8.7 presents the circuit diagram. A balanced topology is used with a cross-coupled differential pair and two emitter followers as a feedback amplifier.

Figure 8.8 shows the simulated $\mathcal{L}(100\,\text{kHz})$ versus $|H_{ol}(j\omega_{osc})|$ of the LC oscillator. The results of the ACPN method are compared in this graph with SpectreRF. Philips' in-house circuit simulator Pstar was used, utilizing only standard AC and transient analysis. An HP 9000 system/model 804, on which the operating system B.10.20A runs, was used to perform the simulations. On this hardware the ACPN method typically needed ~30 seconds CPU time (step 1: 25.76 sec., step 3: 1.33 sec. and step 4: 2.52 sec.).

> Comparison of ACPN with SpectreRF shows an excellent match for open-loop gains up to 1.5. At $|H_{ol}(j\omega_{osc})|$ equal to 1.8, the AC method is 1 dB more pessimistic compared to the SpectreRF phase noise simulations (see Figure 8.8).

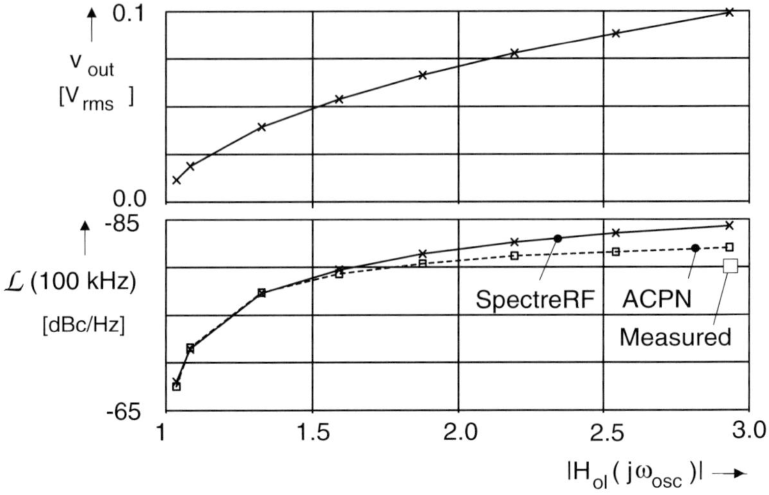

Figure 8.8 V_{out} and $\mathcal{L}(f_m)(100\,\text{kHz})$ versus $|H_{ol}(j\omega_{osc})|$.

This demonstrates that for the presented type of oscillator the ACPN method is an accurate approximation for open-loop gains up to 1.8. If the designer wants to simulate oscillators with larger open-loop gains the ACPN method can be used for coarse optimization, taking advantage of its speed and excellent convergence . Compared to measurements, both the ACPN and the SpectreRF method are too optimistic (respectively 2 dB and 4 dB) because external noise generators, like for example power supply noise, were not included in the simulations[10]. Note that the SpectreRF simulations for $|H_{ol}(j\omega_{osc})| = 2.9$ in Figure 8.8 reveal the effect of noise modulation that we already encountered in Table 6.3 on p. 160.

For higher open-loop gains the oscillator starts switching and the relative contribution of collector shot noise to the total phase noise reduces, and hence $\mathcal{L}(f_m)(100\,\text{kHz})$ improves.

For $|H_{ol}(j\omega_{osc})|$ equal to 1.03 the contribution of four noise sources (Q_1, Q_2, R_1 and R_2), see Figure 8.7) are listed in Table 8.1. This table illustrates that both qualitative (ranking of noise sources) and quantitative $\mathcal{L}(f_m)$ results are simulated correctly using the AC method for small open-loop gain values. Insight into the ranking of the noise sources speeds up the optimization of the oscillator under design, since it points out the most dominant contributors to the phase noise.

[10] If this was the case, SpectreRF would give the most accurate results.

$\mathcal{L}(100\,\text{kHz})$	ACPN method	SpectreRF
Total $\mathcal{L}(100\,\text{kHz})$	-67.9	-68.1
Q_1 (total)	-74.2	-74.1
Q_1 (I_C shot noise)	-77.8	-77.4
Q_1 (R_{base} noise)	-76.9	-77.2
Q_2 (total)	-76.5	-76.2
Q_2 (I_C shot noise)	-79.5	-79.4
Q_2 (R_{base} noise)	-79.7	-79.3
R_1	-80.0	-79.9
R_2	-83.1	-83.0

Table 8.1 ACPN tool and SpectreRF phase noise simulation results of the SOA oscillator for an open-loop gain of 1.03. The four dominant noise contributors are listed.

> The ACPN simulation examples in this section and in Section 8.3 concerned both LC oscillators. However, the ACPN analysis method can equally well be applied to ring oscillators.

The two-integrator oscillator presented in Figure 6.12 on p. 126, can be designed for linear, weakly nonlinear and strongly nonlinear operation. In an RF modulator IC in production, this ring oscillator is used in linear mode (with an AGC) as a sound modulator operating from 4.5MHz to 6.5 MHz [161]. The presented ACPN method was used throughout the design of the sound modulator to optimize $\mathcal{L}(f_m)$. Measurements on the final design resulted in a $\mathcal{L}(100\,\text{kHz})$ of -115 dBc/Hz with a carrier frequency of 5.5 MHz carrier. ACPN simulations predicted these measurements with 1 dB accuracy.

8.5 Summary

CAD-tools play a very important role in the design process of a high-frequency integrated oscillator. Advanced transistor models, models of passive components, combined with accurate model parameters can accurately predict oscillator performance, including $\mathcal{L}(f_m)$. This is crucial for oscillator dimensioning and optimization. Accurate phase noise predictions of most practical oscillators require a circuit simulator that has specialized analysis types, such as PSS and pnoise analysis in the commercial simulator SpectreRF, to capture all nonlinear and time-varying issues, for example noise folding and the modulation of noise sources.

In this chapter, an AC phase noise simulation tool is proposed that can be used to obtain a reasonable (for strongly nonlinear oscillators) or good (for weakly nonlinear oscillators) estimate of $\mathcal{L}(f_m)$. The ACPN tool uses standard AC noise analysis that is available in most circuit simulators, including the freeware simulator Spice. As ACPN simulation is incapable of capturing time-varying and nonlinear phase noise aspects of practical oscillators, it can not compete with the accuracy of sophisticated commercial simulators such as SpectreRF. ACPN simulations only provide the $\mathcal{L}(f_m)$ resulting from noise shaping of noise sources in an LC or ring oscillator. Although the scope of the ACPN tool is limited, several features justify its coexistence with more accurate phase noise simulation tools. Among others, these features are an excellent speed and convergence, a high performance price (the price is practically zero) ratio, and it is ideally suited to verify linear phase noise theories.

In an ACPN simulation, the open-loop gain in an oscillator is reduced to a value $(1-\varepsilon)$. When the oscillator is simulated using AC noise analysis for $\varepsilon \ll |\Delta\omega| Q_p / \omega_{osc}$, the simulated output noise spectrum has a Lorentzian shape. By taking a noise value reading at an offset frequency f_m on the -6 dB per octave part of this curve, dividing it by the carrier and subtracting 3 dB to remove the AM noise contribution, a value for $\mathcal{L}(f_m)$ is obtained. The carrier amplitude can be obtained with one short transient simulation run. In a practical oscillator the open-loop gain is much higher (e.g. 3–5) and needs to be reduced for an ACPN simulation to $(1-\varepsilon)$ without affecting the bias conditions, which in turn affect the amount of generated noise. This can be realized by inserting current sources and giving them the appropriate attenuation factor. These attenuators can be seen as blocks that model the amplitude stabilization mechanism of a practical oscillator in the AC model of this oscillator.

Two examples have illustrated the use of the ACPN tool in this chapter. First, the phase noise generated by a differential behavioral model of a bipolar cross-coupled LC oscillator has been simulated. As expected, the simulated $\mathcal{L}(f_m)$ perfectly matches the $\mathcal{L}(f_m)$-equation, $\mathcal{L}_{bipo}(f_m)$, that was derived for this oscillator in Section 6.3. Second, the phase noise of a realized monolithic 3.6 GHz SOA LC oscillator was simulated with the ACPN tool. The $\mathcal{L}(f_m)$ value obtained with the ACPN tool predicts the measured $\mathcal{L}(100\,\text{kHz})$ value of -80 dBc/Hz with an error smaller than 3 dB. The application of the ACPN method on the design of a two-integrator oscillator was mentioned, as well. Because this oscillator works in linear mode (it was used in an RF modulator), ACPN predictions closely matched the measurements.

9

Design examples

INTEGRATED transceivers are used in many applications. As we have seen in several tables in Chapter 4, the specifications of the oscillators in these transceivers vary considerably when comparing different applications. For example, the phase noise requirements of wireless systems often are much more stringent compared to wired transceivers.

> This chapter covers four oscillator design examples, and illustrates the design of LC oscillators as well as ring oscillators. The high-frequency oscillators are designed for three different applications.

First, the design of a local oscillator for FM radio receivers is highlighted in Section 9.1 [53]. This LC oscillator features a very low power consumption (0.4 mW) while having an excellent spectral purity ($\mathcal{L}(100\,\text{kHz}) \approx -100\,\text{dBc/Hz}$). The LC oscillator is designed in the IC technology Silicon on Anything (SOA).

Second, a 0.9-2.2 GHz ring oscillator [52] and a VHF LC oscillator [54, 162] for a digital satellite receiver is described in Section 9.2 and Section 9.3. Both oscillators are part of a double-loop tuning-system for this satellite receiver. The double-loop tuning-system combines the quadrature property of a two-integrator oscillator with the low-phase-noise property of an LC oscillator [40]. Pulling of the LO and self-reception in the zero-IF architecture of the satellite receiver are minimized, as the two-integrator oscillator is fully integrated and inductorless.

Third, the design of a 10 GHz ring oscillator for the Data Clock Recovery (DCR) block in an optical front-end is highlighted in Section 9.4 [124, 125]. One of the

Specification	Nom. value	Units
VCC	5	V
Supply current	<100	μA
Minimum frequency	785	MHz
Maximum frequency	950	MHz
$\mathcal{L}(100\,\text{kHz})$	-101	dBc/Hz
Process	SOA	

Table 9.1 Short list of nominal FM LO specifications.

challenges for this design was to find a suitable topology that would meet the specifications in a BiCMOS process with a 30 GHz f_T and a power budget of 100 mW. Starting off with a two-integrator, the evolution in ring oscillator topologies resulted in a new circuit with active inductive loads.

9.1 A 670–830 MHz LC oscillator for FM radio in SOA

The FM radio standard is quite an ancient standard compared to modern digital transceiver standards. However, its use is still extremely wide spread. Increasing the level of integration of FM radio receivers is of interest, since many modern electronic consumer products have a built-in FM receiver as an additional selling feature. Examples are mobile phones, MP3 players and PC extension cards. Especially for portable equipment, this FM receiver should be extremely cheap, low power and be fully integrated.

> The first oscillator is a monolithic ultra-low-power LC oscillator running at eight times the FM radio band [53].

9.1.1 Specifications

The nominal target specifications for the FM local oscillator (LO) is listed in Table 9.1. As we know from Table 4.1 on p. 69, the FM radio band starts at 87.5 MHz and extends to 108 MHz. The radio receiver architecture is conventional in the sense that it has a 10.7 MHz IF frequency. In the next section we will see that the quality factor of on-chip inductors is rather low around 100 MHz (the real part is high compared to the imaginary part of the inductor). Therefore, the LO frequency was chosen 8 times higher than the FM radio band plus IF-frequency, and thus ranges from 785 to 950

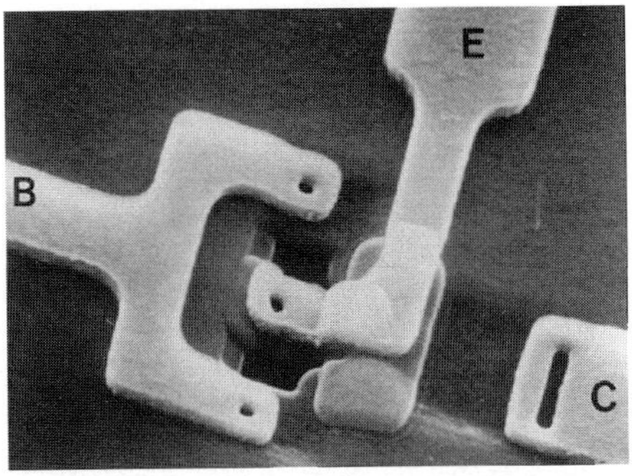

Figure 9.1 Die micro-graph of a SOA NPN transistor.

MHz. This LO frequency is divided down (by 8) via a frequency divider to obtain the LO-signal at $f_{channel} + f_{IF}$.

For FM portable radios a SNR of 60–65 dB is acceptable. The $\mathcal{L}(100\,\text{kHz})$ for FM radio can be calculated using (4.9) on p. 76 with $\delta f = 11.6$ Hz rms (assuming a peak frequency deviation of 22.5 kHz) and $NBW = 5\,\text{kHz}$. It turns out that $\mathcal{L}(100\,\text{kHz})$ must be smaller than -119 dBc/Hz approximately. As is eloquently shown in [3], the phase noise of an oscillator improves 6 dB per division by two of its frequency[1]. Intuitively this is understandable as the period of the waveform is increased by a factor two after division and the noise is ideally unaffected. Since we have chosen a division factor of 8, the $\mathcal{L}(100\,\text{kHz})$ specification of the oscillator around 900 MHz becomes $-119\,\text{dBc/Hz} + 20\log(8) = -101\,\text{dBc/Hz}$.

> To demonstrate the suitability of SOA technology[2] for low-power oscillators, a challenging power budget of 0.5 mW was chosen. SOA is quite a different technology compared to standard bipolar technologies and is briefly introduced in the next section.

9.1.2 SOA technology

SOA technology combines low-power active devices with high Q passives [208]. The on-chip inductors show Q-factors up to 60 @ 2.7 GHz for 1.2 nH inductance. Therefore SOA is well suited for integrated front-ends including RF filters. Varactors for tuning and compensation of the process spread with sufficiently high Q-factors (33 @ 1.9 GHz) also are available. The 13 masks and the 1 μm lithography keep the process costs low enough, so that even complex filters can be integrated economically.

The low power consumption of circuits realized in SOA technology is mainly the result of the removal of the silicon substrate. After processing the wafer is flipped and

[1] That is provided the divider is well designed and does not add phase noise. A typical noise floor of a divider for this application is -140 dBc/Hz at 10 kHz offset.

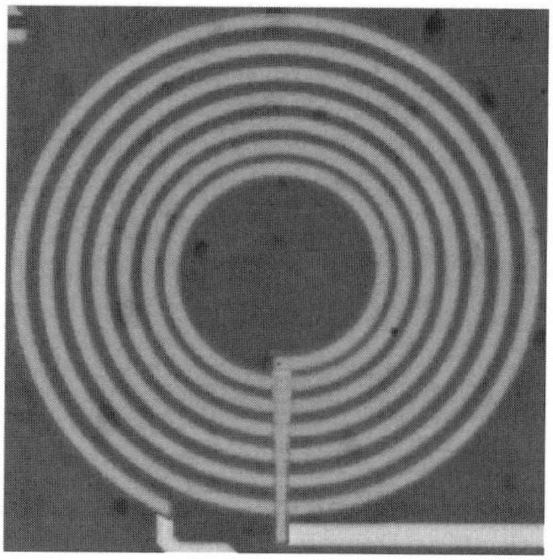

Figure 9.2 Micro-graph of an optimized SOA inductor.

glued to an isolating substrate (normally glass). The silicon substrate then is completely removed by etching. Therefore substrate losses and parasitics are minimized, resulting, among other benefits, in the high-Q factor for inductors.

The NPN transistor in SOA has a lateral current flow (see Figure 9.1 for a photo of the device [88]). In the vertically oriented emitter area one dimension is defined by the epi-layer thickness (0.2 μm). This allows emitter base (and collector base) junctions of 0.2 μm^2, resulting in very small junction capacitances (0.8 fF and 0.22 fF, respectively).

The PNP in SOA is a standard lateral transistor with a minimum emitter area of 0.2 μm^2 too (junction capacitances of 0.22 fF). To obtain these minimal parasitics, the silicon substrate has to be removed and replaced by a glass substrate. As a result, the transistor output bandwidth f_{OUT} is much closer to transition frequency f_T compared to silicon bulk processes.

The maximum supply voltage of the process goes up to 12 V and can be designed by the width of the collector drift region, which is a geometry parameter in the transistors. The process also offers a bipolar I^2L logic running down to 0.7 V, having a normalized gate density of 5000 gates/mm^2 and a gate delay of down to 2 ns. A digital design flow environment is available to support mixed-signal design.

Figure 9.2 shows a micro-graph of the inductor used in the FM LO design. The distance between the curls is increased for each subsequent curl, to tune each curl to approximately the same frequency. Note that this scaling depends on the initial distance d_0 between the inner curls and the inner radius R_0. The bridge also deviates from a constant width; the outer crossings are smaller to shift the poles together, formed of the inductance of all inner curls and the crossing capacitance. For larger inductors (higher number of turns) this compensation becomes less effective. Hence, the impact of these measures on the Q-factor is between 5% and 12% depending on the inductor's geometry parameters. Measured Q-factors of three inductors are shown in Figure 9.3.

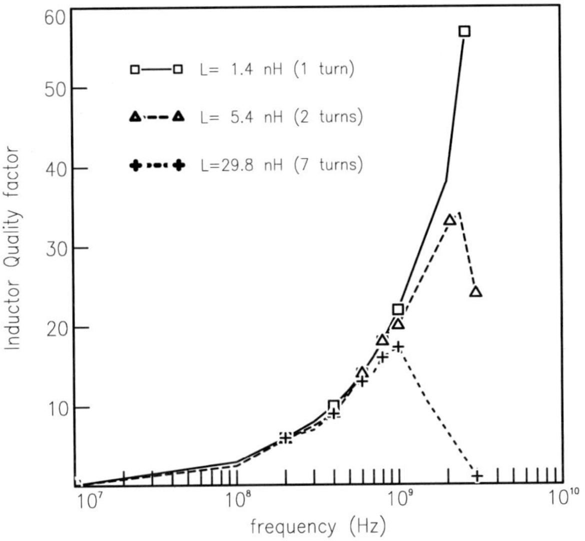

Figure 9.3 Q-factor measurement results of three inductors.

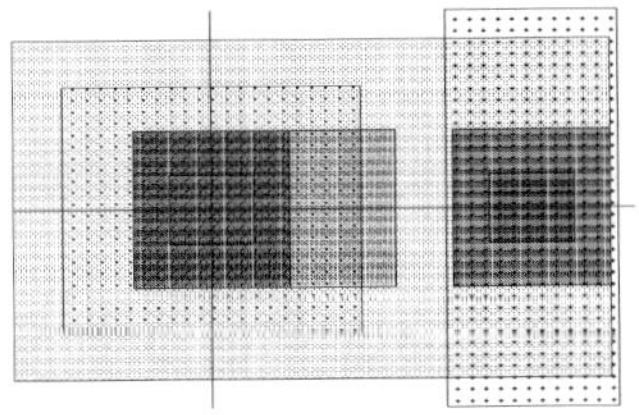

Figure 9.4 Layout of the utilized varactor (one cell).

A single varactor structure is shown in Figure 9.4. It is similar to the emitter-base complex of the NPN transistor and allows the exchange of Q versus C_{max}/C_{min} by design parameters. These basic cells are put together by a varactor layout generator to build up larger arrays. The maximum Q-factors mentioned can be obtained in narrow band systems. In the varactor cell used in the FM LO the length of the low-doped drift region (dr) is 1.2 μm. After realization of the VCO design, new layout design rules allowed a minimum drift region of 0.8 μm. This results in higher Q-factors as can be seen in Figure 9.5.

9.1.3 Oscillator design

Given the combination of the specified $\mathcal{L}(100\,\text{kHz})$ value in Table 9.1 and the power budget of 0.5 mW, it is clear that we can only achieve this with an LC oscillator.

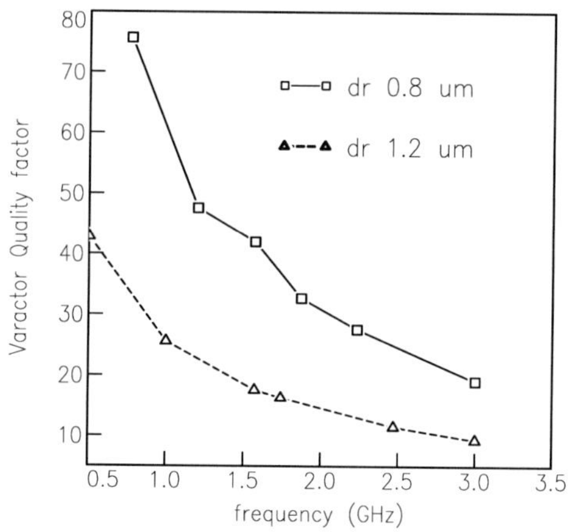

Figure 9.5 Q-factor measurement results of two varactors with drift regions (dr) 0.8 μm and 1.2 μm.

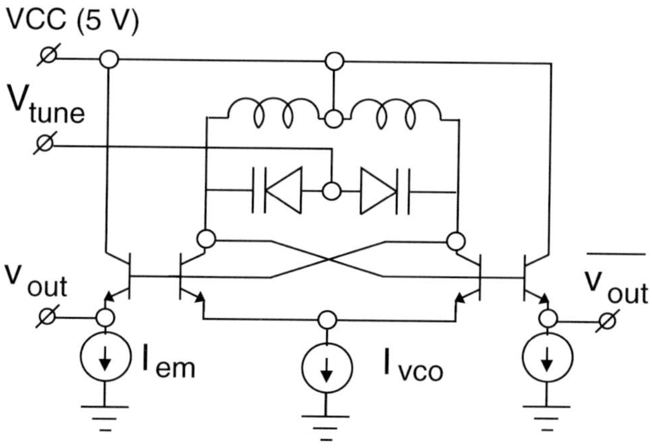

Figure 9.6 Simplified schematic of the LC oscillator.

Calculation of $FOM_{ring-ODE}$ (see (7.14) on p. 197) with this specification for a two-stage ring oscillator yields a value of +7.3 dB. As discussed on p. 198, values beyond -15 dB are challenging and positive values for $FOM_{ring-ODE}$ are extremely unlikely if not impossible.

A balanced oscillator topology was chosen for the oscillator design, to maximize rejection of common mode disturbances. The design is shown in Figure 9.6. Common mode rejection is particularly important since the VCO is integrated in an FM radio receiver. The disadvantage of a balanced LC oscillator design with integrated coils is a substantial increase in the chip area. As we will see the inductor area dominates the total chip area in this design.

The value of the on-chip inductors was chosen to be 29.8 nH with a Q_{max} of $\sim$ 17.3 around 1 GHz. The total parasitic capacitance of this inductor is 150 fF. The inductance value of 29.8 nH is the largest of the three inductors plotted in Figure 9.3. The quality factor of the inductors in this figure is quite similar around this frequency. Given the maximum specified tail current of 100 μA, we need a large effective parallel resistance (R_p) and α_{ol} to benefit from shot-noise reduction due to noise modulation (see Table 6.3 on p. 160). The inductor of 29.8 nH yields[3] an R_p equal to 6.3kΩ around 1 GHz. This gives an open-loop gain α_{ol} of 6 and a peak voltage swing almost 1 V. According to Figure 6.51 on p. 180 this voltage swing is too high: the base-collector junctions will go into forward and the oscillator will operate in the voltage-limited region. However, the effective parallel resistance R_p is reduced by the finite quality factor of the varactor, and we can check the operation region again when we know the total resonator quality factor.

Seven varactor sections in parallel, each of 240 fF (zero bias) are used. At the design time the SOA technology was still in its development stage and the accuracy of the modeling of parasitics including interconnect was an issue. To accommodate this risk, the center frequency of the oscillator was designed somewhat lower than the target center frequency. Each of the seven sections was placed with sufficient surrounding space to make laser trimming possible. Obviously, in production this is not an option for cost reasons, but for a feasibility study laser trimming is viable option.

The capacitance ratio C_{max}/C_{min} of the varactor for a bias voltage $V_{reverse}$ ranging from 0 to 5 volt, is approximately 1.7. The quality factor of the PN-junction varactor is around 26 at 1 GHz (see Figure 9.5 for dr is 1.2 μm). The finite varactor quality factor at 1 GHz reduces R_p to 3.6 kΩ and the maximum peak voltage[4] across the cross-coupled bipolar differential pair is below 0.6 V. This means that the oscillator is working well within the current-limited region. The tail current I_{VCO} in Figure 9.6 was set slightly lower than the maximum value to 80 μA. The emitter followers that implement the first buffering stage each have a current I_{em} of 40 μA.

Dissipation of the VCO can be lowered even further by lowering VCC at the expense of a slightly reduced tuning range. The tuning design FOM, $FOM_{tuning-LC}$ (see (7.4) on p. 188) comes to 110% for the nominal frequency specification in Table 9.1, taking for C_{par} only the parasitic inductor capacitance of 150 fF, for $C_{max} = 7 \cdot 240$fF and finally for $C_{min} = C_{max}/1.7$. Since 100% (or 0 dB) indicates that the specification is exactly met, we have some tuning range margin. However, if we assume 10% spread on the center frequency, $FOM_{tuning-LC}$ drops below 100% and calculates to 75%. In this particular design we can employ laser trimming to compensate for process spread. For a production VCO, the tuning design FOM clearly shows us that we

[3] When calculating $2Q_{max}\omega_{osc}L = 2 \cdot 17.3 \cdot 2\pi\, 10^9 \cdot 29.8\, 10^{-9}$.

[4] Calculated as $\pi/2 \cdot 100\,\mu\text{A} \cdot 3.6\,\text{k}\Omega$.

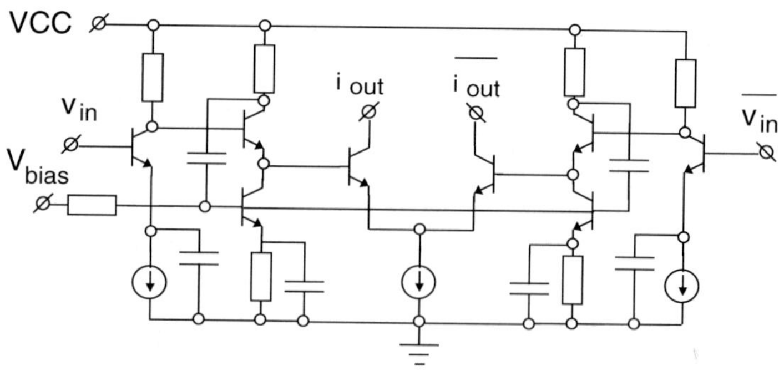

Figure 9.7 Circuit diagram of the output buffer.

would have to use a varactor with a larger capacitance ratio or alternatively implement band-switching or oscillator-switching.

An open-collector output buffer is implemented, in cascade with the VCO and emitter followers, delivering -30 dBm in 50 Ω. Figure 9.7 shows the schematic of the differential buffer. Without special measures in the output buffer, a multi-oscillation would occur (this phenomenon is discussed in Section 2.4.1). In this VCO design, one of the two simultaneous oscillations was introduced by loading of the VCO core due to the buffer. This resulted in a severely distorted VCO output signal.

The feedback in the second stage of the buffer in Figure 9.7 substantially reduces the loading on the VCO circuitry, and eliminates the multi-oscillation phenomenon. The total dissipation of the output buffer is 7.2 mW.

9.1.4 Experimental results

A micrograph of the monolithic SOA LC oscillator is shown in Figure 9.8. The two inductors occupy a dominant part of the 1940 μm × 1280 μm active chip area. On the left two output buffers can be seen. One extra (stand-alone) buffer was used to verify the buffer characteristics.

Frequency and $\mathcal{L}(f_m)$ measurements were performed using an HP8562E spectrum analyzer with pre-amplifier, as presented in Figure 9.9. The measured tuning range is 668 MHz to 830 MHz when V_{tune} is varied from 5 to 0 volt, which is more than 21%. Phase noise measurements are also plotted in Figure 9.9. Laser trimming is needed to change the measured center frequency of 750 MHz to the specified 870 MHz center frequency. For the tuning voltage range from 5 to 0 volt, $\mathcal{L}(100\,\text{kHz})$ varies between -98.1 and -100.4 dBc/Hz. The power spectrum of the oscillator with V_{tune} equal to 0 V is shown in Figure 9.10.

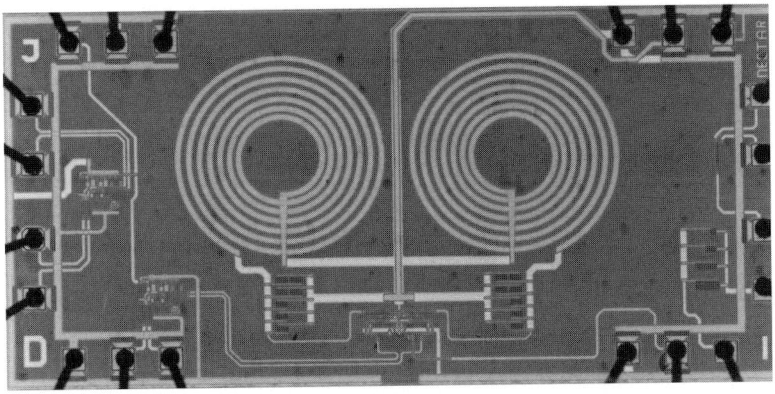

Figure 9.8 Micrograph of the LC oscillator.

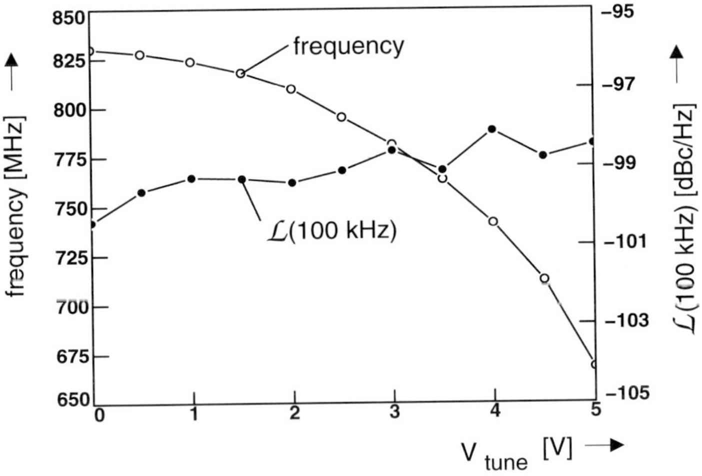

Figure 9.9 Frequency and $\mathcal{L}(100\,\text{kHz})$ versus V_{tune}.

9.1.5 Benchmarking

In Chapter 7 we have introduced several benchmark FOMS. Among other things, the FOM normalized phase noise (Nor-PN) was introduced (see (7.12) on p. 193). Instead of plotting FOM_{Nor-PN} versus frequency, as was done in Figure 7.3, FOM_{Nor-PN} is plotted versus power dissipation in Figure 9.11 for the single-phase LC oscillators in Appendix G.

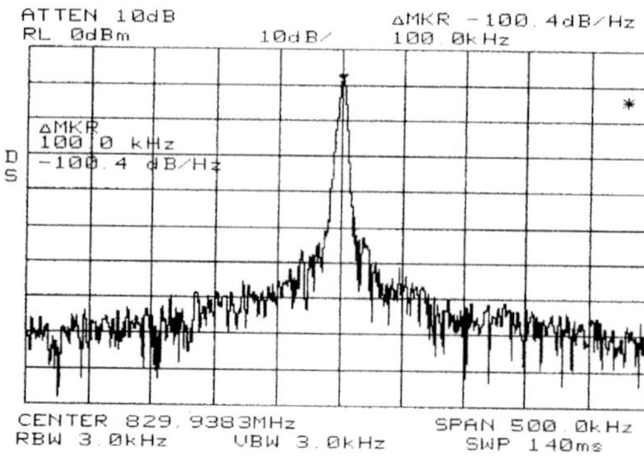

Figure 9.10 Power spectrum of the SOA oscillator.

Benchmark-FOM_{Nor-PN} for the SOA LC oscillator amounts to -182.8 dBc/Hz, with $\mathcal{L}(100\,\text{kHz}) = -100.4\,\text{dBc/Hz}$, a carrier frequency of 830 MHz and a VCO core dissipation of 0.4 mW.

As shown in Figure 9.11 the SOA LC described in this section uses the least amount of power and ranks seventh according to benchmark FOM $Nor - PN$ in the top ten of the plotted oscillators. In this top ten, most oscillators were published after the publication of this work [200].

The benchmark FOM $LC - ODE$ of this SOA LC oscillator is around -14 dB, which is similar to the average encountered ODEs, as we have seen in Figure 7.4 on p. 196.

9.1.6 Conclusion

A 0.4 mW monolithic LC oscillator has been realized in SOA technology. The tuning range is 21.6 % and ranges from 668 MHz to 830 MHz. Measured $\mathcal{L}(100\,\text{kHz})$ is -98.1 dBc/Hz at worst and 100.4 dBc/Hz at best. The power dissipation of the oscillator is the lowest of all LC oscillators listed in Appendix G, while having an excellent benchmark FOM $Nor - PN$ of -182.8 dBc/Hz. The target application for this oscillator is FM radio. To achieve the specification from Table 9.1, the $\mathcal{L}(100\,\text{kHz})$ needs at least 3 dB improvement, which can be obtained with more power dissipation. The nominal tuning range specification is achieved. However, to handle process spread a redesign with a higher varactor ratio is needed or, alternatively, a redesign that implements an oscillator with multiple frequency bands (band-switching).

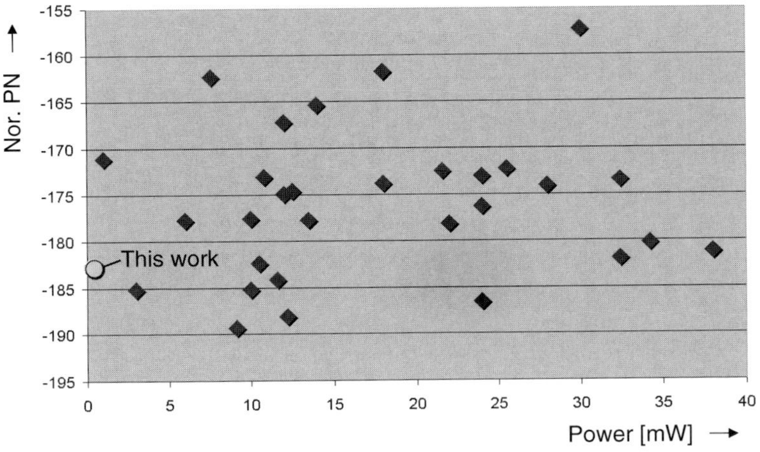

Figure 9.11 The benchmark FOM $Nor - PN$ versus power dissipation. All single-phase LC oscillators from Appendix G with a power dissipation below 40 mW are listed.

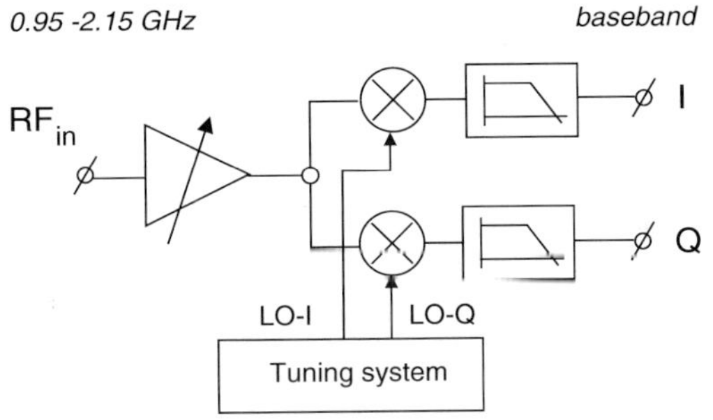

Figure 9.12 Simplified zero-IF satellite receiver.

9.2 A 0.9–2.2 GHz two-integrator oscillator for satellite receivers

Consumer market trends, especially cost and size reduction, have resulted in digital satellite receivers using zero-IF architectures [163–165]. Digital satellite systems use a QPSK modulation scheme. A simplified block diagram of a zero-IF satellite receiver is shown in Figure 9.12. The architecture allows a high degree of integration, especially because expensive IF filters are replaced by low-pass filters, which are relatively easy to integrate. The Local Oscillator (LO) in the tuning system of Figure 9.12 is tuned to a wanted input satellite channel and this channel is directly down-converted to baseband. Although the zero-IF architecture allows for almost full

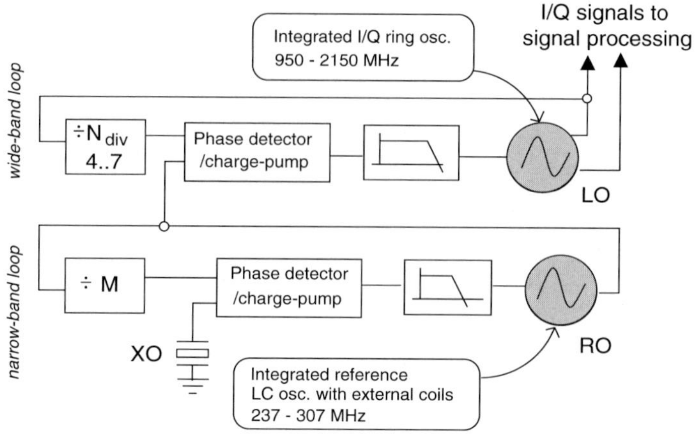

Figure 9.13 Double-loop tuning system for digital satellite receiver front-ends.

integration, it also demands additional complexity of the tuning system compared to a super-heterodyne receiver. Zero-IF receivers require quadrature (I/Q) LO signals covering the total input frequency range [97]. In addition, the LO and the tuning system should be constructed in such a way that LO self-reception [166] is minimized, which is caused by LO leakage [97] to the antenna.

A fundamental solution to combine a large tuning range, I/Q matching and good phase noise is the double-loop architecture shown in Figure 9.13 [3, 40]. In this tuning system a fully integrated ring oscillator provides the required I/Q signals in the satellite frequency band [52].

This ring oscillator can meet the tuning range requirements and minimizes self-reception, as it is fully integrated and has no coils (no radiation to the antenna). The phase noise level of ring oscillators typically is 20 dB higher than the noise level required for reception of the QPSK modulated signals with sufficient BER. Therefore, the ring oscillator is wide-band locked (1-2 MHz) to a low-noise tunable reference oscillator (RO) as shown in Figure 9.13. Without a double-loop tuning system, use of an LC type oscillator would be mandatory to meet the QPSK jitter requirements of the satellite system. The RO oscillates at a much lower frequency than the desired satellite channel and cannot cause or contribute to pulling or self-reception problems. To select a satellite channel, the RO is part of a second PLL and locked to a crystal oscillator reference (XO). The typical bandwidth of this second loop is 10 kHz.

In the presented architecture, the main divider N_{div} (Figure 9.13) is programmable in four steps from 4 to 7. The integrated ring oscillator has four frequency bands, each successive band corresponding to a successive division ratio. The advantage of having

Specification	Nom. value	Units
VCC	5	V
Supply current	20	mA
Minimum frequency	0.95	GHz
Maximum frequency	2.15	GHz
$\mathcal{L}(2\,\text{MHz})$	-106	dBc/Hz
Output signals	I/Q	
Tuning input	high ohmic	
Tuning bandwidth	> 10	MHz
Process	BiCMOS 10 GHz f_T	

Table 9.2 Short list of the I/Q LO specifications.

x bands is a reduction of the needed tuning range of the RO by roughly a factor x. In practice this reduction factor is somewhat smaller than x, because the bands need some overlap. Furthermore, the gain constant (in MHz/V) of the ring oscillator is reduced by the same factor, making it more robust against on-chip disturbances. The consumer satellite band starts at 950 MHz and extends to 2150 MHz. With four LO frequency bands the minimum tuning range of the RO is 237.5 MHz (950 MHz/4) to 307.2 MHz (2150 MHz/7).

In this section, the design of the 0.9–2.2 GHz ring oscillator that implements the LO in the wide-band loop in Figure 9.13 is highlighted. The design of the low phase-noise RO in the narrow-band loop will be described in Section 9.3.

9.2.1 Specifications

Table 9.2 summerizes the most important specification of the quadrature ring oscillator. The $\mathcal{L}(f_m)$ specification is derived after optimization of the double-loop tuning-system on behavioral level [3]. Provided the RO has a $\mathcal{L}(10\,\text{kHz})$ better than -85 dBc/Hz and the loop bandwidth of the wide-band loop in Figure 9.13 is sufficiently large (1-2 MHz), the jitter specification[5] for digital satellite QPSK reception of $< 2.8^\circ$ rms [167] can be achieved with an LO having a $\mathcal{L}(2\,\text{MHz})$ equal to $-106\,\text{dBc/Hz}$.

[5] As discussed on p. 77 (see (4.11)) a high and low frequency limit for f_m are needed to convert $\mathcal{L}(f_m)$ into a jitter specification. f_H is taken to be 100 MHz assuming a symbol rate of 100 Mbits/s. f_L depends on the satellite receiver's carrier-recovery loop-bandwidth being 10, 20, or 30 kHz for example.

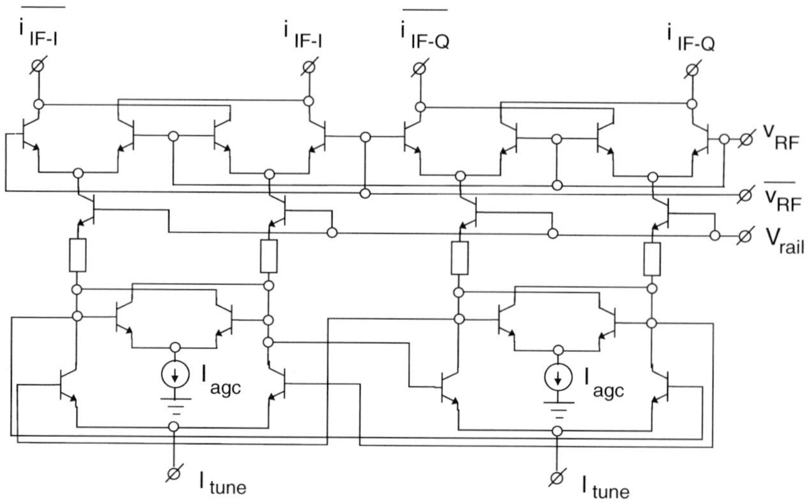

Figure 9.14 Circuit implementation of the quadrature mixer-oscillator.

A relatively large power budget of 100mW is allocated to the LO in the tuning system to allow the use of a ring oscillator. The tuning specification is more than an octave, which is relatively easy to achieve with a ring oscillator. Furthermore, a ring oscillator is compact in chip area and an even-stage implementation provides the required I/Q outputs. LO-leakage and self-reception are expected to be low, as the ring oscillator is inductorless and fully integrated. The LO is part of the wide-band loop and controlled by the output voltage of a loop-filter. It needs to have a high-ohmic voltage tuning-input, otherwise leakage of charge in the loop filter would result in reference-breakthrough signals [94]. In addition, the bandwidth of the tuning input circuitry must be sufficiently high (e.g. ten times the loop bandwidth), so that the stability of the wide-band loop is unaffected by this pole. A low-cost standard BiCMOS technology is specified with a f_T of 10 GHz and f_{MAX} of 11 GHz [168].

9.2.2 Oscillator design

From Section 6.3 we know that a ring oscillator with the minimum number of stages of $N = 2$ yields optimum $\mathcal{L}(f_m)$ given a certain power budget.

For the design of the LO the two-integrator oscillator was chosen. The two-integrator has a minimum number of stages and provides I/Q signals "correct-by-construction". $FOM_{freq-two-int}$ (see p. 187) equals 130% indicating that the specified BiCMOS technology is sufficient to achieve the maximum oscillation frequency of 2.15 GHz with enough margin.

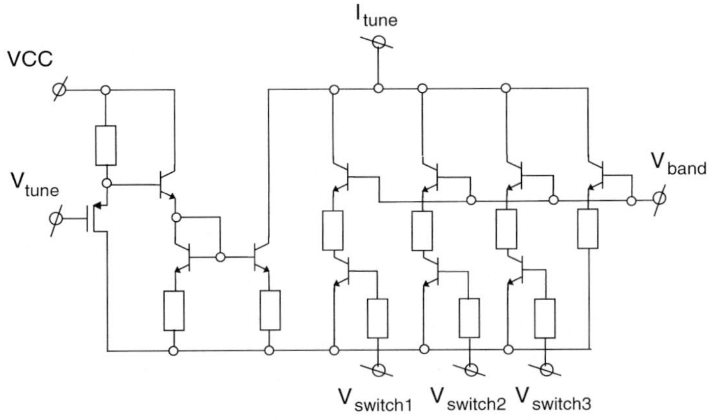

Figure 9.15 Circuit diagram of the V/I-converter and the band-switch circuit.

The circuit diagram of the realized two-integrator is shown in Figure 9.14. The operation of the two-integrator oscillator has already been discussed in Chapters 5 and 6. On top of the two-integrator oscillator, a double-balanced quadrature mixer is stacked that is used to down-convert the desired RF signal to I and Q baseband signals (see Figure 9.12). A cascode stage provides isolation between the oscillator core and the mixer stage. The presence of the mixers on-chip also allows for the accurate measurement of the I/Q matching at baseband frequencies. At 2 GHz the phase relation between the output signals of the two-integrator oscillator would be impossible to measure as a tiny error in bond-wire lengths represent a large phase error at such a high frequency.

> Note that the mixer-oscillator combination in Figure 9.14 is very power efficient: the oscillator current is reused in the mixer.

The cascode bias voltage V_{rail} is 3.8 V nominal. Mixer IF-outputs are implemented as open-collector outputs.

The current I_{agc} is set to 2 mA. At 2 GHz this resulted in a voltage swing of approximately 250 mV. The nonlinear operation of the two-integrator oscillator, i.e. the switching of the transistors in oscillator core, reduced $\mathcal{L}(f_m)$ compared to linear operation. For this particular design, transient simulations were performed to estimate the phase noise [52]. A $\mathcal{L}(2\,\text{MHz})$ of -108 dBc/Hz was simulated.

One of the LO specifications was the need for a high-ohmic tuning input in the voltage domain. The implementation of a V/I-converter with MOS-input, together with band-switching circuitry is shown in Figure 9.15. The V/I converters turns the two-integrator oscillator, which fundamentally is a CCO, into a VCO.

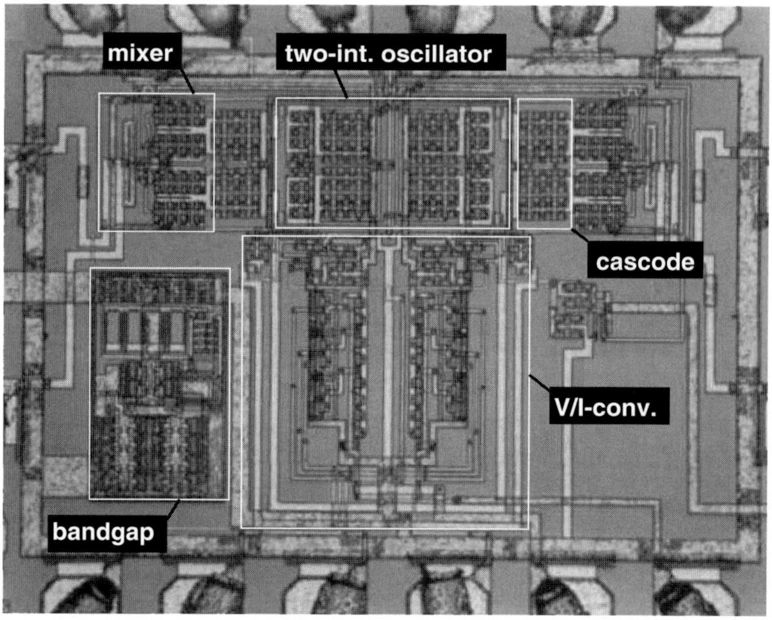

Figure 9.16 Micrograph of the quadrature mixer-oscillator.

> As mentioned several times in Chapter 6, a high VCO gain results in a high sensitivity to noise on the tuning input of an oscillator. To reduce the VCO gain, the tuning range of the oscillator is divided into four bands.

The three band-switches are implemented as switched current sources. When the three current switches are off the minimum I_{tune} is 3.5 mA. The highest oscillation frequency is obtained when all V_{switch} terminals are 5 V and when V_{tune} is 3.5 V. Nominal I_{tune} is in this case 7.5 mA. Note that when bipolar switches are used, as we have done here, a base resistor is important to limit the base current (the "forced" current gain of a bipolar transistor can be very low) and that substrate contacts around each switch are needed to prevent latch-up. The MOS-input of the V/I-converter assures a low leakage current of the loop-filter in the wide-band loop.

9.2.3 Experimental results

A micrograph of the mixer-oscillator circuit is shown in Figure 9.16. The active chip area is less than 0.5 mm^2. As we know from Section 4.7 on p. 81, good I/Q matching is extremely important for a high IRR. Therefore, a lot of effort was put into making the layout as symmetrical as possible. Dummy metal tracks were applied where needed, to make the parasitic capacitance of every track the same.

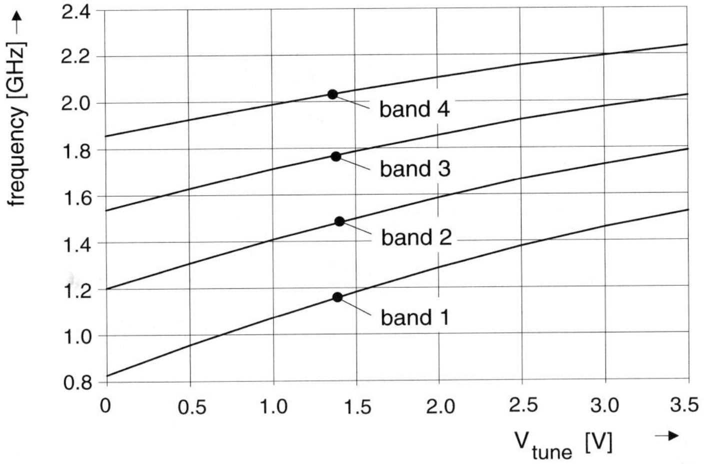

Figure 9.17 Measured frequency versus V_{tune}.

Parasitic extraction of the inter-connect capacitance is a useful effort, not only for verification of the electrical behavior, but also to check whether all matching tracks are truly identical.

In Figure 9.17, the four measured frequency bands versus tuning voltage V_{tune} are plotted. The tuning voltage goes up to 3.5 V. It is instructive to realize that a conventional varactor tuned satellite oscillator needs a second power supply, because its discrete varactor requires 30 V tuning voltage [169]. So compared to 30 V, this design reduces the required tuning voltage by a factor 8 and eliminates the need for a second voltage supply. The tuning range of 80% covers the consumer satellite band with a VCO gain of 150 MHz/V $\pm$ 35%. This relative constant VCO gain facilitates optimization of the wide-band PLL parameters.

Phase noise optimization resulted in a $\mathcal{L}(2\,\text{MHz})$ equal to or better than -106 dBc/Hz across the whole frequency band of interest. Measurements were performed with a spectrum analyzer and with HP 3048 phase noise measurement equipment. In Figure 9.18, the power spectrum of the oscillator is shown, at 2 GHz oscillation frequency.

I/Q matching was measured with a spectrum analyzer and with a HP 5372A frequency and time analyzer. To calibrate for phase and amplitude errors introduced by measurements cables and amplifiers, two measurements per frequency point were taken, results were added and divided by two[6]. The result of the I/Q matching mea-

[6]Suppose the phase error in the I measurement path is ϕ_1 and in the Q measurement path ϕ_2. A first measurement of the phase difference $I-Q$ of the oscillator will yield $I+\phi_1-Q-\phi_2$. Now all wires and amplifiers are exchanged between the I and Q path and the phase measurement is performed again. This will yield $I+\phi_2-Q-\phi_1$. After addition we have removed the error of the measurement set-up and gain $2I-2Q$.

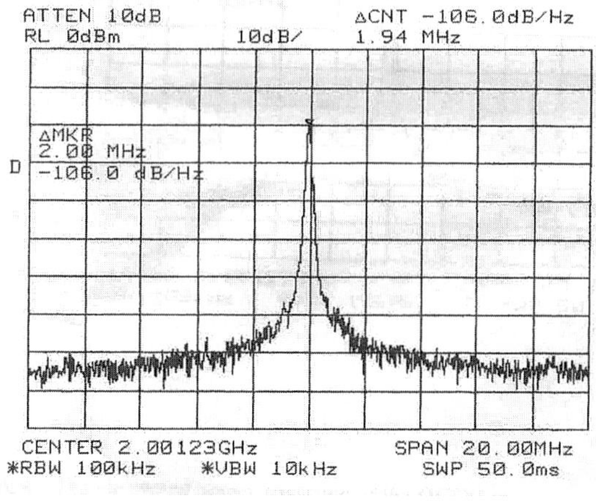

Figure 9.18 Power spectrum of the two-integrator oscillator at 2 GHz.

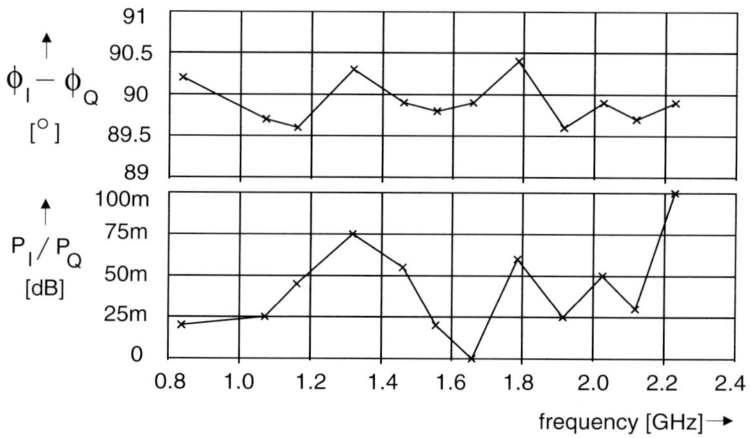

Figure 9.19 Measured phase and amplitude matching of the I/Q signals versus frequency.

surements are shown in Figure 9.19. Phase measurements were performed at an IF frequency of 1 MHz. A careful layout resulted in a phase matching better than 0.5° and an amplitude matching better than 0.1 dB. As shown in Figure 7.5 on p. 198 (see the diamond with reference [241] next to it) the ring oscillator design efficiency (benchmark FOM *ring* − *ODE*) of the oscillator is around -34 dB at 1.55 GHz, which leaves room for further optimization.

Specification	Nom. value	Units
VCC	5	V
Supply current	4	mA
Minimum frequency	237	MHz
Maximum frequency	307	MHz
$\mathcal{L}(10\,\text{kHz})$	<-85	dBc/Hz
Process	Bipolar 11 GHz f_T	

Table 9.3 Short list of RO specifications.

9.2.4 Conclusion

The design of an integrated two-integrator oscillator has been discussed, realized in a standard 10 GHz f_T BiCMOS process. Its tuning range covers the complete satellite band and is equal to 80%. Three band-switched current sources divide the frequency range of 0.9–2.2 GHz into four bands. In this way, the VCO gain constant has been kept low: 150 MHz/V $\pm$ 35%. $\mathcal{L}(2\,\text{MHz})$ measurements across the whole band are equal to or better than -106 dBc/Hz. Combined with a double-loop tuning system, the two-integrator oscillator can be used in digital satellite receivers. The measured I/Q matching corresponds to an IRR better than 40 dB. Nominal supply voltage is 5 V and the power dissipation is 100mW.

9.3 A 225–310 MHz LC oscillator with PMOS varactors

The ring oscillator from the preceding section is part of the wide-band loop shown in Figure 9.13 on p. 226.

Phase noise of the ring oscillator is "cleaned-up" by locking it to a LC reference oscillator (RO) with a much better phase noise performance. In this section the design of this RO is described, which is part of the narrow-band loop in the double-loop tuning system.

9.3.1 Specifications

The most important RO specifications are listed in Table 9.3. The challenge of the RO design is to meet both $\mathcal{L}(10\,\text{kHz})$ and the tuning range requirements. Next to $\mathcal{L}(f_m)$ and tuning range, power dissipation is a third important functional specification that follows from system considerations. The DC current of the RO must be limited to a few mA to fit into the total power budget (500mW) of the satellite front-end.

For digital satellite QPSK reception, the jitter specification is 2.8° rms [167]. This translates to a $\mathcal{L}(10\,\mathrm{kHz})$ for the RO ranging from -85 dBc/Hz to -95 dBc/Hz, depending on the integrated ring oscillator phase noise level, the symbol rate of the satellite system and the loop bandwidth of the clock recovery of the QPSK demodulation back-end [40].

The tuning range in Table 9.3 follows from the chosen double-loop architecture. The main divider ratio N_{div} in Figure 9.13, divides the LO signal by either 4, 5, 6 or 7. Given the satellite frequency band that we need to cover, this results in a nominal RO tuning range of 950 MHz/4 to 2150 MHz/7.

9.3.2 Resonator design

One possible implementation of the RO is an off-the-shelf VHF oscillator with a completely external resonator circuit. However, to comply with the low-cost demands of the satellite consumer market, a higher degree of integration is desirable.

> In order to derive the specification of the RO's resonator, the oscillator's $\mathcal{L}(f_m)$ and tuning requirements have to be translated into resonator specifications.

We can derive a first order estimate for the minimum quality factor we need by solving the equation $FOM_{\mathcal{L}(f_m),LC-bipo} = 1$ for Q_p (see (7.10) on p. 189 for the expression of this design FOM). Choosing L_p=30 nH, $\alpha_{ol} = 3$ and substituting the specification of Table 9.3 in the $\mathcal{L}(f_m)$ design FOM, yields a minimum (loaded) quality factor of 4. This rules out the use of integrated coils in the bipolar technology in which the RO needs to be realized, because around 300 MHz the inductor quality factor is much lower. A cost-effective solution is an off-chip inductor, for example, an air coil or printed PCB inductor. Both air coils and printed PCB-inductors can be realized with quality factors of 60 or better at 300 MHz. Hence the quality factor of the varactor is expected to dominate the Q_p value of the RO resonator. The conclusion is that the varactor quality factor must be significantly[7] larger than 4 to meet the RO $\mathcal{L}(f_m)$ specification.

Tuning design FOM, $FOM_{tuning-LC}$ (see (7.4) on p. 188) can be used to derive a first order estimation of the required varactor ratio $\alpha_{varactor}$. Substituting C_{par}, C_{min} and $C_{max} = \alpha_{varactor} C_{min}$ in (7.4), and solving $FOM_{tuning-LC} = 1$ for $\alpha_{varactor}$ yields the minimum required $\alpha_{varactor}$. In Section 9.3.3, the active oscillator circuit will be described and this circuit will contribute 3.8 pF at 307 MHz to C_{par}. Capacitance C_{min} is 5.2 pF if we choose a resonator inductance of 30 nH. Solving $FOM_{tuning-LC} = 1$ for these values results in a minimum $\alpha_{varactor}$ of 2.16. The derived minimum varactor quality factor $Q_{varactor}$ (of 4) and the minimum capacitance ratio $\alpha_{varactor}$ (of

[7] This is because the design FOM neglects all nonlinear phase noise generation mechanisms.

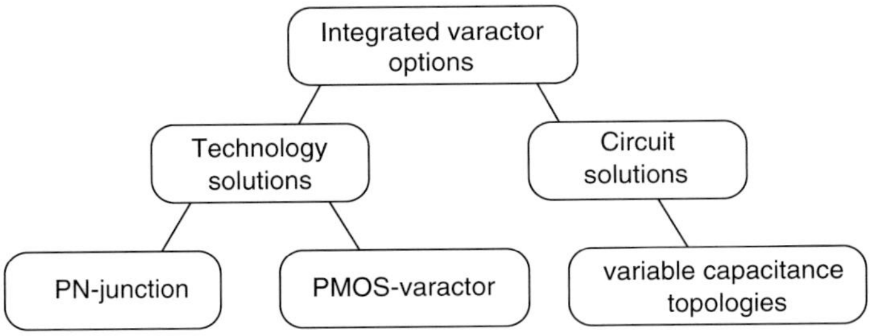

Figure 9.20 Integrated varactor options in the specified IC technology.

2.16) can now be used to benchmark the available varactor options in the specified IC technology.

Three varactor options are available in the IC technology used for the RO design. In this bipolar technology a PMOS device is available too, which is generally used for switches and high-ohmic inputs. The three alternatives are presented in Fig.9.20. Passive options are the PN-junction varactor and a PMOS varactor. Using active devices, circuit solutions can be realized that implement a variable capacitance.

PN-junction varactor

In a bipolar process[8] a PN-junction type varactor is readily available by, for example, connecting the emitter to the base of a transistor and using the collector-base junction reversed biased (see p. 136).

On p. 137 in Figure 6.23 we already demonstrated the simulated capacitance curve of 20 NPNs in parallel with base and emitter short circuited. This varactor was considered for use in the RO's resonator. The simulated capacitance mainly consists of the sum of C_μ and C_{js}. Each transistor has an emitter area of $1.6\,\mu\text{m} \cdot 200\,\mu\text{m}$. The zero reverse-voltage capacitance of this PN-junction type varactor is 12.4 pF. The simulated capacitance ratio is 1.9 with reverse bias voltage V_{tune} varying from 0 V to 5 V. However, since the RO will be incorporated in the double-loop tuning system, zero reverse voltage is unrealistic, since voltage head room is needed for the charge-pump in the tuning system. Furthermore, the oscillator voltage swing limits the usable tuning range. The capacitance ratio with 0.5 V to 5 V reverse bias is reduced to 1.6 for example, which is less than the minimum $\alpha_{varactor}$.

Unloaded quality factors of PN-junctions are measured in the range of 75 and higher at 1 GHz [53]. Therefore varactor quality factors in excess of 75 can be expected at 300 MHz, and phase noise performance is expected to meet the requirements of -85 dBc/Hz at 10 kHz offset of the carrier.

[8] In a standard CMOS process available parasitic PN-junctions could be used, although they are not well characterized generally [223].

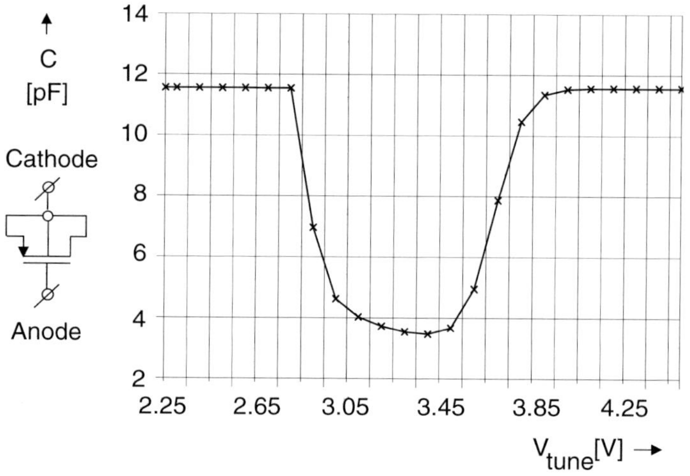

Figure 9.21 Simulated capacitance curve of the PMOS varactor.

Variable capacitance topologies

The active variable capacitance topology in Figure 6.30 on p. 145 was considered. As discussed on p. 144, active varactor topologies can have excellent tuning ranges compared to passive varactors. The simulated capacitance curve of the variable capacitance topology of Figure 6.30 is shown in Figure 6.31 on p. 146 (solid curve). The simulated C_{max}/C_{min} ratio (5.3 pF/0.7 pF) is 7.6, which clearly meets the specifications.

However, calculation of the equivalent varactor quality $Q_{C_{active}}$ (defined on p. 145) results in a value below 1 in the middle of the tuning range. This does not meet the specification.

PMOS varactor

In the case of a CMOS implementation of the RO both NMOS and PMOS devices can be used to implement a MOS-type varactor. NMOS-type varactors yield higher quality factors due to higher carrier mobility. For MOS-type varactors C_{max}/C_{min} ratios of more than 3 have been demonstrated and quality factors larger than 20 at 1 GHz have been measured (see Table 6.2 on p. 139). However, the required RO phase noise levels have not been demonstrated with MOS varactors. In the technology available for the RO design, only PMOS devices are present.

Figure 9.21 shows the simulated capacitance curve of the PMOS-varactor. The simulated device has a W/L of 200 μm / 1.2 μm and is constructed with 20 devices in parallel. Hence the total width of the varactor is 4000 μm. The bulk (or more

Varactor-option	$Q_{varactor}$	$\alpha_{varactor}$	usable?
Specification	> 4	> 2.16	
PN-junction	> 75	< 1.9	no
Active varactor	1	7.6	no
PMOS	?	3.1	possibly

Table 9.4 Benchmarking of the varactor options.

accurately the N-well) is connected to the drain and source contact and forms the varactor cathode. The gate-terminal acts as the anode. MOS-model level 9 was applied for the modeling of the device characteristics [170]. In the simulation, the anode was given a fixed reference voltage of 2.5 V to be able to evaluate the accumulation region and inversion region with positive tuning voltages. Figure 9.21 reveals a minimum and maximum capacitance of 3.6 pF and 11.3 pF, respectively. Therefore, $\alpha_{varactor}$ is equal to 3.1. In the applied MOS model, source, drain and channel resistance are not modeled, which results in unrealistic quality factor predictions.

Table 9.4 summarizes the evaluation of the varactor options. The quality of the varactor and the capacitance ratios are compared with the varactor specifications derived in Section 9.3.1. The PMOS-type varactor is the only option for the RO design with integrated varactors, but its quality factor is an uncertainty. The simulation results and expectation of the PMOS varactor must be experimentally verified. A robust active oscillator circuit is needed for the RO with PMOS varactors. In the following section the design of this active circuit will be discussed, followed by a presentation of the experimental results in Section 9.3.4.

9.3.3 Active oscillator design

In Figure 9.22, a simplified diagram of the RO is presented. A fully balanced design is chosen in order to minimize interference via supply lines in the mixed-signal satellite receiver IC. The design consists of a cross-coupled pair that implements the negative transconductance needed. The tank is connected to the differential pair via two emitter followers. The open-collector differential pair facilitates measurements and the collectors can be directly connected to 50 Ω measurement equipment. The power dissipation of the cross-coupled pair and the emitter followers (the VCO core) is 14 mW when the internal supply voltage V_{stabi} is 3.5 V.

During the design, simulations revealed potential parasitic oscillations in the frequency range of 1.5 GHz - 2.5 GHz caused by package and PCB parasitics. To eliminate these oscillations, which would obstruct proper characterization, base-collector (C_{bc} in Figure 9.22) capacitances are added. This reduces oscillator activity at high frequencies.

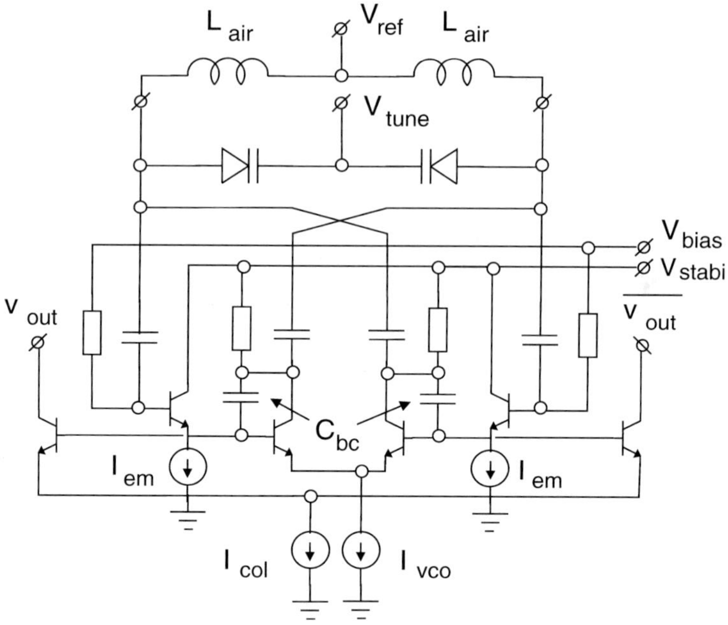

Figure 9.22 Reference oscillator circuit diagram.

The tank circuit is capacitively coupled to the active RO part by high Q (> 100 at 300 MHz) on-chip capacitors. In this RO design, air coils are used, having quality factors better than 100 at 300 MHz. The capacitance curve of the PMOS-varactor versus V_{tune} can be mapped to the desired DC range with V_{ref}. V_{ref} effectively changes the threshold voltage V_{TH} of the PMOS varactor. For V_{ref} equal to zero, part of the tuning range in the accumulation area has a negative V_{tune}. By making V_{ref} equal to 2.5 V for example, the complete tuning range can be covered with positive tuning voltages (as shown in Figure 9.21).

9.3.4 Experimental results

The RO test-IC with PMOS varactor has been realized, together with a 5 V voltage stabilizer that generates the 3.5 V voltage V_{stabi} and the 2.7 V bias voltage V_{bias}. The chip micrograph of the RO with PMOS-varactors is shown in Figure 9.23. The active IC area is about 0.26 mm^2. An eight-pin plastic Shrink Small Outline (SSO) package is used for packaging.

Simulated and measured oscillation frequency versus V_{tune} are shown in Figure 9.24. As can be expected from the MOS capacitance characteristic (Figure 9.21), the frequency curve is not monotonic. The frequency curve in the inversion region is well predicted by the applied MOS model, but the modeling needs improvement to predict

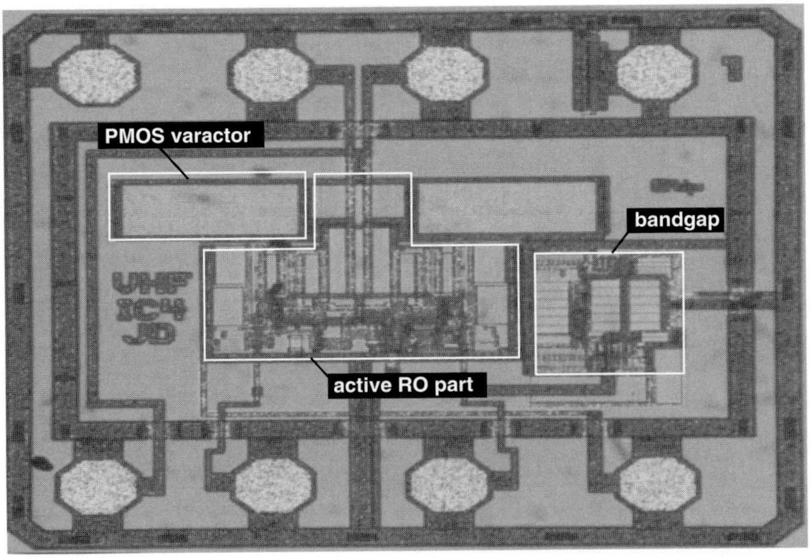

Figure 9.23 Die photo of the RO with PMOS varactors.

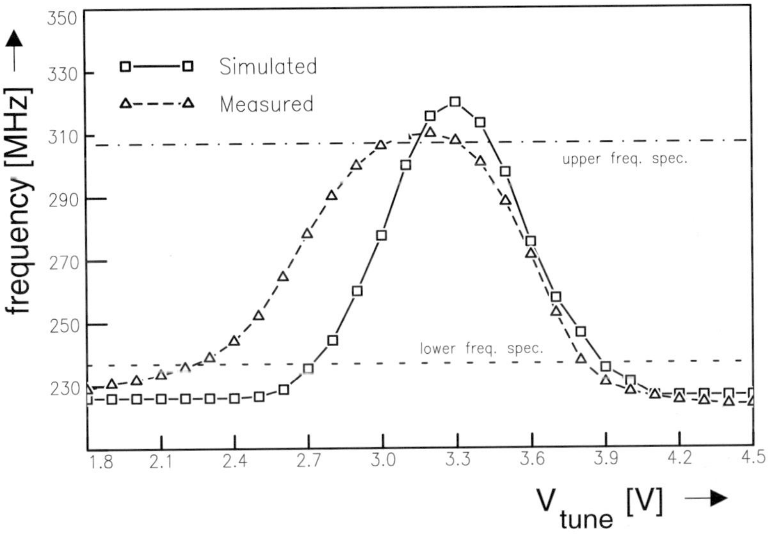

Figure 9.24 Frequency versus V_{tune}.

the accumulation region accurately. Minimum and maximum frequency of the RO are 225 MHz and 310 MHz, respectively. The complete tuning range needed for the RO is covered and a tuning range of more than 31% is measured.

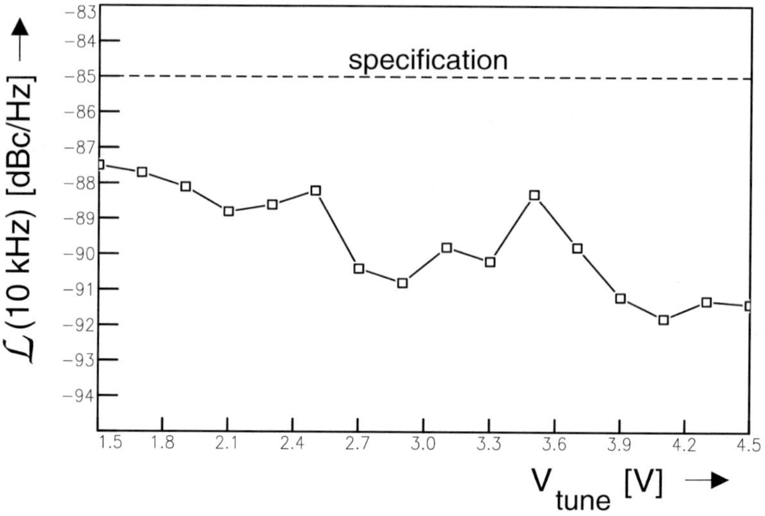

Figure 9.25 $\mathcal{L}(10\,\text{kHz})$ versus V_{tune}.

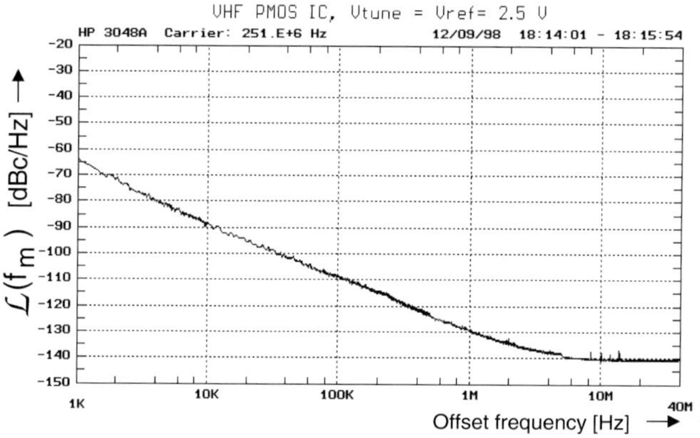

Figure 9.26 $\mathcal{L}(f_m)$ versus f_m for V_{tune} equal to 2.5 V ($f_{osc} = 251\,\text{MHz}$).

The measured phase noise at 10 kHz offset of the carrier over the complete tuning range is presented in Figure 9.25.

> Over the entire RO tuning range $\mathcal{L}(10\,\text{kHz})$ is better than -87 dBc/Hz and the specification is met. $\mathcal{L}(f_m)$ versus offset frequency is shown in Figure 9.26 for V_{tune} is 2.5 V.

9.3.5 Discussion

Characterization of the RO design with integrated PMOS varactors shows that the PMOS solution meets both phase noise requirements and tuning range requirements. However, for a robust product, additional effort is needed. Measurements were performed with air coils and these have to be replaced by printed PCB coils for two reasons. Firstly, this removes an alignment point since the PCB coils are much more reproducible than air coils. Secondly, air coils are susceptible to microphonics, which can lead to cycle slips in the satellite receivers. To cope with process spread more tuning range is needed. This can be achieved by a redesign of the active RO circuitry and optimizing for minimum C_{par}. Furthermore, if the minimum varactor capacitance C_{min} is increased, the (relative) influence of parasitic capacitance C_{par} reduces and the tuning range increases.

The tuning behavior of the RO with PMOS varactors is not monotonic. In the test-IC, the well of the PMOS is connected to the drain and source contact. This allowed evaluation of both accumulation region and inversion region of the varactor. Since the RO will be part of the double-loop tuning system, the tuning curve needs to be made monotonic otherwise the double-loop tuning system could latch up. This can be realized by connecting the well-contact to V_{stabi}. Depending on whether the gate is connected as an anode or as a cathode, the accumulation region or inversion region can be selected for operation. Figure 9.24 shows that the accumulation region (left region) has a lower VCO gain (MHz/V) which leads to lower spurious components resulting from noise sources present at the tuning input.

> The quality factor of the PMOS varactor can be improved through careful optimization of the layout.

In the presented design standard layout rules are applied. Using the design guidelines suggested in [116] the layout of PMOS varactor can be optimized to maximize the quality factor and thus the phase noise performance of the RO.

9.3.6 Conclusion

PMOS integrated varactors are successfully used to realize a reference oscillator for a double-loop tuning system, which can be used in digital satellite receiver architectures. State-of-the-art phase noise performance of -87 dBc/Hz at 10 kHz offset is experimentally demonstrated for a highly tunable oscillator with integrated PMOS varactors. The measured tuning range is 225 MHz to 310 MHz, which is larger than 31%. Power dissipation is 14 mW with a supply voltage of 3.5 V.

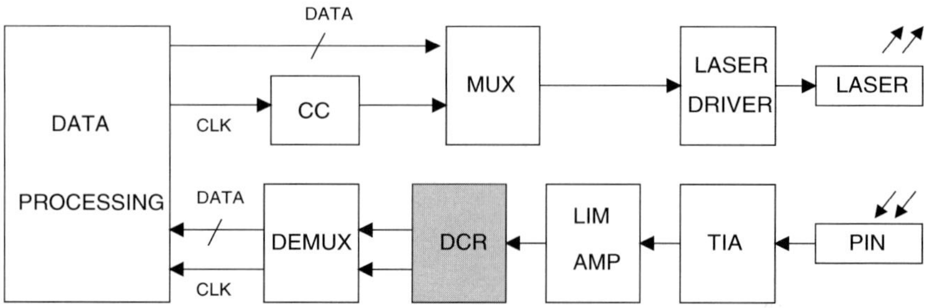

Figure 9.27 Block diagram of an optical transceiver.

9.4 A 10 GHz I/Q ring oscillator for optical receivers

In data transmission over optical fibers, one of the key functions of the receiver front-end is Data Clock Recovery (DCR). In networks following the Synchronous Optical Network (SONET) standard or the Synchronous Digital Hierarchy (SDH) standard, non-return-to-zero (NRZ) data signals are used. A block diagram of an optical transceiver is shown in Figure 9.27. At the receiver side, the optical signal from the glass fiber is transfered to the electrical domain by a PIN-diode. Next, the output current of this diode is amplified and converted into a voltage by a low-noise Trans-Impedance Amplifier (TIA). A limiting amplifier further amplifies the signal. The task of the DCR circuit is to extract the clock information from the NRZ data. This means that the DCR circuit must be able to acquire phase lock with the clock signal from the random data.

Many advanced integrated DCR circuits are PLL based [109]. Since the free-running frequency of the oscillator in the PLL is never exactly the same as the incoming data rate, the DCR circuit must obtain frequency lock prior to phase lock. In practice this means that every PLL-based DCR circuit needs some type of frequency acquisition aid [109]. One option is to use a crystal oscillator to keep the oscillator frequency within the acquisition range of the PLL, but this solution requires an expensive external crystal and an IC pin. Fully integrated solutions have been realized in which the DCR architecture has a frequency discriminator as an integral part of the architecture [231]. Often, this requires an oscillator that provides quadrature (I/Q) signals. The availability of quadrature signals also allows for the construction of half-rate DCR architectures [171]. Half-rate DCR circuits operate at half the frequency of the incoming data rate.

> This section presents an investigation of three I/Q ring oscillator topologies for 10 Gb/s DCR circuits that require quadrature signals.

The SONET standard OC-192 (equivalent to SDH STM-64) requires clock extraction at a bit-rate of 9953.28 Mb/s ($\sim$ 10 Gb/s).

Specification	Value	Unit
Technology	BiCMOS 30 GHz f_T	
Frequency	9.953280	GHz
$\mathcal{L}(2\,\text{MHz})$	-95	dBc/Hz
Power budget	100	mW
Power supply voltage	2.7	V
Tuning input	Differential	
Output(s)	In-phase / Quadrature	

Table 9.5 A short list of the specifications of the DCR ring oscillator.

The specifications for the I/Q ring oscillator will be reviewed in Section 9.4.1. Three two-stage ring oscillators are discussed at the circuit level in Section 9.4.2. The most promising ring oscillator implementation employs stacked active inductances. Section 9.4.4 explains how the active inductances in this ring oscillator can be used to improve the isolation between the oscillator core and cascaded circuits. The experimental results of the quadrature oscillator with stacked active inductances are discussed in Section 9.4.5, and compared with other reported ring oscillators in section 9.4.6.

9.4.1 Specifications

The target specifications for the quadrature oscillator are shown in Table 9.5 and have been derived from DCR system considerations and simulation. The power budget of 100 mW is based on the knowledge that a ring oscillator will be used rather than an LC oscillator. Compared to a LC oscillator, a ring oscillator needs a higher level of power dissipation to meet the carrier to noise ratio $\mathcal{L}(f_m)$ of -95 dBc/Hz at 2 MHz offset[9]. However, the $\mathcal{L}(f_m)$ specification required for DCR circuits is several orders of magnitude lower compared to, for example, local oscillator requirements in wireless front-ends, which makes the use of ring oscillators feasible in DCR systems. Especially the compact chip area and the, in general, large tuning range of a ring oscillator, make it a good candidate for use in DCR circuits.

A key aspect of the presented ring oscillator topology study is the technology used to investigate the performance of the oscillators. As shown in Table 9.5, a BiCMOS technology is specified with a 30 GHz f_T [141]. Constructing a quadrature ring oscil-

[9]This frequency domain specification is derived from the time domain OC-192 specification assuming the oscillator is part of a PLL with a loop bandwidth of 10 to 12 MHz (see Table 4.3 on p. 78).

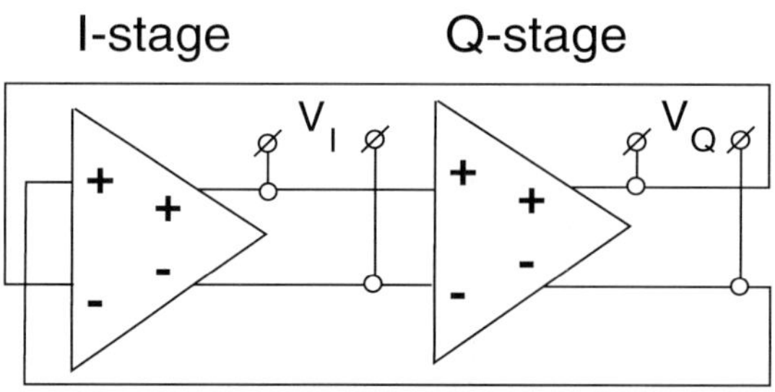

Figure 9.28 Block diagram of a two-stage quadrature oscillator.

lator with a $\mathcal{L}(2\,\mathrm{MHz})$ of -95 dBc/Hz oscillating at 1/3 of the f_T with less than 100 mW dissipation, is a design challenge. The combination of the oscillation frequency, $\mathcal{L}(2\,\mathrm{MHz})$, dissipation target and technology has led to an evolution of ring oscillator circuits, presented below.

> The influence of the parasitics is dominant at 10 GHz, so that circuit complexity must be low. Every additional device adds device and interconnect parasitics, which reduce the oscillation frequency and the carrier level.

9.4.2 Two-stage ring oscillator topologies

The oscillation frequency and $\mathcal{L}(2\,\mathrm{MHz})$ specification are the primary objectives for the DCR ring oscillator design. The large signal oscillation frequency in a ring oscillator is equal to $(2N\tau_{delay})^{-1}$, in which N is the number of stages and τ_{delay} the large signal delay of a stage. A two-stage ring oscillator therefore is most interesting as a starting point for the DCR oscillator design, since it has the highest frequency and provides I/Q signals.

> Two stages is the optimum number of stages also for differential ring oscillators with respect to phase noise minimization, as we have seen in Chapter 6.

A block diagram of a two-stage ring oscillator is shown in Figure 9.28. The model consists of two identical sections and one inversion. The phase condition for oscillation dictates a 90° phase shift per section. Therefore, a two-stage oscillator with identical stages provides "correct-by-construction" I/Q signals on a behavioral level. Practical I/Q matching is limited by device matching and layout symmetry.

Below, three implementations of the stages in Figure 9.28 are discussed. Oscillation frequency maximization is used as a criterion to select the most promising topology. The circuit implementation with the maximum oscillation frequency (above 10

GHz) will have the highest gain and signal swing at the (lower) target frequency of 10 GHz, which will result in better $\mathcal{L}(f_m)$ figures. The qualitative discussion in this section is followed by a quantitative analysis of the three oscillator implementations in Section 9.4.3, in which the maximum oscillation frequency of the topologies in a 30 GHz f_T BiCMOS process is determined by an automatic circuit optimizer.

A candidate for the realization of the DCR oscillator is the two-integrator oscillator [241]. For convenience (we already encountered the circuit a number of times) the circuit diagram of one stage of this oscillator is redrawn in Figure 9.29(a). Maximum oscillation frequency is obtained when lumped capacitor C_1 is omitted. In such a case the integration capacitance in each section consists completely of parasitic capacitance. In Section 9.2 we saw that this oscillator, implemented in a 10 GHz f_T BiCMOS process, performed well in the satellite band. However, simulations of this circuit in the specified 30 GHz f_T BiCMOS process, revealed poor $\mathcal{L}(2\,\text{MHz})$ figures at 10 GHz. Furthermore, $FOM_{freq-two-int}$ amounts[10] to a value slightly below 100%, indicating that it is very difficult to achieve the target frequency with sufficient margin (the FOM does not include inter-connect capacitance, etcetera). A topological change is needed to achieve a higher maximum oscillation frequency and better $\mathcal{L}(2\,\text{MHz})$.

A promising option to extend the oscillation frequency and improve $\mathcal{L}(f_m)$ is to alter the circuit in Figure 9.29(a), such that the parasitic phase shift of transistors Q_{T1},Q_{T2} is partly compensated. Interestingly, only a small topological change in Figure 9.29(a) is needed to realize this. If the transistors Q_{A1},Q_{A2} in Figure 9.29(a) are connected with shorted base and collector to the collectors of Q_{T1},Q_{T2}, the oscillator circuit in Figure 9.29(b) is obtained [238].

> For high frequencies, a transistor with shorted base and collector implements an active inductance. It effectively implements inductive peaking and thus extends the oscillation frequency of the oscillator (see p. 150) [238].

This circuit will be referred to as the ring oscillator with folded active inductances. Tuning is realized by varying the bias current I_{tune} of transistors Q_{L1},Q_{L2}, which changes the inductance value.

The oscillation frequency of the ring oscillator with folded active inductances is maximized, if the parasitic capacitance seen at the collectors of transistors Q_{A1},Q_{A2} in Figure 9.29(b) is minimized. In the oscillator stage Figure 9.29(b), the collector substrate capacitance of transistors Q_{L1},Q_{L2} adds to the total parasitic capacitance seen at the collectors of Q_{A1},Q_{A2}. This contribution is eliminated when stacking the active inductances as shown in Figure 9.29(c). However, the current through transistors Q_{A1},Q_{A2} in Figure 9.29(c) ($I_{level}/2$ in balanced condition) is now reused in transistors Q_{L1},Q_{L2}. Therefore, a means of frequency control is needed to set the carrier level and frequency independently. This is realized with variable resistors R_{tune1},R_{tune2}, which control the inductance value of Q_{L1},Q_{L2} [235].

[10] See Figure 6.13 on p. 126: in this figure the oscillation frequency of a two-integrator oscillator is plotted, implemented in the technology specified in Table 9.5. $FOM_{freq-two-int}$ is defined on p. 187.

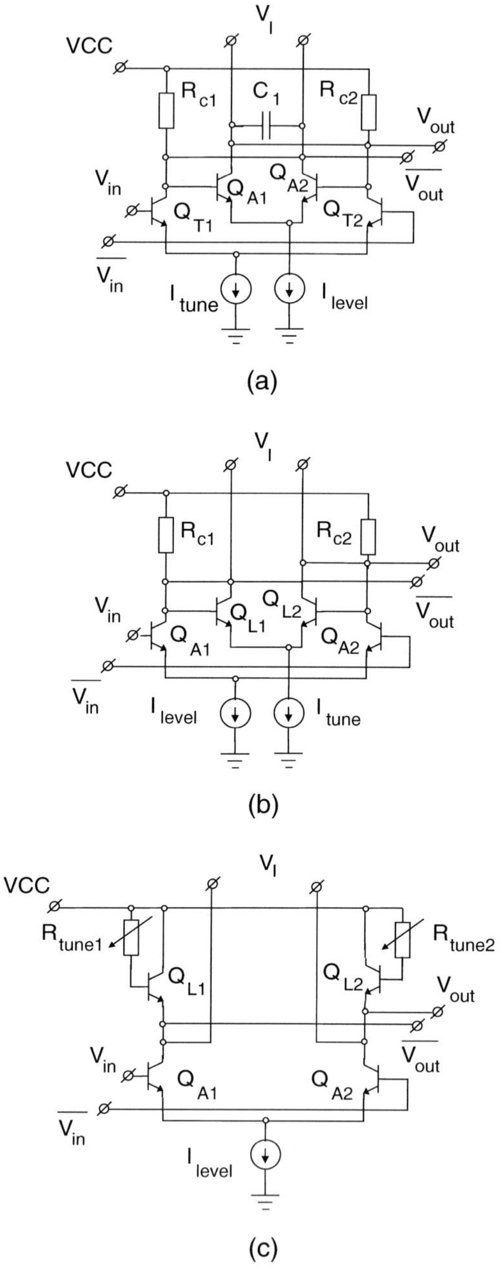

Figure 9.29 Evolution in oscillator stage topologies. A stage of a two-integrator oscillator implementation (a), a stage with folded active inductances (b), and a stage with stacked active inductances (c).

Oscillator	Figure	$f_{osc\|max}$ [GHz]	$\frac{f_{osc\|max}}{f_T}$
Two-integrator	Fig. 9.29(a)	10.3	0.34
Folded inductances	Fig. 9.29(b)	13.5	0.45
Stacked inductances	Fig. 9.29(c)	20.2	0.67

Table 9.6 Maximum oscillation frequency $f_{osc|max}$ of the three I/Q ring oscillator topologies.

9.4.3 Simulation of the maximum oscillation frequency

In order to quantify the maximum oscillation frequency of the three ring oscillator topologies, an automatic circuit optimizer was used. All transistors where modeled with the MEXTRAM transistor model [172] to include all high-frequency parasitics effects. The optimization goal was simply to maximize the oscillation frequency. All currents and resistor values where given as design parameters. The oscillation frequency was simulated using transient analysis and the value of the oscillation frequency was fed back to the optimizer, which adjusted the design parameters until a maximum value of oscillation frequency $f_{osc|max}$ was reached.

The results of the optimization are shown in Table 9.6. As expected, the ring oscillator with stacked active inductances achieves the highest oscillation frequency. In practice, $f_{osc|max}$ is lower since interconnect capacitance is not taken into account.

The simulation results listed in Table 9.6 are obtained without loading the ring oscillators with buffers, for example.

> For the realization of the ring oscillator with stacked active inductances, loading of the circuit by cascaded circuits has to be addressed. Minimization of the loading effect by buffers will be considered next.

9.4.4 Adding buffered outputs

The quadrature ring oscillator is used in a DCR architecture. Therefore, buffers, flip-flops or other circuits load the ring oscillator. If these circuits are connected to the collectors of transistors Q_{A1},Q_{A2} in Figure 9.29(c), their contribution to the total parasitic capacitance can significantly lower the oscillation frequency. This loading effect can be reduced by providing alternative output terminals, which are isolated from the collectors of Q_{A1},Q_{A2}.

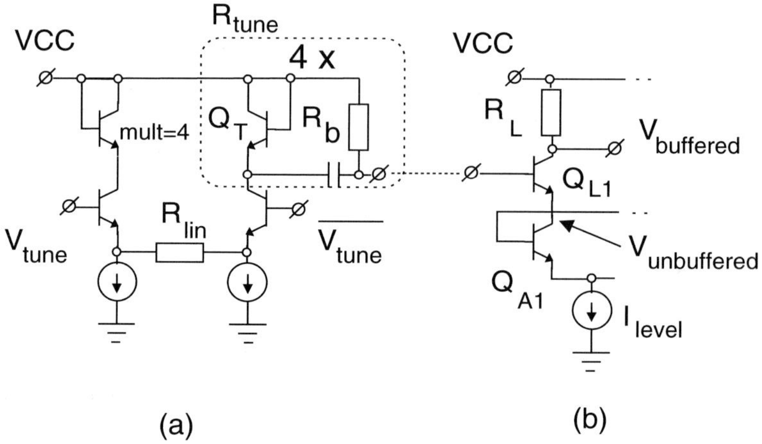

Figure 9.30 Circuit implementation of the V/I-converter with differential tuning inputs (a) and the half-circuit of one oscillator section from Figure 9.29(c), with the active inductance (Q_{L1}) used as cascode stage (b).

Figure 9.30(b) shows the half circuit of Figure 9.29(c).

> The collector of transistor Q_{L1} is connected to VCC with resistor R_L. This creates a buffered output node.

For small values of R_L the inductance of Q_{L1} remains practically unchanged when this resistor is added. Transistor Q_{L1} is now used for two functions. First of all it implements the active inductance and secondly, it provides cascode buffering.

The simulated buffered and unbuffered output signals ($V_{buffered}$ and $V_{unbuffered}$ in Figure 9.30(b)) of the oscillator in Figure 9.29(c), with collector resistors R_L inserted between the collectors of Q_{L1}, Q_{L2} and VCC, are shown in Figure 9.31. The tail current I_{level} (see Figure 9.29(c)) was set to 12 mA, which resulted in a large-signal oscillation frequency around 10 GHz. Resistor R_L was set to 20 Ω. The simulation results in Figure 9.31 are obtained without loading the oscillator and performed at a maximum oscillation frequency of 11.6 GHz. The influence of loading effects was investigated by connecting two differential pairs with 12 mA tail current (the same current level as the oscillator) to the collectors of Q_{A1}, Q_{A2} in Figure 9.29(c). The simulated frequency dropped from 11.6 GHz to 9.2 GHz and the output voltage, which was 216 mV_{peak} in Figure 9.31, dropped to 101 mV_{peak}. On the other hand, the internal signal swing of the oscillator was 9 mV lower, and the oscillation frequency changed only 100 MHz, if the differential pairs were connected to the collectors of the stacked active inductances (Q_{L1}, Q_{L2} in Figure 9.29(c)) with added collector resistors R_L. These simulation results illustrate the effectiveness of the output configuration shown in Figure 9.30(b).

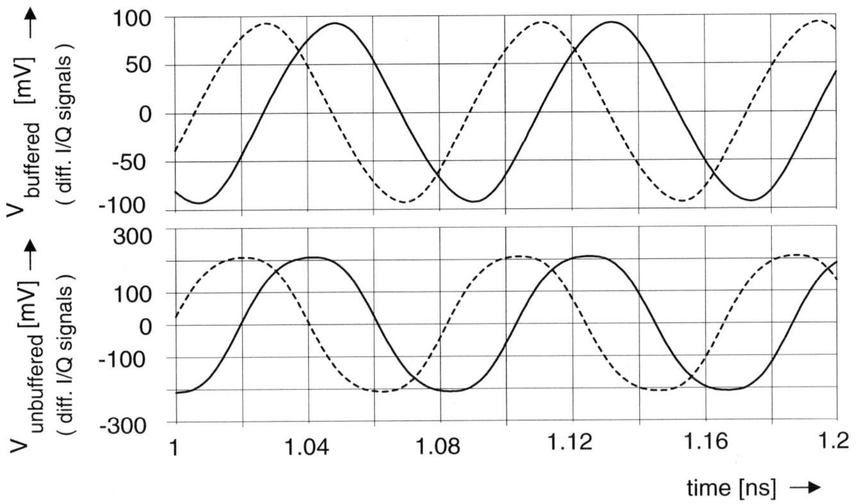

Figure 9.31 Simulated differential quadrature signals at the buffered and unbuffered output terminals.

Figure 9.30(a) shows the implementation of R_{tune}. To realize the variable resistance, diode connected transistor Q_T is used, which is AC-coupled to the base of the active inductance. Resistor R_b provides the base current for Q_{L1}. The current of transistor Q_T is controlled by a linearized differential pair. This implements differential tuning inputs.

The $\mathcal{L}(2\,\text{MHz})$ of the quadrature oscillator constructed with the stage shown in Figure 9.29(c), complete with the V/I-converter shown in Figure 9.30(a) and with R_L added to have buffered outputs, was simulated with SpectreRF. At 10 GHz the simulated $\mathcal{L}(2\,\text{MHz})$ is -95.3 dBc/Hz. The noise contribution of transistor Q_A (four in total, since there are two stages) is a dominant source (4×7.25 %), followed by the active inductance Q_L (4×5.6 %). Other contributors are the implementation of tail current source I_{level} (2×3 %), the base resistance of Q_A (4×2.2 %) and a multitude of other small noise sources. The tunable resistors R_{tune}, implemented by the boxed circuit Figure 9.30(a), each contribute less than 0.3 %. The same is true for the contribution of R_L, hence the value of these resistors is not critical concerning the $\mathcal{L}(f_m)$ of the ring oscillator with stacked active inductances.

9.4.5 Experimental results

The ring oscillator with stacked active inductances and buffered outputs was realized in the specified BiCMOS process [235]. The micrograph of the IC is shown in Figure 9.32. The active chip area of the oscillator with V/I-converter is less than 0.13 mm^2.

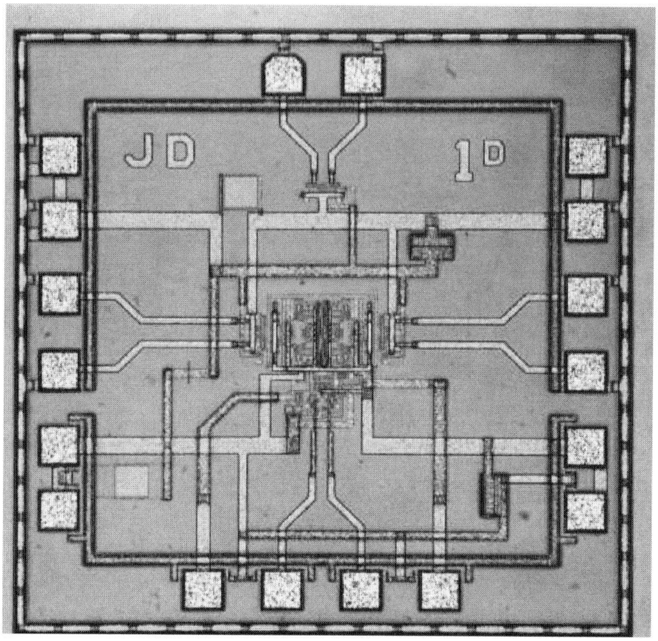

Figure 9.32 Micrograph of the realized ring oscillator.

Total chip area including bond pads is 1.5x1.5 mm^2. The power dissipation of the total IC is 230 mW of which 75 mW is dissipated by the VCO core. The power supply voltage is 2.7 V. All measurements were performed on packaged samples (16 pins HTSSOP package). On-chip 50 Ω I/Q buffers provided the quadrature output signals with -20 dBm output power.

Measured frequency and $\mathcal{L}(2\,\text{MHz})$ curves versus differential tuning voltage V_{tune} are shown in Figure 9.33. The tuning range is 16 % and ranges from 9.8 GHz to 11.5 GHz. The $\mathcal{L}(2\,\text{MHz})$ was measured with a spectrum analyzer and results were verified with HP3048 phase noise measurement equipment, which has an accuracy of ±2 dB.

Measured $\mathcal{L}(2\,\text{MHz})$ offset is better than -94 dBc/Hz over the complete tuning range. Best case $\mathcal{L}(2\,\text{MHz})$ is -98 dBc/Hz at 9.8 GHz. At higher frequencies, the carrier is somewhat smaller resulting in a worst case $\mathcal{L}(2\,\text{MHz})$ of -94 dBc/Hz.

The power spectrum of the ring oscillator at an oscillation frequency of 11.5 GHz is shown in Figure 9.34, measured with a resolution bandwidth of 100 kHz.

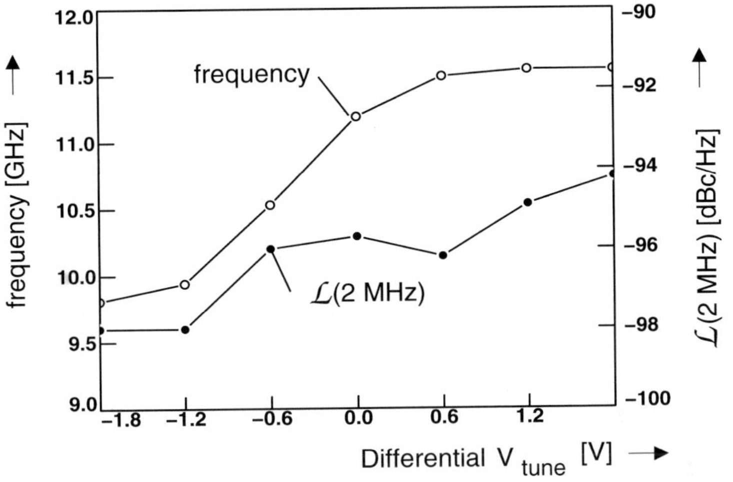

Figure 9.33 Frequency and $\mathcal{L}(2\text{ MHz})$ versus differential tuning voltage V_{tune}.

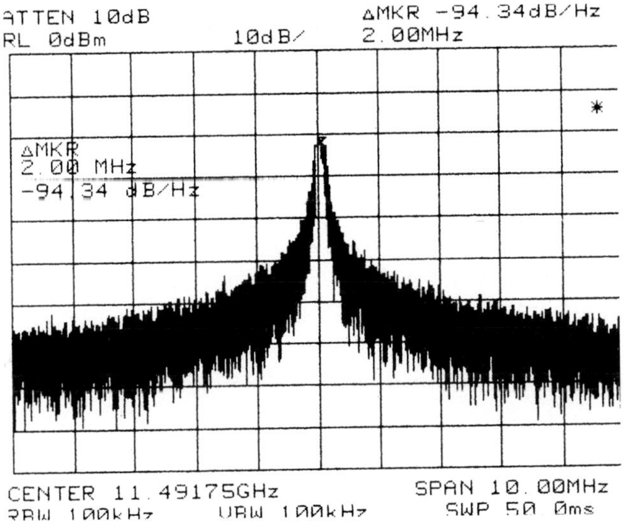

Figure 9.34 Power spectrum of the oscillator at 11.5 GHz.

9.4.6 Benchmarking

Ring oscillators reported in literature are realized in a variety of IC technologies, ranging from CMOS, BiCMOS and SiGe to InP and GaAs implementations. A number of reported ring oscillators are compared with the presented quadrature oscillator design.

This comparison is shown in Table 9.7. The benchmark FOM $Nor - PN$ in Table 9.7 is defined on p. 193. Note that parameter P_{DC} is the total power dissipation of the oscillator, excluding buffers.

> The realized quadrature oscillator achieves a state-of-the-art f_{osc}/f_T ratio in comparison with the other listed ring oscillators.

The ring oscillator in [231] with f_{osc}/f_T equal to 0.36 comes close to the achieved record of 0.38 but has a normalized $\mathcal{L}(f_m)$ that is 20 dB lower compared to the ring oscillator circuit with stacked active inductances.

9.4.7 Conclusion

In this section, a two-stage ring oscillator with stacked active inductances was presented. The oscillator can be tuned between 9.8 GHz and 11.5 GHz and is suitable for use in DCR circuits of optical receivers that require quadrature signals. The active inductances isolate the oscillator core from cascaded circuits, which makes the circuit relatively insensitive to loading effects. The quadrature oscillator is realized in a 30 GHz f_T BiCMOS technology and achieves an oscillation frequency over f_T ratio of 0.38. The $\mathcal{L}(2\,\text{MHz})$ is -94 dBc/Hz or better, which is realized with 75 mW dissipation and a 2.7 V supply voltage.

Ref.	$\mathcal{L}(f_m)$ [dBc/Hz]	f_m MHz	f_{osc} GHz	P_{DC} mW	Techn./ f_T – / GHz	f_{osc}/f_T	FOM Nor – PN [dBc/Hz]
This Work	-94.3	2	11.5	75	BiCMOS/ 30	0.38	-150.4
[233]	-97	1	10	250	InP HBT/ 53	0.19	-153
[243]	–	–	6.5	–	AlGaAs, GaAS/22	0.3	–
[231]	-100	10	9	850	Si Bipo./ 25	0.36	-130
[240]	-75	1	2.2	1.3	BiCMOS/ 18	0.13	-140.7
[232]	-114	10	10	–	SiGe/ 45	0.23	
[234]	-94	1	2.2	11.8	0.5 μm CMOS/ $\sim$ 15	$\sim$ 0.15	-150.2
[230]	-98.5	1	5.43	80	0.25 μm CMOS/ $\sim$ 40	$\sim$ 0.14	-154.2

Table 9.7 Ring oscillator benchmarking.

A

Resonator quality factor

RESONATORS with a high quality factor ("high-Q") are an essential ingredient of low-noise oscillators. In this appendix, the example of a parallel LCR resonator is used to get "a feel" for the energy storage capability of a high-Q resonator.

The most general formula for the quality factor of a resonator is given by [56],

$$Q_{res} = 2\pi \frac{\text{Total energy stored in the resonator}}{\text{Energy lost per cycle from the resonator}}. \quad \text{(A.1)}$$

Note that Q_{res} is a constant, characteristic of a resonator. For low-Q resonators the rate of energy loss is not uniform. Furthermore, for low-Q resonators the loss per cycle is comparable to the total stored energy. In this case, the numerator of (2.16) should be taken as the *average* of the energy during one oscillation period [56].

Figure A.1 illustrates the importance of a high quality factor for the amount of energy stored in the resonator. A certain amount of energy is provided to three parallel LCR resonators, with the same resonance frequency but with different quality factors. A $t = 0$ the energy source is removed and a transient is initiated. It can be shown that the voltage transient across the parallel resonator decreases exponentially. Specifically, the amplitude ratio V_0 at a certain time to $V_{n_{cycle}}$ at a time n_{cycle} cycles later, can be expressed as [56]

$$\frac{V_0}{V_{n_{cycle}}} = e^{\left(n_{cycle}\frac{\pi}{Q_{res}}\right)}, \quad \text{(A.2)}$$

which can be solved for n_{cycle}, resulting in

$$n_{cycle} = \frac{Q_{res}}{\pi} \ln\left(\frac{V_0}{V_{n_{cycle}}}\right). \quad \text{(A.3)}$$

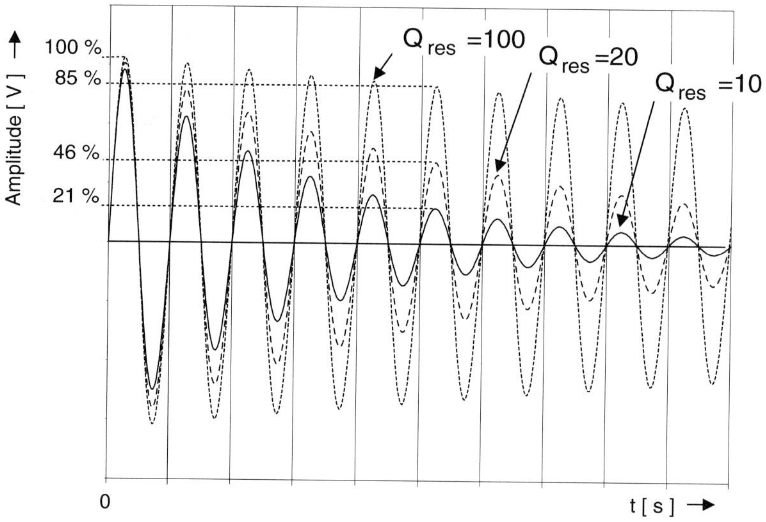

Figure A.1 Damping transients of a parallel LCR resonator with Q_{res} is 100, 20, and 10.

The term $\ln(V_0/V_{n_{cycle}})$ in (A.3) is called logarithmic decrement. The approximated percentage of which remains of V_0 after 5 cycles is calculated with (A.2) and denoted in Figure A.1. With $Q_{res} = 100$, $V_{n_{cycle}=5}$ still is 85% of V_0. Only 21% remains of V_0 after 5 cycles if the resonator quality factor is 10.

B

Behavioral modeling building blocks

BEHAVIORAL modeling is a powerful tool for modeling first-order oscillator behavior and reducing the complexity of a real integrated oscillator. In a behavioral model the passive elements are considered ideal. This in contrast to for example a resistor symbol in a circuit diagram. Here it represents a physical resistor. In addition to passive elements, a number of other building blocks are used in this work to construct behavioral models. In this appendix the symbols of these blocks are presented and described.

The ideal representation of a bipolar and a MOS transistor is an ideal transconductor. Hence it comes as no surprise that important active behavioral building blocks are transconductors. Four transconductor models are show in Figure B.1. Figure B.1(a) represents an ideal single-ended transconductor. Apart from the ideal transconductor in Figure B.1(a), two other single-ended models are used. The transconductor model in Figure B.1(b) is a single-pole model with time constant τ. This model is useful for studying the effects of parasitic delays that occur in transistors at high frequencies. A nonlinear transconductor is modeled by Figure B.1(c). When used, the function $f(V_{in})$ will be specified. For example, $f(V_{in})$ can be the transfer function of a differential bipolar or MOS pair (see also Appendix C).

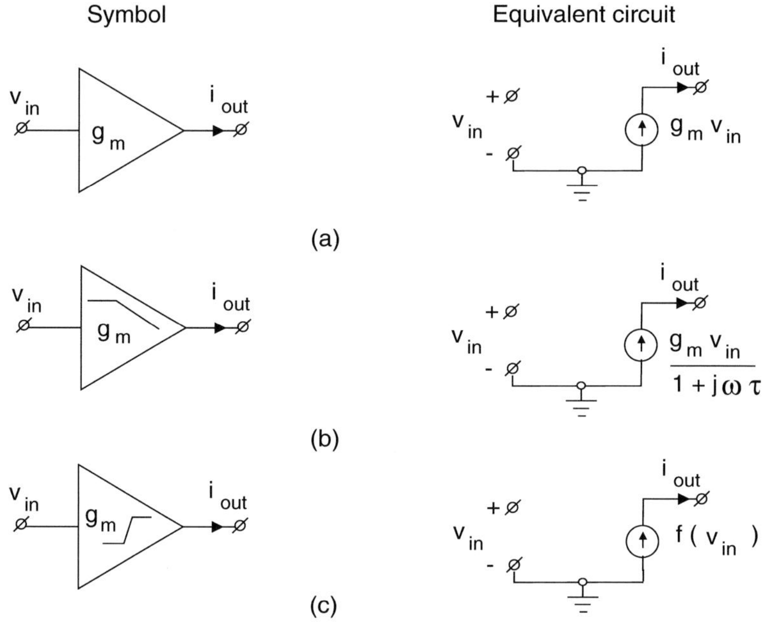

Figure B.1 Behavioral transconductor building blocks.

In Figure B.2 additional behavioral building blocks are shown. Each element is briefly described below:

(a) Phase shifter. This block provides a phase shift between input and output voltage of $\theta°$.

(b) Inverter. Inverts the input voltage. By convention there is no isolation between input and output of this block (since in a differential circuit it is implemented by simply cross-coupling differential wires).

(c) Switch. This symbol represents an ideal, lossless switch.

(d) Noise current source. Noise in behavioral models of oscillators will be represented by one or more noise current sources. The component name of noise current sources is i_n, possibly with additional subscript letters.

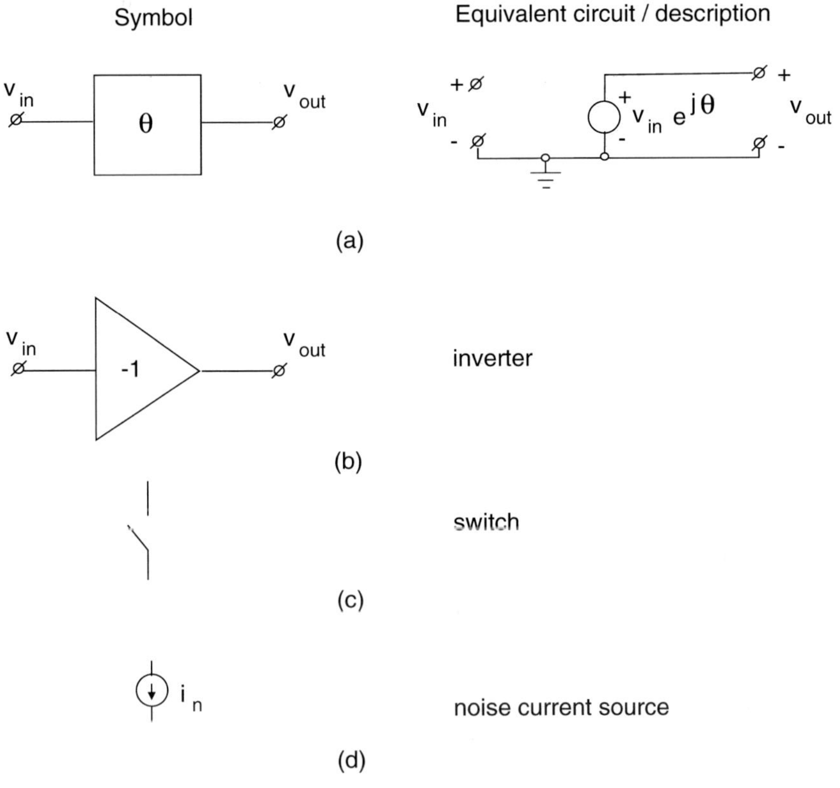

Figure B.2 Miscellaneous behavioral building blocks.

C

The ideal limiter and two implementations

BIPOLAR and MOS differential pairs have a limiting transfer characteristic. The differential pair is an important building block often utilized in high-frequency oscillators. The limiting transfer curve is used in many oscillators designs, where generation of some harmonics is allowable for amplitude stabilization (self-limiting). In behavioral modeling an approximation of the $I_{out}(V_{in})$-curve of a differential pair is useful, as it can simplify discussions and calculations considerably. In this appendix, the DC transfer characteristics of a MOS and a bipolar differential pair are briefly reviewed. In addition, these characteristics are approximated by a piece-wise linear limiting characteristic (see Figure 5.3 on p. 93) described by (C.1). In other words, the relation between parameters I_{lim}, g_m and V_{lim} of (C.1), and the parameters of a bipolar and MOS differential pair are derived.

$$I_{out}(V_{in}) = \begin{cases} I_{lim} = g_m V_{lim} & \text{if } V_{in} \geq V_{lim}, \\ g_m V_{in} & \text{if } -V_{lim} < V_{in} < V_{lim}, \\ -I_{lim} = -g_m V_{lim} & \text{if } V_{in} \leq -V_{lim}. \end{cases} \tag{C.1}$$

C.1 DC transfer characteristics of a MOS differential pair

Consider an NMOS differential pair with tail current I_{tail}. Assuming that the transistors operate in the saturation region ($V_{ds} > V_{gs} - V_{TH}$, where V_{ds}, V_{gs} and V_{TH} are the drain-source voltage, gate-source voltage and the threshold voltage, respectively), a first order approximation for the differential output current $I_{out_{MOS}}$ versus the differential input voltage V_{in} is given by [89,99]

$$I_{out_{MOS}}(V_{in}) = \begin{cases} I_{tail} & \text{if } V_{in} \geq \sqrt{2}V_{TMOS}, \\ \frac{I_{tail}}{V_{TMOS}} V_{in} \sqrt{1 - \frac{V_{in}^2}{4V_{TMOS}^2}} & \text{if } -\sqrt{2}V_{TMOS} < V_{in} < \sqrt{2}V_{TMOS}, \\ -I_{tail} & \text{if } V_{in} \leq -\sqrt{2}V_{TMOS}, \end{cases} \quad \text{(C.2)}$$

with

$$V_{TMOS} = \sqrt{\frac{I_{tail}}{2K_n}} = V_{gs} - V_{TH}, \quad \text{(C.3)}$$

in which K_n is $1/2 \cdot \mu_n C_{ox} \cdot W/L$ with μ_n, C_{ox}, W and L, the channel mobility, the oxide per unit area, the MOS width and the effective length of the MOS, respectively[1]. It follows that the small-signal transconductance of an NMOS differential pair is

$$g_{m_{MOS}} = \frac{I_{tail}}{V_{TMOS}}, \quad \text{(C.4)}$$

which is identical to the transconductance of one transistor biased at current $I_{tail/2}$.

Equations (C.2), (C.3), and (C.4) show how the parameters of an ideal limiter should be set to approximate a MOS differential pair. The transconductance g_m in (C.1) is set to $g_{m_{MOS}}$ and V_{lim} should be made equal to V_{TMOS}. Since the tail current is the maximum current the differential pair can supply, I_{lim} should be set to I_{tail}.

C.2 DC transfer characteristics of a bipolar differential pair

Consider an NPN bipolar differential pair with a tail current I_{tail}. Given the well-known exponential relation of a bipolar transistor the differential output current $I_{out_{bipo}}$ versus differential input voltage V_{in} can be expressed as [87]

$$I_{out_{bipo}}(V_{in}) = I_{tail} \tanh(\frac{V_{in}}{2V_T}), \quad \text{(C.5)}$$

with V_T equal to kT/q in which k is Boltzmann's constant, T the absolute temperature and q the charge of an electron. The transconductance of a bipolar differential pair is equal to that of one transistor biased at $I_{tail}/2$, which is

$$g_{m_{bipo}} = \frac{I_{tail}}{2V_T}. \quad \text{(C.6)}$$

Given the previous equations on the bipolar differential pair, it is straightforward to set the parameters of an ideal limiter to approximate the transfer curve of a bipolar differential pair. Similar as what was done for the MOS differential pair, g_m should be set to $g_{m_{bipo}}$. The voltage V_{lim} should be set to $2V_T$, and I_{lim} to I_{tail}.

[1] K_n needs to be adjusted to a smaller value than the value predicted by $1/2 \cdot \mu_n C_{ox} \cdot W/L$ even for a long channel NMOS, to take into account the finite output impedance of a MOS. In [99] this is realized by introducing the parameter $K_n^{'} < K_n$.

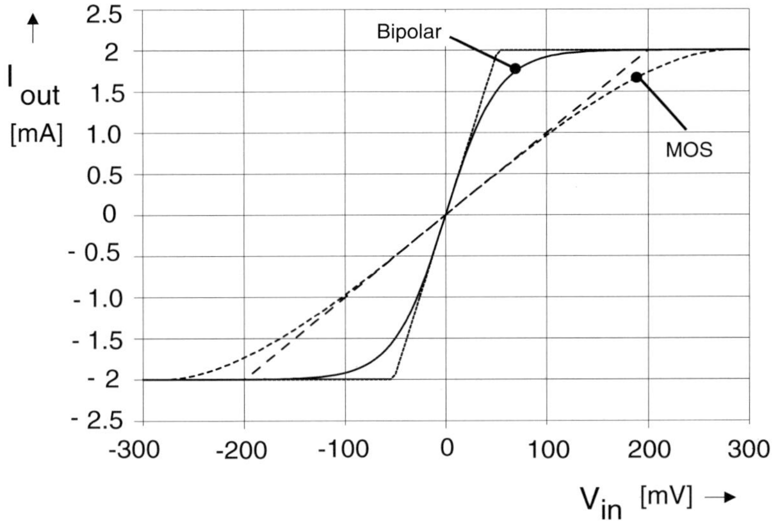

Figure C.1 The DC transfer characteristics of a bipolar and MOS differential pair approximated by the curve of an ideal limiter. I_{lim} is set 2 mA, $V_{lim} = 2V_T$ (bipolar), and $V_{lim} = 0.2V$ (MOS).

C.3 Graphical example

Figure C.1 shows the DC transfer plots of a MOS and a bipolar differential pair plus the transfer curves of two ideal limiters set to represent an approximation of the transfer curves. Current I_{lim} is made 2 mA. The voltage V_{lim} is set to $V_{lim} = 2V_T$ for the limiting curve which mimics a bipolar differential pair, and 0.2 V for the limiter that fits the MOS DC transfer characteristic. Note that the limiting voltage of a bipolar is fixed to $2V_T$, but that V_{TMOS} can be varied, because the W/L ratio of a MOS transistor offers an additional degree of freedom compared to the bipolar transistor. In this example V_{TMOS} is made equal to 0.2 V, which is a good setting for high gain amplifiers. In general, for high speed, $V_{TMOS} = V_{gs} - V_{TH}$ is made larger than 0.2 V, to have a high transconductance at a high output current level [99].

D

Quadrature signal generation implementation examples

MANY modern transceiver architectures require a signal source that provides I/Q output signals. These I/Q signals can be generated in many ways. Nine methods are listed in Table D.1 with their qualitative properties. Advantages are marked +, and disadvantages are marked −. In the third column of Table D.1, one or more references are given that describe the implementation example in more detail.

Implementation example	Properties	Ref.
Even-stage LC oscillator	+ excellent $\mathcal{L}(f_m)$, + low power, + low harmonic content, + I/Q correct by construction, - relatively large chip area, - limited tuning range.	[184, 185, 221, 222, 224, 225]
Even-stage ring oscillator	+ large tuning range, + compact chip area, + I/Q correct by construction - poor $\mathcal{L}(f_m)$, - large power consumption.	[235, 241]
continued on next page		

Implementation example	Properties	Ref.
Oscillator at f_{osc} and RC-CR network	+ simple circuit, + good $\mathcal{L}(f_m)$ in combination with LC oscillator, - limiters required, - undistorted waveform needed, - bandwidth limited if LC oscillator is used.	[173]
Oscillator at f_{osc} and poly-phase filter	+ good $\mathcal{L}(f_m)$ in combination with LC oscillator,- high noise floor, - high insertion loss, - bandwidth limited.	[174]
Oscillator at 2 (or 4) $\times$ f_{osc} with divider	+ good $\mathcal{L}(f_m)$ in combination with LC oscillator, - In case of division by 2, 50 % duty cycle of oscillator required, - oscillator must constructed at 2 or 4 $\times$ f_{osc} resulting in more power dissipation.	[175, 176]
Double PLL loop: ring oscillator at f_{osc} locked to LC oscillator	+ good $\mathcal{L}(f_m)$, + wide-band, - high complexity, - high power dissipation.	[40]
Four-stage oscillator at $1/2 \cdot f_{osc}$ with mixers or addition of phases	+ oscillator required at half the desired frequency, + wide-band if a ring oscillator is used, - large chip area if a (four-stage) LC oscillator is used, - poor $\mathcal{L}(f_m)$ if a ring oscillator is used.	[229]
One of above techniques with calibration technique added	properties of one of above methods, + improved amplitude and phase matching, + can have improved bandwidth, - high complexity, - high power dissipation (depending on technique).	[177–179]
Digital implementation, for example utilizing lookup table and reference clock	+ accurate, + wide-band, - A/D-converter needed if signals are needed in analog domain - low frequency.	[180]

Table D.1: Nine ways to generate quadrature signals at a frequency f_{osc}.

E

The frequency of a switching N-stage ring oscillator

DEPENDING on the frequency of operation and the degree of nonlinearity in a ring oscillator, this oscillator can work in linear or nonlinear mode. In this appendix the oscillation frequency of an N-stage ring oscillator ($N \geq 3$) is derived for a ring oscillator operating in strongly nonlinear mode.

Consider the behavioral in Figure E.1 in which the transconductors are implemented with the ideal limiting characteristic described by (C.1) on p. 261, and assume that the output voltage of the ring oscillator is such that the transconductors are completely switching. The latter assumption implies that V_{out} is much larger than V_{lim}, the

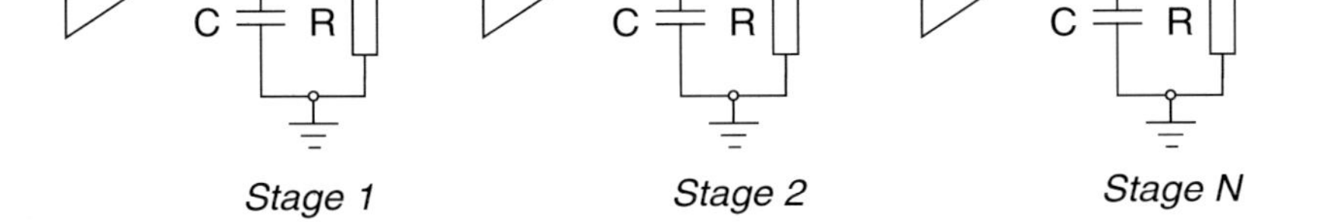

Figure E.1 A nonlinear N-stage ring oscillator behavioral model.

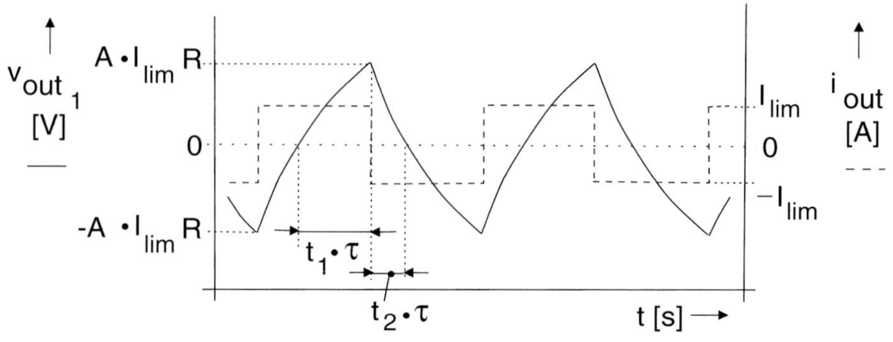

Figure E.2 The output voltage and output current waveform of a 3-stage ring oscillator (modeled by Figure E.1), when the transconductors are fully switching.

voltage where the transconductors start limiting. Since the transconductors in Figure E.1 are completely switching, current i_{out}, is either I_{lim}, or $-I_{lim}$ (see (C.1) on p. 261).

The waveforms of v_{out_1} and i_{out} of the ring oscillator model (for $N = 3$) are shown in Figure E.2. On the left Y-axis is v_{out_1} is set out. Output voltage of the ring oscillator is indicated and written as a fraction A of the maximum output voltage $I_{lim}R$. The X-axis represents the time. One half of the signal period is $(t_1 + t_2)\tau$ in which $\tau = RC$. The phase condition for oscillation dictates that $t_1 + t_2$ must be equal to Nt_2. By making use of the well-known transient response of circuits containing one reactive element[1], the following set of equations allows us to solve the unknown variables of interest, A and t_2,

$$\begin{aligned} t_1 &= -\ln(-A+1), \\ t_2 &= -\ln(\frac{1}{A+1}), \\ t_1 + t_2 &= Nt_2. \end{aligned} \tag{E.1}$$

The set of equations (E.1) can be solved numerically, and the results are shown in Table E.1 for $N = 3 \ldots 8$. For a large number of stages, the voltage on the parallel RC circuit in each ring oscillator stage practically reaches its end value, and A practically becomes unity.

Given the values for t_2 from Table E.1, the large signal oscillation frequency of an N-stage ring oscillator can be calculated with (see (5.18) on p. 98),

$$\omega_{Nring,large-signal} = \frac{2\pi}{2N\tau_{delay}}, \tag{E.2}$$

[1] $f(t) = f_{lim} + (f_0 - f_{lim})e^{(-t/\tau)}$, where is f_{lim} the steady-state final value, f_0 the initial value and τ the circuit time constant [181].

N	A	t_2	$\frac{\omega_{Nring,large-signal}}{\omega_{Nring}}$
3	0.618	0.481	1.26
4	0.839	0.609	1.29
5	0.927	0.656	1.32
6	0.965	0.675	1.34
7	0.983	0.684	1.36
8	0.991	0.689	1.38
...	...	...	...
∞	1	$\ln(2)$	≈ 1.44

Table E.1 The rounded numerical solutions for A, t_2 and $\omega_{Nring,large-signal}/\omega_{Nring}$ for $N = 3\ldots8$, and their limiting values for $N \to \infty$.

where $\tau_{delay} = t_2 \cdot \tau$. For large N, t_2 approximately becomes equal to $\ln(2)$, and (E.2) can be written as

$$\omega_{Nring,large-signal} \approx \frac{2\pi}{2NRC\ln(2)}. \tag{E.3}$$

The third column in Table E.1 lists the oscillation frequency given by (E.2) divided by the small-signal oscillation frequency $(\tan(\pi/N)/RC)$, showing that the nonlinear operation of the N-stage ring oscillator increases the frequency of this type of oscillators.

F

Bipolar and MOS small-signal calculation model

BIPOLAR and MOS transistors are complex devices and especially at high frequencies accurate models are of prime importance for "first-time-right" oscillator design. A number of models were already mentioned in Table 3.1 on p. 64. These models, with parameter counts ranging from 50 to over 300, are not very inviting for back-of-the-envelope calculations. Since these calculations provide considerable design insight and a feeling for the dominant parameters in a design, simplified models are normally used for this purpose [87, 89, 99, 139]. In this appendix a generic small-signal transistor model is presented and its corresponding bipolar and MOS parameters, suitable for first order calculations. Furthermore, equations are given for the main parameters and technology FOMs such as f_T and f_{MAX}.

F.1 Generic transistor model

A simple generic transistor model is shown in Figure F.1. The simplified small-signal models of the bipolar and MOS transistor are almost identical and represented by this model. All components are present in both technologies, except for R_{in}. This resistance, which models the base current for bipolar transistors, must be set to infinite for MOS technology. The generic parameters of the small-signal model and their bipolar and MOS interpretation are listed in Table F.1. The small-signal model and Table F.1 show that once calculations are performed using the bipolar parameters, results can easily be extended to a MOS implementation. Simply by replacing the parameters of the second column of Table F.1 with parameters of the third column. It should be stressed once more that the model in Figure F.1 is a simplified model, to be used only for first order estimates. For example, the emitter resistance, collector

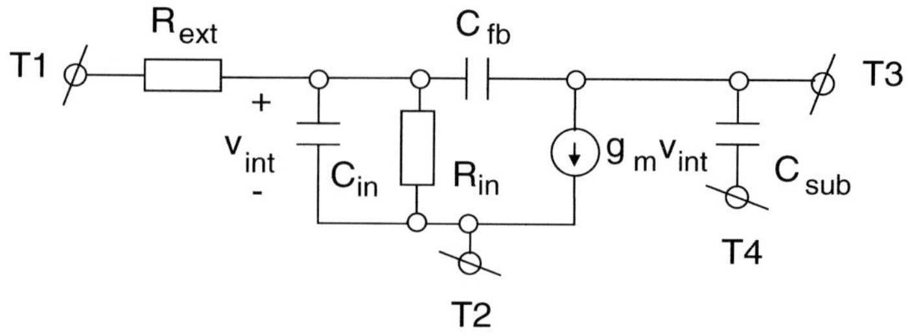

Figure F.1 First-order generic transistor small-signal model.

Generic parameter	Bipolar parameter	MOS parameter
T1	Base (b)	Gate (g)
T2	Emitter (e)	Source (s)
T3	Collector (c)	Drain (d)
T4	Substrate (s)	Bulk (b)
R_{ext}	R_{base}	R_{gate}
C_{in}	C_{π}	C_{gs}
R_{in}	R_{π}	∞
C_{fb}	C_{μ}	C_{gd}
g_m	$g_{m_{bipo}}$	$g_{m_{MOS}}$
C_{sub}	C_{js}	C_{db}

Table F.1 Parameters of the generic transistor model and their bipolar and MOS interpretation.

resistance, output resistance and distributed nature of the base in bipolar transistors are not modeled. For MOS transistors, the model neglects, among other effects, the bulk network, source, drain and output resistance and channel resistance (non-quasi static effects) [88].

F.2 Bipolar and MOS parameter values

Most parameters in Table F.1 are simply numbers that characterize particular aspects of a specific bipolar or MOS device in an IC technology. Others, like the transconductance, can be expressed in a useful formula for first order calculations. Most common expressions, technology FOMs, and underlying variables for bipolar and MOS transistors are tabulated in Table F.2 and F.3, respectively.

Bipolar parameter	Value	Description
$g_{m_{bipo}}$	$\frac{I_c}{V_T}$	small-signal transconductance
C_π	$\tau_F g_{m_{bipo}} + C_{je}$	input capacitance
R_π	$\frac{\beta_O}{g_{m_{bipo}}}$	
f_T	$\frac{g_{m_{bipo}}}{2\pi(C_\pi+C_\mu)}$	transition frequency
f_{MAX}	$\sqrt{\frac{f_T}{8\pi R_{base} C_\mu}}$	max. oscillation frequency
τ_F		forward transit time
C_{je}		emitter junction capacitance
β_O		common-emitter current gain
I_c		collector current
V_T		thermal voltage

Table F.2 Small-signal bipolar parameter values and technology FOMs

MOS parameter	Value	Description
$g_{m_{MOS}}$	$\mu_n C_{ox} \frac{W}{L}(V_{gs} - V_{TH})$	small-signal transconductance (saturation, long-channel)
	$\sqrt{2\mu_n C_{ox} \frac{W}{L} I_d}$	idem
$g_{m_{MOS}}$	$\frac{\mu_n C_{ox}}{2} W E_{crit}$	small-signal transconductance (saturation, short-channel)
f_T	$\frac{g_{m_{MOS}}}{2\pi(C_{gs}+C_{gd})}$	transition frequency
f_{MAX}	$\sqrt{\frac{f_T}{8\pi R_{gate} C_{gd}}}$	max. oscillation frequency
W		channel width
L		effective channel length
μ_n		(NMOS) mobility
C_{ox}		oxide capacitance per unit area
I_d		drain current
V_{TH}		threshold voltage
E_{crit}		critical field strength

Table F.3 Small-signal MOS parameter values and technology FOMs

G

Performance overview of LC oscillator designs

OPEN literature provides an abundance of interesting integrated LC oscillators ranging from low performance to the state of the art. A number of LC oscillator designs and their key performance parameters are gathered and listed below in Table G.1 (Single-phase) and Table G.2 (I/Q). Blank positions indicate that a particular performance figure for this oscillator was not reported. Some tabulated performance figures are self-explanatory, except perhaps for the following:

- TR, tuning range in percentage.
- P_{DC}, Power dissipation of LC oscillator core (without buffers).
- V_{DC}, Supply voltage of LC oscillator core.
- Q_p, unloaded quality factor of total LC resonator, or only of the inductor(s) if the varactor quality factor is not reported.
- $\mathcal{L}(2\text{MHz})$, $\mathcal{L}(f_m)$ extrapolated to $f_m = 2\,\text{MHz}$.
- Tech., stands for technology: B (bipolar), BC (BiCMOS), C (CMOS), SG (SiGe Bipolar) or S (Silicon-on-Anything). The number indicates the f_T in case of B, BC, SG or S technology (e.g. B30 means a bipolar technology with a 30 GHz f_T). For CMOS technology the minimum channel length is given (e.g. C018 means a 0.18μm process).

Ref.	f_{osc} GHz	TR %	$\mathcal{L}(f_m)$ dBc/Hz	f_m MHz	P_{DC} mW	V_{DC} V	Q_p	$\mathcal{L}(2MHz)$ dBc/Hz	Tech.
[208]	3.6	0.7	-106	2	1	3.6	13.4	-106	S9
[195]	6	15	-116	1	22	3	7.5	-122	SG45
[197]	4	9	-106	1	12	3	3	-112	C0.5[1]
[216]	1.55	13	-102	0.1	21.6	2.7	7.6	-128	B15
[196]	4.7	4.3	-110	1	10.8	2.7	2.4	-116	C035
[218]	2	7.5	-136	4.7	32.4	2.7	3.3	-128.6	B25
[213]	9.8	3	-115	1	11.6	2.7	3	-121	C035
[211][2]	6.29	16.8	-98.4	1	18	1.5	4	-104.4	C035
[206]	1.5	10	-105	0.1	28	3.6	5	-131	BC11
[203][3]	1.8	11.4	-88	0.1	70	5	5	-114	BC10
[209]	1.5	4.7	-129	0.6	38	1.9	7.1	-139	BC35
[199][4]	1.8	4.5	-85	0.01	24	3		-131	C07
[202]	1.8	6.7	-100	0.5	7.6	3.3		-112	C06
[201]	2.4		-92	0.1	54	3.6	9.3	-118	BC12
[219]	1.8	14	-116	0.6	6	1.5	4.8	-126.4	C07
[215][5]	1.84	26	-101	0.1	25.5	3	2.9	-127	C06
[214]	1.8		-125.7	0.6	48	3	7.5	-136.2	C025
[183]	2.03	26	-117	0.6	10	2.5		-127.5	C035[6]
[226]	1.9	13.2	-100	0.1	12.5	2.5		-126	C025
[207][7]	1.4	17	-107	0.1	3	3		-133	C05
[194]	11	5	-106	1	24	3	5	-112	SG45
[205]	1.3	28	-119	0.6	12	2	3	-129.5	C035
[188][8]	2	30	-86	0.1	30	2		-112	BC12
[204][9]	2	38	-128.5	0.6	12.2	2.7		-139	BC25
[223][10]	1.8	15.6	-130	2	13.5	2.7		-130	C08
[217]	2	11	-125.1	0.6	34.2	1.8	11	-135.6	C065
[187][11]	1	15	-148.5	3	9.1	2.5	10	-145	C[12]
[212]	17	8.6	-108	1	10.5	1.4		-114	C025

continued on next page

[1]BiCMOS ($f_T = 12\,\text{GHz}$) but only CMOS is used.
[2]I/Q LC oscillator but the two outputs are summed.
[3]Uses two LC tuned circuits with different resonance frequencies.
[4]Utilizes bond-wires and an enhanced LC tank.
[5]Switched tuning; different oscillators or capacitors are switched.
[6]BiCMOS but only CMOS is used.
[7]Uses tapped bond-wire inductances.
[8]The varactor is an (active) variable-impedance converter. Bond-wire inductors.
[9]Two-band oscillator (one band-switch). Bond-wire inductors.
[10]MOS varactors, and bond-wire inductors.
[11]Tail current noise is filtered. It also uses a switched binary weighted capacitor bank.
[12]BiCMOS process but only CMOS is used, minimum channel length not specified.

Ref.	f_{osc} GHz	TR %	$\mathcal{L}(f_m)$ dBc/Hz	f_m MHz	P_{DC} mW	V_{DC} V	Q_p	$\mathcal{L}(2MHz)$ dBc/Hz	Tech.
[210]	1.8	28	-127.5	0.6	32.4	1.8	8.7	-138	C025
[191]	3.6	19.4	-145	3	24	3	5	-141.5	C025
[198]	2.5	40	-128	3	18	3.3	17	-124.5	BC30
[192][13]	2.12	30.5	-138.2	3	10	2.5	6	-135	C035
[200]	0.83	21.6	-100.4	0.1	0.4	5	10.4	-126.4	S9
[189][14]	0.310	30	-87	0.01	14	3.5		-133	B11

Table G.1: Single-phase LC oscillator designs and their key parameters.

Ref.	f_{osc} GHz	TR %	$\mathcal{L}(f_m)$ dBc/Hz	f_m MHz	P_{DC} mW	V_{DC} V	Q_p	$\mathcal{L}(2MHz)$ dBc/Hz	Tech.
[184]	0.9	17	-85	0.1	30	3	5	-111	C1
[221]	10	12.4	-106	1	45	1.5	10	-112	C025
[227][15]	2.6	8.9	-110	5	13	2.5		-102	C035
[222]	1.9	6.3	-123	0.6	43.2	2.7	6.5	-133.4	B40
[185]	1	10	-126	0.6	13	2.5		-136.5	C06
[190]	1.57	24	-130.5	0.6	30	2	20	-141	C025
[228]	13	37	-112.3	10	39	3.3		-98	SG55
[220]	5	6.4	-113	2	21.2	2.7	4	-113	BC30
[193]	1.8	15.6	-143	3	20	2.5	8	-139.5	C025
[186]	1.8	18	-140	3	50	2	6	-136.5	C035

Table G.2: I/Q (two-stage) LC oscillator designs and their key parameters.

[13]Differential Colpitts oscillator with switching current source.

[14]External coils, integrated PMOS varactors.

[15]Common-source nodes used as single-phase 5.2 GHz output.

H

Performance overview of ring oscillator designs

RING oscillators are often used in optical transceiver front-ends and are less common in portable wireless applications, since they require much more dissipation given a certain $\mathcal{L}(f_m)$ specification. A number of ring oscillator designs and their key performance parameters are gathered and listed below in Table H.1. Blank positions indicate that a particular performance figure for this oscillator was not reported. The abbreviations of the tabulated performance figures are explained in Appendix G. Compared to the columns in Appendix G, there is one change: Q_p has been replaced by the number of stages N. All even-stage oscillators are I/Q oscillators. In addition to the technologies listed in Appendix G, abbreviations G (GaAs) and I (Inp) are also used.

Ref.	f_{osc} GHz	TR %	$\mathcal{L}(f_m)$ dBc/Hz	f_m MHz	P_{DC} mW	V_{DC} V	N	$\mathcal{L}$(2MHz) dBc/Hz	Tech.
[241]	1.55	80	-106	2	100	5	2	-106	BC11
[235]	10	17	-94.3	2	75	2.7	2	-94.3	BC30
[233]	10		-97	1	250	5	3	-103	I53
[231]	8	87.5	-100	10	850	5	2	-86	B25

continued on next page

Ref.	f_{osc} GHz	TR %	$\mathcal{L}(f_m)$ dBc/Hz	f_m MHz	P_{DC} mW	V_{DC} V	N	$\mathcal{L}(2MHz)$ dBc/Hz	Tech.
[243][1]	6	28.3					4		G22
[240]	2	17.5	-75	1	1.3	3	2 [2]	-81	BC18
[232][3]	10		-114	10		3.3	2	-100	SG45
[238]	8	18.8	-96	2	76	2.7	2	-96	BC30
[234]	2.2		-94	1	11.8	5	3	-100	C05
[230]	5.43	25	-98.5	1	80	2.5	4	-114.5	C025
[236]	0.9	68	-105.5	0.6	15.5	2.5	2	-116	C05
[242]	1.63	88	-95	10	78	5	4	-81	C08
[237]	0.9	50	-117[4]	0.6	30	3	4	-127.5	C06
[239]	2.5	32	-90	5	10	3.3	2	-82	C04
[244]	1.25	61.5			100	3.3	8[5]		C035

Table H.1: Ring oscillator designs and their key parameters.

[1] Two mixers are used for frequency doubling.
[2] Stages are not identical. Inductive peaking used.
[3] Tuning is accomplished with delay interpolation.
[4] Power spectrum plot indicates $\approx$ -95 dBc/Hz at 100 kHz.
[5] Sub-feedback loops used to achieve high speed operation.

I

Q and $\mathcal{L}(f_m)$ of linear LC oscillators

NOISE shaping of white noise in oscillators gives rise to oscillator sidebands with a $1/f^2$ slope (-6 dB per octave). For linear[1] oscillators, this phase noise generating mechanism is the only one present[2]. As shown in Section 3.4.1 on p. 58, noise shaping calculations can be derived at a system level. Using the theory outlined in that section, the quality factor and $\mathcal{L}(f_m)$ of single-phase and multi-phase LC oscillators is derived in this appendix (Section I.2). Although the theory of Section 3.4.1 can be applied to single-phase oscillators, it is instructive to show that $\mathcal{L}(f_m)$ can be derived directly from a behavioral model (as long as the model is not too complex). In the following section $\mathcal{L}_{LC}(f_m)$ of a single-phase oscillator is calculated.

I.1 Single-phase LC oscillators

The linear single-phase LC oscillator model discussed in Chapter 6 is repeated for convenience, and shown in Figure I.1. To derive $\mathcal{L}(f_m)$ of this oscillator, we have to find an expression for noise output current $\overline{i_{n_{out}}^2}$. As defined in (4.3) on p. 71, division of $\overline{i_{n_{out}}^2}$ by the carrier signal (in this case a current) will give us an expression for $\mathcal{L}_{LC}(f_m)$.

[1]The assumption here is that an AGC control is present that stabilizes the oscillation amplitude in the linear region of the devices in the oscillator. If the AGC is relatively slow, the oscillator can be considered a Linear Time Invariant (LTI) system for a large number of cycles.

[2]That is, in absence of noise on modulation inputs such as the tuning input of an oscillator.

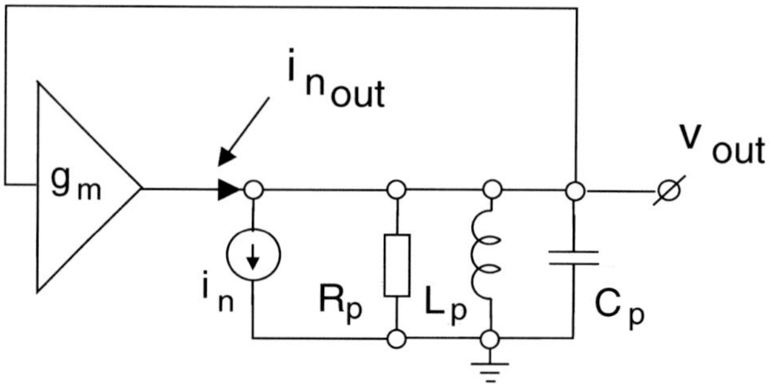

Figure I.1 Linear behavioral model of a single-stage LC oscillator.

Note that for the parallel circuit of R_p, L_p and C_p we can write (see p. 23)

$$Z_p(j\omega) = \frac{R_p}{1 + jvQ_p}, \tag{I.1}$$

with

$$v = \frac{\omega}{\omega_{osc}} - \frac{\omega_{osc}}{\omega}, \tag{I.2}$$

$$Q_p = R_p\sqrt{\frac{C_p}{L_p}} = \frac{R_p}{\omega L_p}. \tag{I.3}$$

In terms of impedance Z_p, we can write the following for the closed-loop gain $H_{cl}(j\omega)$ of Figure I.1,

$$H_{cl}(j\omega) = \frac{g_m Z_p(j\omega)}{1 - g_m Z_p(j\omega)}. \tag{I.4}$$

Substitution of (I.1) in (I.4) yields,

$$H_{cl}(j\omega) = \frac{g_m R_p}{1 + jvQ_p - g_m R_p}. \tag{I.5}$$

Assuming we are only interested in the behavior of the oscillator near the oscillation frequency $f_{osc} = \omega_{osc}/(2\pi)$ at an offset frequency Δf, a good approximation of v can be derived. If $|\Delta f| \ll f_{osc}$, v simplifies to $(2\Delta f)/f_{osc}$. Substitution of the gain condition of oscillation, $g_m R_p = 1$, and $v = (2\Delta f)/f_{osc}$ in (I.5) leads to a simplified expression for $H_{cl}(j(f_{osc} + \Delta f))$,

$$|H_{cl}(j(f_{osc} + \Delta f))| = \frac{1}{\frac{2\Delta f}{f_{osc}} Q_p}. \tag{I.6}$$

The output noise current $\overline{i_{n_{out}}^2}$ is the result of noise current $\overline{i_n^2}$, amplified by the

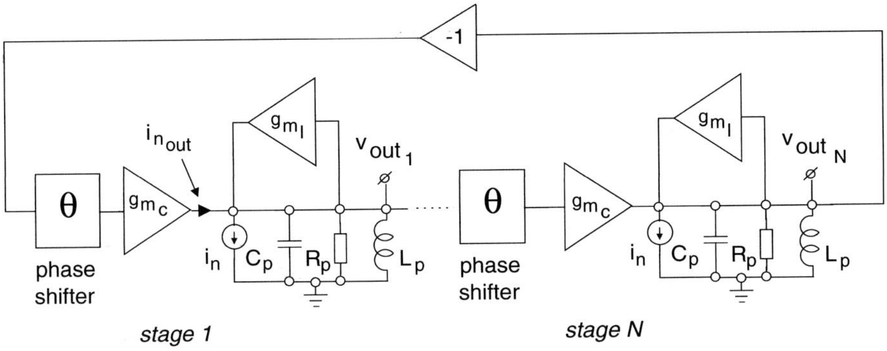

Figure I.2 Linear behavioral model of an N-stage LC oscillator.

squared closed-loop gain. Therefore, $\overline{i^2_{n_{out}}}$ can be calculated with

$$\overline{i^2_{n_{out}}}(j(f_{osc}+\Delta f)) = \frac{\overline{i^2_n}}{4Q_p^2}\left(\frac{f_{osc}}{\Delta f}\right)^2. \tag{I.7}$$

In accordance with the definition of $\mathcal{L}(f_m)$ in (4.3), the phase noise to carrier ratio of a linear single-phase LC oscillator can be written as

$$\mathcal{L}_{LC}(f_m) = \frac{1}{2}\cdot\frac{1}{4Q_p^2}\left(\frac{f_{osc}}{f_m}\right)^2\frac{\overline{i^2_n}}{i^2_{carrier}}, \tag{I.8}$$

in which f_m $(=\Delta f)$ is the offset frequency from the carrier and $i^2_{carrier}$ the squared rms carrier current. Note the factor $1/2$ in (I.8). In accordance with the convention adopted in this book (see p. 72), $\mathcal{L}(f_m)$ only takes into account the phase noise at the oscillator output, whereas $\overline{i^2_n}$ is responsible for both phase and amplitude noise [21].

I.2 Multi-phase LC oscillators

The quality factor and $\mathcal{L}(f_m)$ of an N-stage LC oscillator can be calculated using the oscillator model shown in Figure I.2. This model was introduced and discussed in Chapter 6. Application of the system level theory discussed in Section 3.4.1, and in particular application of (3.6)[3] on p. 59 leads to the following expression for $\mathcal{L}(f_m)$ of a multi-phase oscillator with N-stages

$$\mathcal{L}_{NLC}(f_m) = \frac{1}{2}\cdot\frac{1}{4\,Q_{NLC}^2}\left(\frac{f_{osc}}{f_m}\right)^2\cdot\frac{N\cdot\overline{i^2_n}}{i^2_{carrier}}, \tag{I.9}$$

[3]With Y^2_{out} replaced by $\overline{i^2_{n_{out}}}$ and $\overline{i^2_n}$ substituted for X^2_{in}.

where $i^2_{carrier}$ is the squared rms carrier current and Q_{NLC} is defined as (see p. 61)

$$Q_{NLC}{}_{|\omega=\omega_{osc}} = N \cdot \frac{\omega_{osc}}{2} \sqrt{\left(|H(j\omega)|^{N-1} \cdot \frac{d|H(j\omega)|}{d\omega}\right)^2 + \left(\frac{d\arg(H(j\omega))}{d\omega}\right)^2}, \qquad (I.10)$$

in which $H(j\omega)$ is the transfer function of one stage of the N-stage LC oscillator in Figure I.2. Note the extra factor $1/2$ in (I.9), which was explained in the previous section, and the factor N in the numerator, resulting from the presence of N noise sources.

In order to derive Q_{NLC} as a function of N and the model parameters of one LC stage, $|H(j\omega)|$ and the derivatives of $|H(j\omega)|$ and $\arg(H(j\omega))$ need to be calculated. In Figure I.2, several feedback loops can be identified. The main feedback is the connection between the stages by the transconductances g_{m_c}. In addition, a local feedback loop is present in each stage, implemented by transconductance g_{m_l}. Transconductance g_{m_l} effectively implements a negative resistance in parallel with R_p. If g_{m_l} is equal to $1/R_p$, the "electronic quality factor" of each parallel LCR circuit in Figure I.2 is infinity (the power dissipated in the resistor is completely supplied by g_{m_l}). In order for Q_{NLC} to be a measure of energy preservation in a multi-phase LC oscillator, and model the phase noise dependency on the quality factor accurately, g_{m_l} must be set to zero (removing the internal energy source in each stage).

With transconductance g_{m_l} set to zero, the absolute value of the transfer function of one stage of the model in Figure I.2 becomes

$$|H(j\omega)| = \frac{g_{m_c} R_p}{\sqrt{1+\nu^2 Q_p^2}}, \qquad (I.11)$$

with ν and Q_p defined in (I.2) and (I.3), respectively. The argument, $\arg(H(j\omega))$, then is

$$\arg(H(j\omega)) = -\arctan(\nu Q_p) + \theta, \qquad (I.12)$$

with θ the phase shift of the phase shifter in a stage. The derivatives of (I.11) and (I.12) are

$$\frac{d|H(j\omega)|}{d\omega} = \frac{g_{m_c} R_p Q_p^2 (\omega^4 - \omega_{LC}^4)}{\omega(\omega^2 \omega_{LC}^2 + Q_p^2 (\omega^2 - \omega_{osc_{LC}}^2)^2)} \times \frac{1}{\sqrt{1+Q_p^2(-2+(\omega/\omega_{LC})^2+(\omega_{LC}/\omega)^2)}}, \qquad (I.13)$$

and

$$\frac{d\arg(H(j\omega))}{d\omega} = \frac{Q_p\, \omega_{LC}(\omega^2+\omega_{LC}^2)}{\omega^2\, \omega_{LC}^2 + Q_p^2 (\omega^2 - \omega_{LC}^2)^2}, \qquad (I.14)$$

with $\omega_{LC} = 1/\sqrt{C_p L_p}$.

Substitution of the amplitude condition for oscillation $g_{m_c} = 1/(\cos(\phi_{res}) R_p)$, ω_{osc} equals ω_{NLC} (see (5.11) on p. 95), (I.11), (I.13) and (I.14) in (I.10), followed by computer assisted simplification, yields the following expression (continued on the next page) for Q_{NLC} as a function of the resonator phase shift ϕ_{res},

$$Q_{NLC}(\phi_{res}) = \frac{N\cos(\phi_{res})\left(\tan(\phi_{res}) + \sqrt{4{Q_p}^2 + \tan(\phi_{res})^2}\right)\sqrt{\left(4{Q_p}^2 + \tan(\phi_{res})^2\right)}}{4\sqrt{2}{Q_p}^2}$$

$$\times \sqrt{\left(2{Q_p}^2 + \tan(\phi_{res})^2 - \tan(\phi_{res})\sqrt{4{Q_p}^2 + \tan(\phi_{res})^2}\right)}, \qquad \text{(I.15)}$$

which by good approximation is equal to,

$$Q_{NLC}(\phi_{res}) \approx N \cdot Qp \cdot \cos(\phi_{res}). \qquad \text{(I.16)}$$

Equation (I.16) has an error less than 1% of (I.15) for a resonator phase shift ϕ_{res} ranging from 0° to 70°. At 80° and 89° the error is 3.8% and 70% respectively. Note that for zero resonator phase shift, the approximation and the exact Q_{NLC} reduce to the maximum of Q_{NLC}, which is NQ_p.

In conclusion, $\mathcal{L}(f_m)$ of a linear N-stage LC oscillator is approximately equal to

$$\mathcal{L}_{NLC}(f_m) \approx \frac{1}{2} \cdot \frac{1}{4N^2 Q_p^2 \cos^2(\phi_{res})} \left(\frac{f_{osc}}{f_m}\right)^2 \cdot \frac{N \cdot \overline{i_n^2}}{i_{carrier}^2}. \qquad \text{(I.17)}$$

J

Q and $\mathcal{L}(f_m)$ of linear ring oscillators

RING oscillators, can work in linear mode and, theoretically, generate pure sine waves (see Chapter 5). Under the same conditions[1] as were mentioned for LC oscillators in Appendix I, the phase noise[2] of in linear ring oscillator models can be described by the theory presented in Section 3.4.1 on p. 58. Application of this theory leads to the formulation of the quality factor and $\mathcal{L}(f_m)$ of N-stage ring oscillators in Section J.2 of this appendix. The expression for the closed-loop gain in two-integrator oscillators is transparent enough to derive $\mathcal{L}(f_m)$, without making use of the system level approach presented in Section 3.4.1, as will be shown in the next section.

J.1 The two-integrator oscillator

Consider the behavioral model of the two-integrator oscillator that is shown in Figure J.1, and will be used to calculate $\mathcal{L}(f_m)$ of two-integrators. If transconductance g_{m_a} is made equal to $1/R$, g_{m_a} exactly compensates the losses modeled by R, and a true two-integrator oscillator is obtained.

The oscillation frequency of the two-integrator oscillator was derived in Chapter 5 and is

$$\omega_{two-int} = \frac{g_{m_t}}{C}. \tag{J.1}$$

The absolute value of closed-loop gain of the two-integrator comes to

$$|H_{cl}(j\omega)| = \frac{g_{m_t}^2}{g_{m_t}^2 - (\omega C)^2}. \tag{J.2}$$

[1]An AGC is present, and the oscillator can be considered to be an LTI system.

[2]That is, the $1/f^2$ part of the phase noise sidebands.

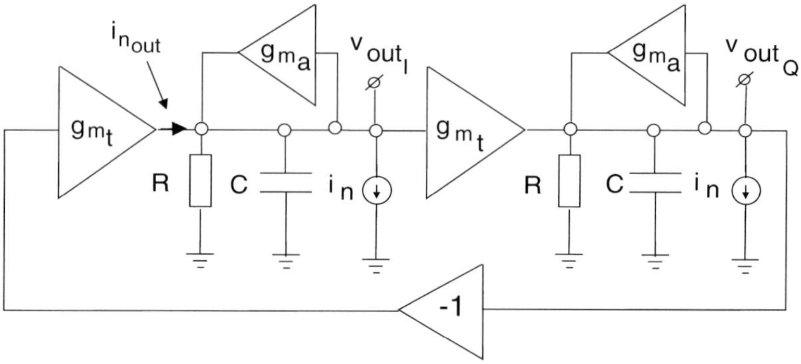

Figure J.1 Linear behavioral model of the two-integrator oscillator.

For phase-noise-to-carrier ratio calculations, the transfer function close to the carrier ω_{osc} is of interest. Therefore ω in (J.2) is replaced with $\omega_{osc} + \Delta\omega$,

$$|H_{cl}(\omega_{osc}+\Delta\omega)| = \frac{g_{m_t}^2}{g_{m_t}^2 - (\omega_{osc}C)^2 - 2\omega_{osc}\Delta\omega C^2 - (\Delta\omega C)^2}. \tag{J.3}$$

Since ω_{osc} is equal to $\omega_{two-int} = gm_t/C$, (J.3) simplifies to

$$|H_{cl}(\omega_{osc}+\Delta\omega)| = \frac{g_{m_t}^2}{-2\omega_{osc}\Delta\omega C^2 - (\Delta\omega C)^2}. \tag{J.4}$$

Provided that $(\Delta\omega C)^2 \ll 2\omega_{osc}\Delta\omega C^2$, (J.4) can be approximated by

$$|H_{cl}(\omega_{osc}+\Delta\omega)| = \frac{g_{m_t}^2}{-2\omega_{osc}\Delta\omega C^2}. \tag{J.5}$$

To define the quality factor of a two-integrator oscillator, (J.5) is rewritten as

$$|H_{cl}(\omega_{osc}+\Delta\omega)| = \frac{g_{m_t}^2}{2\omega_{osc}^2 C^2}\frac{-\omega_{osc}}{\Delta\omega}. \tag{J.6}$$

With $Q_{two-int}$ defined as $\omega_{osc}^2 C^2/g_{m_t}^2$, the squared output noise current density $\overline{i_{n_{out}}^2}$ of the two-integrator oscillator becomes

$$\overline{i_{n_{out}}^2}(\omega_{osc}+\Delta\omega) = |H_{cl}|^2 \cdot 2 \cdot \overline{i_n^2} = \frac{1}{4\,Q_{two-int}^2}\left(\frac{\omega_{osc}}{\Delta\omega}\right)^2 \cdot 2 \cdot \overline{i_n^2}. \tag{J.7}$$

The factor two in front of $\overline{i_n^2}$ stems from the two identical noise sources in the oscillator model. When the oscillation frequency (given by (J.1)) is substituted in $Q_{two-int}$, it becomes evident that the quality factor of a two-integrator oscillator is equal to unity.

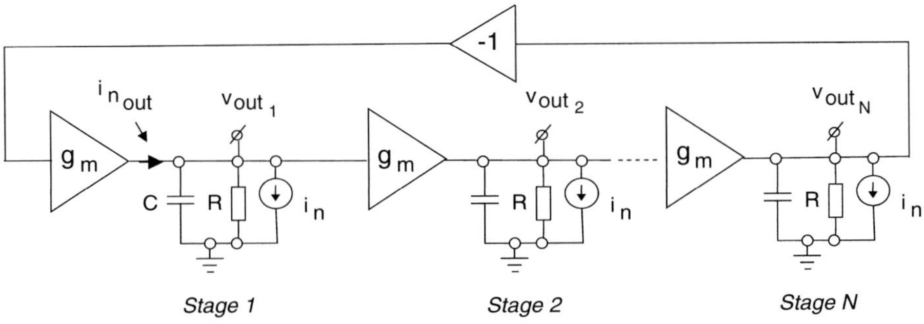

Figure J.2 Linear behavioral model of an N-stage ring oscillator.

Division of (J.7) by the squared rms carrier current $i^2_{carrier}$ yields $\mathcal{L}(f_m)$ of the two-integrator oscillator

$$\mathcal{L}_{two-int}(f_m) = \frac{1}{2} \cdot \frac{1}{4\,Q^2_{two-int}} \left(\frac{f_{osc}}{f_m}\right)^2 \cdot \frac{2 \cdot \overline{i_n^2}}{i^2_{carrier}}, \tag{J.8}$$

where $f_m = \Delta\omega/(2\pi)$. An extra factor 1/2 is included in (J.8) since $\mathcal{L}_{two-int}(f_m)$ only takes into account the phase noise, whereas $\overline{i_n^2}$ is responsible for both phase and amplitude noise.

J.2 N-stage ring oscillators

Consider the N-stage ring oscillator model presented in Figure J.2. The oscillation frequency of this model is defined for $N \geq 3$. For an N-stage ring oscillator the absolute value of the transfer function $|H(j\omega)|$ of one stage, is equal to

$$|H(j\omega)| = A = \frac{g_m R}{\sqrt{1 + (\omega CR)^2}}, \tag{J.9}$$

and the argument $\arg(H(j\omega))$ of a single stage is

$$\arg(H(j\omega)) = \phi = -\arctan(\omega CR). \tag{J.10}$$

To derive an expression for the quality factor of an N-stage ring oscillator, $dA/d\omega$ and $d\phi/d\omega$ need to be calculated (see the quality factor definition (3.12) on p. 61),

$$\frac{dA}{d\omega}\Big|_{\omega=\omega_{osc}} = \frac{-\alpha^2}{(1+\alpha^2)}\frac{1}{\omega_{osc}}, \tag{J.11}$$

$$\frac{d\phi}{d\omega}\Big|_{\omega=\omega_{osc}} = \frac{-\alpha}{(1+\alpha^2)}\frac{1}{\omega_{osc}}, \tag{J.12}$$

with α equal to

$$\alpha = \tan(\frac{\pi}{N}). \tag{J.13}$$

Substitution of (J.9), (J.11), (J.12) and (J.13) in the quality factor definition of (3.12) leads, after simplification, to the quality factor of an N-stage ring oscillator

$$Q_{Nring} = \frac{1}{2} N \sin\left(\frac{\pi}{N}\right). \tag{J.14}$$

When N goes to infinity, Q_{Nring} reaches its maximum: $\pi/2$. It is interesting to note that an effective quality factor for a switching ring oscillator, derived by modeling the ring oscillator with a delay line, was also found to be $\pi/2$ [182]. Application of (3.6), with the same considerations as were made in Section I.2 (p. 283), gives the expression for $\mathcal{L}_{Nring}(f_m)$

$$\mathcal{L}_{Nring}(f_m) = \frac{1}{2} \cdot \frac{1}{4\, Q_{Nring}^2} \left(\frac{f_{osc}}{f_m}\right)^2 \cdot \frac{N \cdot \overline{i_n^2}}{i_{carrier}^2}. \tag{J.15}$$

References

[1] C. Huygens, *Horologium Oscillatorium sive De Motu Pendulorum ad Horologia aptato Demonstrationes Geometricae*, Published in Paris, 1673.

[2] G. Hopkins, *Experimental Science*, Lindsay Publications Inc., 1987.

[3] C. S. Vaucher, *Architectures for RF frequency synthesizers*, Ph.D. thesis, University of Technology Twente, 2001.

[4] R. E. Ziemer and W. H. Tranter, *Principles of Communication*, Houghton Mifflin Company, Boston, 3rd edition, 1990.

[5] J. W. S. Rayleigh, *The Theory of Sound*, Dover, New York, 2nd edition, 1894.

[6] B. van der Pol, "The nonlinear theory of electric oscillations," *Proceedings of the IRE*, vol. 22, no. 9, pp. 1051–1086, Sept. 1934.

[7] H. von Barkhausen, *Lehrbuch der Elektronen-Röhren, Band 3, Rückkopplung*, Verlag S. Hirzel, 1935.

[8] N. Nyquist, "Regeneration Theory," in *Bell Systems Technical Journal*, 1932, pp. 126–147.

[9] J. Groszkowski, *Frequency of Self-Oscillations*, Pergamon Press, New York, 1964.

[10] N. Kryloff and N. Bogoliuboff, *Introduction to Non-linear Mechanics*, Princeton University Press, Princeton, 1943.

[11] H. Bode, *Network Analysis and Feedback Amplifier Design*, D. van Nostrand Company, Inc., 1945.

[12] W. A. Edson, "Noise in oscillators," *Proceedings of the IRE*, vol. 48, no. 8, pp. 1454–1466, Aug. 1960.

[13] E. Hafner, "The Effects of Noise in Oscillators," *Proceedings of the IEEE*, vol. 54, no. 2, pp. 179–198, Feb. 1966.

[14] D. B. Leeson, "A simple model of feedback oscillator noise spectrum," *Proceedings of the IEEE*, pp. 329–330, Feb. 1966.

[15] A. Blaquière, *Nonlinear System Analysis*, Academic Press, New York, 1966.

[16] K. Kurokawa, "Noise in synchronized oscillators," *IEEE Transaction on microwave theory and techniques*, vol. 16, no. 4, pp. 234–240, Apr. 1968.

[17] A. Takaoka and K. Ura, "Noise analysis of nonlinear feedback oscillator with AM-PM conversion coefficient," *IEEE Transaction on microwave theory and techniques*, vol. 28, no. 6, pp. 654–662, June 1980.

[18] A. A. Abidi, *Effects of random and periodic excitations on relaxation oscillators*, Ph.D. thesis, University of California, Berkeley, 1981.

[19] A. Abidi *et al.*, "Noise in relaxation oscillators," *IEEE Journal of Solid-State Circuits*, vol. 18, no. 6, pp. 794–802, June 1983.

[20] B. D. Bates and P. J. Kahn, "Stability of Multi-frequency Negative-Resistance Oscillators," *IEEE Transaction on microwave theory and techniques*, pp. 1310–1318, Oct. 1984.

[21] W. P. Robins, *Phase Noise in Signal Sources*, 9. IEE Telecomm., London, 2nd edition, 1996.

[22] A. N. Riddle, *Oscillator noise: Theory and characterization*, Ph.D. thesis, North Carolina State University, 1986.

[23] E. Vittoz *et al.*, "High-performance crystal oscillator circuits: Theory and Application," *IEEE Journal of Solid-State Circuits*, vol. 23, no. 3, pp. 774–783, Mar. 1988.

[24] C. J. M. Verhoeven, *First-order Oscillators*, Ph.D. thesis, Delft University of Technology, 1990.

[25] C. J. M. Verhoeven, "A New model for regenerative circuits," in *Midwest symposium on circuits and systems*, 1988, pp. 631–634.

[26] B. Razavi, "A study of phase noise in CMOS oscillators," *IEEE Journal of Solid-State Circuits*, vol. 31, no. 3, pp. 331–343, Mar. 1996.

[27] J. A. McNeill, *Jitter in Ring Oscillators*, Ph.D. thesis, Boston University, 1994.

[28] J. McNeill, "Jitter in Ring Oscillators," *IEEE Journal of Solid-State Circuits*, vol. 32, no. 6, pp. 870–879, June 1997.

[29] A. Hajimiri *et al.*, "A General Theory of Phase Noise in Electrical oscillators," *IEEE Journal of Solid-State Circuits*, vol. 33, no. 2, pp. 179–194, Feb. 1998.

[30] A. Hajimiri *et al.*, "Corrections to "A General Theory of Phase Noise in Electrical Oscillators"," *IEEE Journal of Solid-State Circuits*, vol. 33, no. 6, June 1998.

[31] C. Samori *et al.*, "Spectrum folding and phase noise in LC tuned oscillators," *IEEE Transactions on Circuits and Systems II*, vol. 45, no. 7, pp. 781–789, July 1998.

[32] H. Darabi and A. A. Abidi, "Noise in RF-CMOS Mixers: A Simple Physical Model," *IEEE Journal of Solid-State Circuits*, vol. 35, no. 1, pp. 15–25, Jan. 2000.

[33] J. J. Rael and A. A. Abidi, "Physical Processes of Phase Noise in Differential LC Oscillators," in *IEEE Custom Integrated Circuits Conf. (CICC)*, 2000, pp. 569–572.

[34] Q. Huang, "Phase Noise to Carrier Ratio in LC oscillators," *IEEE Transactions on Circuits and Systems I*, vol. 47, no. 7, pp. 965–980, July 2000.

[35] Q. Huang, "On the Exact Design of RF oscillators," in *IEEE Custom Integrated Circuits Conf. (CICC)*, 1998, pp. 41–44.

[36] C. Lin, *Incremental Mixed-Signal Layout Generation Concepts, Theory & Implementation*, Ph.D. thesis, Eindhoven University of Technology, 2002.

[37] F. M. Gardner, *Phase-lock Techniques*, Wiley, New York, 2nd edition, 1979.

[38] U.L. Rohde, *RF and Microwave Digital Frequency Synthesizers*, Wiley, New York, 1997.

[39] B. Razavi, *Monolithic Phase-locked Loops and Clock Recovery Circuits*, IEEE Press, New York, 1996.

[40] C. S. Vaucher and D. Kasperkovitz, "A Wide Band Tuning System for Fully Integrated Satellite Receivers," *IEEE Journal of Solid-State Circuits*, vol. 33, no. 7, pp. 987–997, July 1998.

[41] R. J. Matthys, *Crystal oscillator Circuits*, Krieger Publishing Company, 1991.

[42] M. Soyuer, "Design considerations for high-frequency crystal oscillators," *IEEE Journal of Solid-State Circuits*, vol. 26, no. 6, pp. 889–893, June 1991.

[43] E. H. Nordholt *et al.*, "Single-pin integrated crystal oscillators," *IEEE Transactions on Circuits and Systems*, vol. 37, no. 2, pp. 175–182, Feb. 1990.

[44] J. J. Goedbloed, *Electromagnetic Compatibility*, Prentice Hall International, Hertfordshire, 1992.

[45] U. Rohde, "Designing SAW Resonators and DRO Oscillators Using Nonlinear CAD tools," in *IEEE International Frequency Control Symposium (FCS)*, 1995, pp. 379–395.

[46] H. Wu and A. Hajimiri, "A 10 GHz CMOS Distributed Voltage Controlled Oscillator," in *IEEE Custom Integrated Circuits Conf. (CICC)*, 2000, pp. 581–584.

[47] C. A. M. Boon *et al.*, "Modeling the phase noise of RC multivibrators," in *Midwest symposium on circuits and systems*, 1984, pp. 421–424.

[48] B. M. Fleischer, *Jitter in relaxation oscillators*, Ph.D. thesis, Stanford University, 1989.

[49] C. J. M. Verhoeven, "A high-frequency electronically tunable quadrature oscillator," *IEEE Journal of Solid-State Circuits*, vol. 27, no. 7, pp. 1097–1100, July 1992.

[50] J. Westra, *High-Performance Oscillators and Oscillator Systems*, Ph.D. thesis, Delft University of technology, 1998.

[51] A. H. M. van Roermund, *Noise and Accuracy in Switched Capacitor Modulation Circuits*, Ph.D. thesis, Katholieke Universiteit Leuven, 1987.

[52] J. van der Tang and D. Kasperkovitz, "A 0.9-2.2 GHz Monolithic Quadrature Mixer Oscillator for Direct-Conversion Satellite Receivers," in *IEEE International Solid-State Circuits Conf. (ISSCC)*, 1997, pp. 88–89.

[53] J. van der Tang and S. Hahn, "A Monolithic 0.4 mW SOA LC Voltage-Controlled Oscillator," in *European Solid-State Circuits Conf. (ESSCIRC)*, 1999, pp. 150–153.

[54] J. van der Tang and D. Kasperkovitz, "Low Phase Noise Reference Oscillator with Integrated PMOS Varactors for Digital Satellite Receivers," *IEEE Journal of Solid-State Circuits*, vol. 35, no. 8, pp. 1238–1243, Aug. 2000.

[55] L. O. Chua, *Introduction to Nonlinear Network Theory*, McGraw-Hill Inc., New York, 1969.

[56] W. A. Edson, *Vacuum-Tube Oscillators*, John Wiley & Sons, Inc., New York, 1953.

[57] W. Chen, *Active Network and Feedback Amplifier Theory*, Hemisphere Publishing Corporation, 1980.

[58] N. Nguyen, *Monolithic microwave oscillators and amplifiers*, Ph.D. thesis, University of California, 1991.

[59] N. M. Nguyen and R. G. M. Meyer, "Start-up and Frequency Stability in High-Frequency Oscillators," *IEEE Journal of Solid-State Circuits*, vol. 27, no. 5, pp. 810–820, May 1992.

[60] Y. Sun, *Monolithically Integrated Silicon Bipolar RF Oscillators*, Ph.D. thesis, University of Technology Delft, 1999.

[61] G. Gonzales, *Microwave transistor amplifiers, analysis and design*, Prentice-Hall, 2nd edition, 1997.

[62] A. Zanchi *et al.*, "Impact of AAC Design on Phase Noise Performance of VCOs," *IEEE Transactions on Circuits and Systems I*, vol. 48, no. 6, pp. 537–547, June 2001.

[63] J. Vlach and K. Singhal, *Computer Methods for Circuit Analysis and Design*, Van Nostrand Reinhold, New York, 1983.

[64] R. Airiau *et al.*, *Circuit Synthesis with VHDL*, Kluwer Academic Publishers, Norwell, 1994.

[65] G. van der Plas *et al.*, "AMGIE—A Synthesis Environment for CMOS Analog Integrated Circuits," *IEEE Transactions on Computer-Aided Design of Integrated Circuits and Systems*, vol. 20, no. 9, pp. 1037–1058, Sept. 2001.

[66] M. Ismail and J. Franca, Eds., *Introduction to Analog VLSI Design Automation*, Kluwer Academic Publishers, Norwell, 1990.

[67] G. W. Roberts and A. S. Sedra, *Spice*, The Oxford Series in Electrical and Computer Engineering. Oxford University Press, 2nd edition, 1996.

[68] K. S. Kundert, *The Designer's Guide to SPICE & SPECTRE*, Kluwer Academic Publishers, Norwell, 1995.

[69] B. Martin, "Automation Comes to Analog," *IEEE Spectrum*, pp. 70–75, June 2001.

[70] APLAC, www.aplac.com, *APLAC User's guide Vol. I.*

[71] Philips, ED&T/Analogue Simulation, *Adapt User Manual, for Adapt 3.0*, 1st edition, 2000.

[72] S. Kirkpatrick *et al.*, "Optimization by Simulated Annealing," *Science*, vol. 220, no. 4598, pp. 671–680, May 1983.

[73] Z. Michalewicz, *Genetic Algorithms + Data Structures Evolution Programs*, Springer-Verlag, 1992.

[74] K. Madsen *et al.*, "Efficient Minimax Design of Networks without Using Derivatives," *IEEE Transaction on microwave theory and techniques*, vol. 23, no. 10, pp. 803–809, Oct. 1975.

[75] W. H. Press *et al.*, *Numerical Recipes in C*, Cambridge University Press, 1988.

[76] M. del Mar Hershenson *et al.*, "Design and optimization of LC oscillators," in *International Conference on Computer Aided Design (ICCAD)*, 1999, pp. 65–69.

[77] Cadence Design Systems, *Cadence openbook documentation system, SpectreRF*, 2000.

[78] H. Chang *et al.*, *A Top-down Constraint-driven Design Methodology for Analog Integrated Circuits*, Kluwer Academic Publishers, Norwell, 1997.

[79] L. R. Carley *et al.*, "Synthesis tools for mixed-signal ICs: Progress on front-end and back-end strategies," in *IEEE/ACM Design Automation Conf.*, 1996, pp. 298–303.

[80] D.M.W. Leenaerts, *TOPICS, A Contribution to Analog Design Automation*, Ph.D. thesis, Eindhoven University of Technology, 1992.

[81] F. El-Turky and E. E. Perry, "BLADES: An Artificial Intelligence Approach to Analog Circuit Design," *IEEE Transactions on Computer-Aided Design*, vol. 8, no. 6, pp. 680–692, June 1989.

[82] H. Koh *et al.*, "OPASYN: a compiler for CMOS operational amplifiers"," *IEEE Transactions on Computer-Aided Design*, vol. 9, no. 2, pp. 113–125, Feb. 1990.

[83] K. Lampaert *et al.*, *Analog Layout Generation for Performance and Manufacturability*, Kluwer Academic Publishers, Norwell, 1999.

[84] C. De Ranter *et al.*, "CYCLONE: Automated Design and Layout of RF LC-Oscillators," in *IEEE/ACM Design Automation Conf.*, 2000, pp. 11–14.

[85] A. van Staveren, *Structured electronic design of high-performance low-voltage references*, Ph.D. thesis, Delft University of Technology, 1997.

[86] C. A. Desoer and E. S. Kuh, *Basic Circuit Theory*, McGraw-Hill Book Company, New York, 1969.

[87] P. R. Gray and R. G. Meyer, *Analysis and Design of Analog Integrated Circuits*, John Wiley & Sons, New York, 3rd edition, 1993.

[88] D. Leenaerts, J. van der Tang, and C. Vaucher, *Circuit Design for RF Transceivers*, Kluwer Academic Publishers, Boston, 2001.

[89] B. Razavi, *Design of Analog CMOS Integrated Circuits*, McGraw-Hill, New York, 2001.

[90] F. L. Walls *et al.*, "Extending the Range and Accuracy of Phase Noise Measurements," in *IEEE International Frequency Control Symposium (FCS)*, 1988, pp. 432–441.

[91] S. L. J. Gierkink, *Control Linearity and Jitter of Relaxation Oscillators*, Ph.D. thesis, Twente, University of Technology, 1999.

[92] R. Poore, *Phase Noise and Jitter*, Agilent EEsof EDA, 2001, May.

[93] N. Schmitt and D. Grant, *Jitter Measurements of Agilent Technology OC-48 Optical Transceivers using the OmniBER718*, Agilent Technologies, 2001.

[94] J. van der Tang and C. Vaucher, "Design and Optimization of a Low Jitter Clock-Conversion PLL for SONET/SDH Optical Transmitters," in *IEEE International Conference on Electronics, Circuits and Systems (ICECS)*, 2001, vol. I, pp. 31–34.

[95] Maxim Integrated Products, *Converting between RMS and Peak-to-Peak Jitter at a Specified BER*, 2000, Application Note HFAN-4.0.2.

[96] K. Sam Shanmugam, *Digital and Analog Communication Systems*, John Wiley & Sons, New York, 1985.

[97] A. A. Abidi, "Direct-conversion radio transceivers for digital communications," in *IEEE International Solid-State Circuits Conf. (ISSCC)*, 1995, pp. 186–187.

[98] B. Razavi, "Design Considerations for Direct-Conversion Receivers," *IEEE Transactions on Circuits and Systems II*, vol. 44, no. 6, pp. 428–435, June 1997.

[99] K. R. Laker and W. M. C. Sansen, *Design of Analog Integrated Circuits and Systems*, McGRAW-Hill, Inc., New York, 1994.

[100] C. A. M. Boon, *Design of High-Performance Negative-Feedback Oscillators*, Ph.D. thesis, Delft University of Technology, 1989.

[101] R. J. Baker *et al.*, *CMOS Circuit Design, Layout, and Simulation*, IEEE press, New York, 1998.

[102] C. P. Yue *et al.*, "A physical model for planar spiral inductors on silicon," in *International Electron Devices Meeting (IEDM)*, 1996, pp. 155–158.

[103] H. Wang, "A 50 GHz VCO in 0.25 um CMOS," in *IEEE International Solid-State Circuits Conf. (ISSCC)*, 2001, pp. 372–373.

[104] A. Hajimiri and T. H. Lee, *Low Noise Oscillators*, Kluwer Academic Publishers, 1999.

[105] H. Wang, "Comments on "Design Issues in CMOS Differential LC Oscillators"," *IEEE Journal of Solid-State Circuits*, vol. 35, no. 2, pp. 286–287, Feb. 2000.

[106] P. van de Ven, J. van der Tang, D. Kasperkovitz, and A. van Roermund, "An optimally coupled 5 GHz quadrature LC oscillator," in *Symposium on VLSI Circuits*, 2001, pp. 115–118.

[107] A. Rofougaran *et al.*, "A Single-Chip 900 MHz Spread-Spectrum Wireless Transceiver in 1-um CMOS–Part I: Architecture and Transmitter Design," *IEEE Journal of Solid-State Circuits*, vol. 33, no. 4, pp. 515–534, Apr. 1998.

[108] J. van der Kreek, *Elektrische netwerken*, Educaboek - Stam Technische Boeken, Culemborg, 2nd edition, 1983.

[109] A. Buchwald and K. Martin, *Integrated Fiber-Optic Receivers*, Kluwer Academic Publishers, Norwell, 1995.

[110] M. Alioto and G. Palumbo, "Oscillation Frequency in CML and ESCL Ring Oscillators," *IEEE Transactions on Circuits and Systems I*, vol. 48, no. 2, pp. 210–214, Feb. 2001.

[111] R. Wassenaar, *Analysis of Analog CMOS Circuits*, Ph.D. thesis, Twente University of Technology, 1996.

[112] M. Alioto and G. Palumbo, "CML and ECL: Optimized Design and Comparison," *IEEE Transactions on Circuits and Systems I*, vol. 46, no. 11, pp. 1330–1341, Nov. 1999.

[113] U. L. Rohde and P. Newkirk D, *RF/Microwave Circuit Design for Wireless Applications*, John Wiley & Sons, Inc., New York, 2000.

[114] J. Rabaey, *Digital Integrated Circuit, A Design Perspective*, Prentice-Hall, Englewood Cliffs, 1996.

[115] P. Andreani and S. Mattisson, "On the Use of MOS Varactors in RF VCO's," *IEEE Journal of Solid-State Circuits*, vol. 35, no. 6, pp. 905–910, June 2000.

[116] T. Soorapanth *et al.*, "Analysis and Optimization of Accumulation-Mode Varactor for RF ICs," in *Symposium on VLSI Circuits*, 1998, pp. 32–33.

[117] J. Maget *et al.*, "A Physical Model of CMOS Varactor with High Capacitance Tuning Range and its Application to Simulate a Voltage Controlled Oscillator," in *International Semiconductor Device Research Symposium*, 2001, pp. 609–612.

[118] A. Porret *et al.*, "Design of High-Q Varactors for Low-Power Wireless Applications Using a Standard CMOS Process," *IEEE Journal of Solid-State Circuits*, vol. 35, no. 3, pp. 337–345, Mar. 2000.

[119] R. Casetllo *et al.*, "A ±30% Tuning Range Varactor Compatible with future Scaled Technologies," in *Symposium on VLSI Circuits*, 1998, pp. 34–35.

[120] J. van der Tang, D. Kasperkovitz, H. Rumpt, and A van Roermund, "A Cost-Effective Multi-band LC Oscillator for Low-IF FM Radio Receivers," in *European Solid-State Circuits Conf. (ESSCIRC)*, 2002, pp. 819–822.

[121] N. Sneed, "A 2-GHz CMOS LC-Tuned VCO using Switched-Capacitors to Compensate for Bond Wire Inductance Variation," M.S. thesis, University of California, Berkeley, Dec. 2000.

[122] Q. Huang *et al.*, "Design and Implementation of a CMOS VXCO for FM Stereo Decoders," *IEEE Journal of Solid-State Circuits*, vol. 23, no. 3, pp. 784–793, June 1988.

[123] F. Herzel *et al.*, "A New Approach to Fully Integrated CMOS LC-Oscillators with a Very Large Tuning Range," in *IEEE Custom Integrated Circuits Conf. (CICC)*, 2000, pp. 573–576.

[124] J. van der Tang, D. Kasperkovitz, and A. van Roermund, "A 9.8-11.5 GHz Quadrature Ring Oscillator for Optical Receivers," *IEEE Journal of Solid-State Circuits*, vol. 37, no. 3, pp. 438–442, Mar. 2002.

[125] J. van der Tang, D. Kasperkovitz, and A. van Roermund, "A 9.8-11.5 GHz I/Q Ring Oscillator for Optical Receivers," in *IEEE Custom Integrated Circuits Conf. (CICC)*, 2001, pp. 323–326.

[126] J. van der Tang, D. Kasperkovitz, and A. van Roermund, "A 2.7 V, 8 GHz Monolithic I/Q RC Oscillator with Active Inductive Loads," in *European Solid-State Circuits Conf. (ESSCIRC)*, 2000, pp. 304–307.

[127] J. O. Voorman, *The Gyrator As a Monolithic Circuit in Electronic Systems*, Ph.D. thesis, Katholieke Universiteit Nijmegen, 1977.

[128] A. Thanachayanont and A. Payne, "CMOS floating active inductor and its application to bandpass filter and oscillator designs," *IEE Proc. Circuits, Devices and Systems*, vol. 147, no. 1, pp. 42–48, Feb. 2000.

[129] M. Gröezing *et al.*, "A 2.5 V CMOS Differential Active Inductor with Tunable L and Q For Frequencies up to 5 GHz," in *IEEE Radio Frequency Integrated Circuits Symposium (RFIC)*, 2001, pp. 271–274.

[130] A. Jayaraman *et al.*, "A Fully Integrated Broadband Direct-Conversion Receiver for DBS Applications," in *IEEE International Solid-State Circuits Conf. (ISSCC)*, 2000, pp. 140–141.

[131] A. Kral *et al.*, "RF-CMOS Oscillators with Switched Tuning," in *IEEE Custom Integrated Circuits Conf. (CICC)*, 1998, pp. 555–558.

[132] J. van der Tang, D. Kasperkovitz, and A. Bretveld, "A 65 mW, 0.4-2.3 GHz Bandpass Filter for Satellite Receivers," *Analog Integrated Circuits and Signal Processing, an international journal*, vol. 31, no. 1, pp. 5–18, Apr. 2002.

[133] F. Doorenbosch, *A Monolithically Integrated Wide-Tunable Sine Oscillator*, Ph.D. thesis, Delft University of Technology, 1982.

[134] I. Young *et al.*, "A PLL clock generator with 5 to 110 MHz lock range for microprocessors," in *IEEE International Solid-State Circuits Conf. (ISSCC)*, 1992, pp. 50–51.

[135] L. Sun *et al.*, "A 1.0V GHz Range 0.13 um CMOS Frequency Synthesizer," in *IEEE Custom Integrated Circuits Conf. (CICC)*, 2001, pp. 327–330.

[136] K. A. Kouznetsov and R. G. Meyer, "Phase Noise in LC Oscillators," *IEEE Journal of Solid-State Circuits*, vol. 35, no. 8, pp. 1244–1248, Aug. 2000.

[137] F. Herzel *et al.*, "Phase Noise in a Differential CMOS Voltage-Controlled Oscillator for RF Applications," *IEEE Transactions on Circuits and Systems II*, vol. 47, no. 1, pp. 11–15, Jan. 2000.

[138] D. Triantis *et al.*, "Thermal Noise Modeling for Short-Channel MOSFET's," *IEEE Transactions on Electron Devices*, vol. 43, no. 11, pp. 1950–1955, Nov. 1996.

[139] Y. Tsividis, *Operation and Modeling of the MOS Transistor*, McGraw-Hill, second edition, 1999.

[140] J. van der Tang, P. van de Ven, D. Kasperkovitz, and A. van Roermund, "Analysis and Design of an Optimally Coupled 5 GHz Quadrature LC Oscillator," *IEEE Journal of Solid-State Circuits*, vol. 37, no. 5, pp. 657–661, May 2002.

[141] A. Pruijmboom *et al.*, "QUBiC3: A 0.5um BiCMOS Production Technology, with fT=30 GHz, fmax=60GHz and High-Quality Passive Components for Wireless Telecommunication Applications," in *Bipolar Circuits and Technology Meeting (BCTM)*, 1998, pp. 120–123.

[142] L. Dai and R. Harjani, "Design of Low-Phase-Noise CMOS Ring Oscillators," *IEEE Transactions on Circuits and Systems I*, vol. 49, no. 5, pp. 328–338, May 2002.

[143] L. Dai and R. Harjani, *Design of High-Performance CMOS Voltage Controlled Oscillators*, Kluwer Academic Publishers, Norwell, 2002.

[144] P. Andreani and H. Sjöland, "Tail Current Noise Suppression in RF CMOS VCOs," *IEEE Journal of Solid-State Circuits*, vol. 37, no. 3, pp. 342–348, Mar. 2002.

[145] T. H. Lee and A. Hajimiri, "Oscillator Phase Noise: A Tutorial," *IEEE Journal of Solid-State Circuits*, vol. 35, no. 3, pp. 326–336, Mar. 2000.

[146] T. H. Lee, *The Design of CMOS Radio-Frequency Integrated Circuits*, Cambridge University Press, Cambridge, 1998.

[147] A. Sempel and H. van Nieuwenburg, "A fully-integrated HIFI PLL FM-demodulator," in *IEEE International Solid-State Circuits Conf. (ISSCC)*, 1990, pp. 102–103.

[148] J. van der Tang and D. Kasperkovitz, "Oscillator Design Efficiency: A New Figure of Merit for Oscillator Benchmarking," in *IEEE International Symposium on Circuits and Systems (ISCAS)*, 2000, vol. II, pp. 533–536.

[149] J. van der Tang and D. Kasperkovitz, "Fast Phase Noise Analysis Method for Noise Optimization of Oscillators," in *European Solid-State Circuits Conf. (ESSCIRC)*, 1998, pp. 504–507.

[150] "AC phase noise analysis concept," Conversation with Dr. D. Kasperkovitz, 1996, Philips Research Laboratories, Eindhoven.

[151] B. de Smedt *et al.*, "Accurate Simulation of Phase Noise in Oscillators," in *European Solid-State Circuits Conf. (ESSCIRC)*, 1997, pp. 208–211.

[152] F. Herzel, "An Analytical Model for the Power Spectral Density of a Voltage-Controlled Oscillator and Its Analogy to the Laser Linewidth Theory," *IEEE Transactions on Circuits and Systems I*, vol. 45, no. 9, pp. 904–908, Sept. 1998.

[153] A. Demir *et al.*, "Phase Noise and Timing Jitter in Oscillator," in *IEEE Custom Integrated Circuits Conf. (CICC)*, 1998, pp. 45–48.

[154] M. Randall and T. Hock, "General Oscillator Characterization Using Linear Open-Loop S-Parameters," *IEEE Transaction on microwave theory and techniques*, vol. 49, no. 6, pp. 1094–1100, June 2001.

[155] L. Cascio, "Comments on "General Oscillator Characterization Using Linear Open-Loop S-Parameters"," *IEEE Transaction on microwave theory and techniques*, vol. 50, no. 1, pp. 226–228, Jan. 2002.

[156] A. van Staveren *et al.*, *Structure Electronic Design, High-Performance Harmonic Oscillators and Bandgap References*, Kluwer Academic Publishers, Boston, 2000.

[157] A. van Staveren and C. Verhoeven, "Optimizing a resonator tap for maximizing oscillator carrier-to-noise ratio," in *IEEE International Symposium on Circuits and Systems (ISCAS)*, 2001, vol. 1, pp. 480–483.

[158] J. K. A. Everard, "Low noise oscillators: theory and design.," in *10th International Conf. on European Freq. and Time*, 1996, pp. 436–441.

[159] J. K. A. Everard, "Minimum sideband noise in oscillators," in *IEEE International Frequency Control Symposium (FCS)*, 1986, pp. 336–339.

[160] A. Wagemans *et al.*, "A 3.5 mW 2.5 GHz Diversity Receiver and a 1.2 mW 3.6 GHz VCO in Silicon-On-Anything," in *IEEE International Solid-State Circuits Conf. (ISSCC)*, 1998, pp. 250–251.

[161] Philips Semiconductors, *Universal I^2C-bus programmable RF modulator*, 1997, Datasheet TDA8822.

[162] J. van der Tang and D. Kasperkovitz, "A Low Phase Noise Reference Oscillator with Integrated PMOS Varactors for Digital Satellite Receivers," in *IEEE International Conference on Electronics, Circuits and Systems (ICECS)*, 1999, vol. I, pp. 181–184.

[163] S. Brett and G. Stanton, "A Direct Conversion L-band Tuner for Digital DBS," in *IEEE International Solid-State Circuits Conf. (ISSCC)*, 1998, pp. 126–127.

[164] K. Richter, "Zero IF Satellite Tuners for DBS," in *IEEE Conf. on Consumer Electronics (ICCE)*, 1998, pp. 70–71.

[165] P. Blaud, "A direct conversion IC for digital satellite TV," in *European Solid-State Circuits Conf. (ESSCIRC)*, 1997, pp. 124–127.

[166] J. F. Wilson *et al.*, "A single-chip VHF and UHF receiver for radio paging," *IEEE Journal of Solid-State Circuits*, vol. 26, no. 12, pp. 1944–1950, Dec. 1991.

[167] EUTALSAT, *Earth Station Standard EESS 500*, 3 edition, Aug. 1996.

[168] J. de Jong *et al.*, "Single Polysilicon-Layer Advanced Super High Speed BiCMOS Technology," in *Bipolar Circuits and Technology Meeting (BCTM)*, 1989.

[169] K. Hadjizada *et al.*, "TV and TVSAT Mixer-oscillator PLL ICs," *IEEE Transactions on Consumer Electronics*, vol. 41, no. 3, pp. 942–945, Mar. 1995.

[170] R. Velghe, "Compact MOS modelling for analogue circuit simulation," in *European Solid-State Device Research Conf. (ESSDERC)*, 1994, pp. 833–836.

[171] J. Hauenschild *et al.*, "A plastic packaged 10 Gb/s BiCMOS Clock and Data Recovering 1:4-demultiplexer with external VCO," *IEEE Journal of Solid-State Circuits*, vol. 31, no. 12, pp. 2056–2059, Dec. 1996.

[172] H. de Graaf *et al.*, "Experience with the new compact MEXTRAM model for bipolar transistors," in *Bipolar Circuits and Technology Meeting (BCTM)*, 1989, pp. 246–249.

[173] B. Razavi, *RF Microelectronics*, Prentice Hall, New York, 1998.

[174] J. Crols and M. Steyeart, "A Single-Chip 900 MHz CMOS Receiver Front-End with a High Performance Low-IF Topology," *IEEE Journal of Solid-State Circuits*, vol. 30, no. 12, pp. 1483–1492, Dec. 1995.

[175] J. Fenk *et al.*, "An RF Front-end for Digital Mobile Radio," in *Bipolar Circuits and Technology Meeting (BCTM)*, 1990, pp. 244–247.

[176] F. Behbahani *et al.*, "An Adjustable Bipolar Quadrature LO Generator with an Improved Divide-by-2 stage," in *Bipolar Circuits and Technology Meeting (BCTM)*, 1996, pp. 157–160.

[177] S. Navid *et al.*, "Level-locked loop, A Technique for Broadband Quadrature Signal Generation," in *IEEE Custom Integrated Circuits Conf. (CICC)*, 1997, pp. 411–414.

[178] D. Lovelace and J. Durec, "A Self Calibrating Quadrature Generator with Wide Frequency Range," in *IEEE Radio Frequency Integrated Circuits Symposium*, 1997, pp. 147–150.

[179] C. Park *et al.*, "A 1.8 GHz Self-Calibrating Phase-Locked Loop with Precise I/Q Matching," *IEEE Journal of Solid-State Circuits*, vol. 36, no. 5, pp. 777–783, May 2001.

[180] S. Nahm *et al.*, "A Cordic-Based Digital Quadrature Mixer: Comparison with A ROM-based Architecture," in *IEEE International Symposium on Circuits and Systems (ISCAS)*, 1998, vol. IV, pp. 385–388.

[181] J. H. Bentley and K. M. Hess, *A Programmed Review For Electrical Engineering*, Van Nostrand Reinhold Company, New York, 1978.

[182] A. Abidi, "How Phase Noise Appears in Oscillators," in *Workshop on Advances in Analog Circuit Design (AACD)*, 1997.

Literature on LC oscillator designs

[183] D. Ham and A. Hajimiri, "Concepts and Methods in Optimization of Integrated LC VCOs," *IEEE Journal of Solid-State Circuits*, vol. 36, no. 6, pp. 896–909, June 2001.

[184] A. Rofougaran *et al.*, "A 900 MHz CMOS LC-Oscillator with Quadrature Outputs," in *IEEE International Solid-State Circuits Conf. (ISSCC)*, 1996, pp. 392–393.

[185] J. Kim and B. Kim, "A Low-Phase-Noise CMOS LC Oscillator with a Ring Structure," in *IEEE International Solid-State Circuits Conf. (ISSCC)*, 2000, pp. 430–431.

[186] P. Andreani, "A Low-Phase-Noise Low-Phase-Error 1.8GHz Quadrature CMOS VCO," in *IEEE International Solid-State Circuits Conf. (ISSCC)*, 2002, pp. 290–291.

[187] E. Hegazi *et al.*, "A Filtering Technique to Lower Oscillator Phase Noise," in *IEEE International Solid-State Circuits Conf. (ISSCC)*, 2001, pp. 364–365.

[188] W. Chen and J. Wu, "A 2-V 2-GHz BJT Variable Frequency Oscillator," *IEEE Journal of Solid-State Circuits*, vol. 33, no. 9, pp. 1406–1410, Sept. 1998.

[189] J. van der Tang and D. Kasperkovitz, "Low Phase Noise Reference Oscillator with Integrated PMOS Varactors for Digital Satellite Receivers," *IEEE Journal of Solid-State Circuits*, vol. 35, no. 8, pp. 1238–1243, Aug. 2000.

[190] P. Vancorenland and M. Steyeart, "A 1.57 GHz Fully integrated Very Low Phase Noise Quadrature VCO," in *Symposium on VLSI Circuits*, 2001, pp. 111–114.

[191] J. Kucera and B. Klepser, "3.6 GHz VCOs for multi-band GSM transceivers," in *European Solid-State Circuits Conf. (ESSCIRC)*, 2001.

[192] R. Aparicio and A. Hajimiri, "A CMOS Differential Noise-Shifting Colpitts VCO," in *IEEE International Solid-State Circuits Conf. (ISSCC)*, 2002, pp. 288–289.

[193] M. Tiebout, "Low-Power Low-Phase-Noise Differentially Tuned Quadrature VCO Design in Standard CMOS," *IEEE Journal of Solid-State Circuits*, vol. 36, no. 7, pp. 1018–1024, July 2001.

[194] M. Soyuer *et al.*, "An 11-GHz 3-V siGe Voltage Controlled Oscillator with Integrated Resonator," *IEEE Journal of Solid-State Circuits*, vol. 32, no. 9, pp. 1451–1454, Sept. 1997.

[195] J. Plouchart *et al.*, "Fully-Monolithic 3V SiGe Differential Voltage-Controlled Oscillators for 5GHz and 17GHz Wireless Applications," in *European Solid-State Circuits Conf. (ESSCIRC)*, 1998, pp. 332–335.

[196] P. Kinget, "A Fully Integrated 2.7V 0.35um CMOS VCO for 5 GHz Wireless Applications," in *IEEE International Solid-State Circuits Conf. (ISSCC)*, 1998, pp. 226–227.

[197] M. Soyeur *et al.*, "A 3V 4GHz nMOS Voltage-Controlled Oscillator with Integrated Resonator," in *IEEE International Solid-State Circuits Conf. (ISSCC)*, 1996, pp. 394–395.

[198] M. A. T. Sanduleanu and J. P. Frambach, "1 GHz Tuning Range, Low Phase Noise, LC Oscillator with Replica Biasing Common-Mode Control and Quadrature Outputs," in *European Solid-State Circuits Conf. (ESSCIRC)*, 2001.

[199] J. Craninckx and M. Steyaert, "A CMOS 1.8 GHz Low-Phase-Noise Voltage-Controlled Oscillator with Prescaler," in *IEEE International Solid-State Circuits Conf. (ISSCC)*, 1995, pp. 266–267.

[200] J. van der Tang and S. Hahn, "A Monolithic 0.4 mW SOA LC Voltage-Controlled Oscillator," in *European Solid-State Circuits Conf. (ESSCIRC)*, 1999, pp. 150–153.

[201] M. Soyuer *et al.*, "A 2.4-GHz Silicon Bipolar Oscillator with Integrated Resonator," *IEEE Journal of Solid-State Circuits*, vol. 31, no. 2, pp. 268–270, Feb. 1996.

[202] B. Razavi, "A 1.8GHz CMOS Voltage-Controlled Oscillator," in *IEEE International Solid-State Circuits Conf. (ISSCC)*, 1997, pp. 388–389.

[203] N. M. Nguyen and R. G. Meyer, "A 1.8 GHz Monolithic LC Voltage-Controlled Oscillator," in *IEEE International Solid-State Circuits Conf. (ISSCC)*, 1992, pp. 158–159.

[204] J. J. Kucera, "Wideband BiCMOS VCO for GSM/UMTS Direct Conversion Receivers," in *IEEE International Solid-State Circuits Conf. (ISSCC)*, 2001, pp. 374–375.

[205] F. Svelto *et al.*, "A 1.3GHz CMOS VCO with 28% frequency tuning," in *IEEE Custom Integrated Circuits Conf. (CICC)*, 1999, pp. 645–648.

[206] L. Dauphinee *et al.*, "A Balanced 1.5 GHz Voltage Controlled Oscillator with an Integrated LC Resonator," in *IEEE International Solid-State Circuits Conf. (ISSCC)*, 1997, pp. 390–391.

[207] T. Ahrens and T. Lee, "A 1.4-GHz 3-mW CMOS LC Low Phase Noise VCO Using Tapped Bond Wire Inductances," in *International Symposium on Low-Power Electronics and Design*, 1998, pp. 16–19.

[208] P. G. M. Baltus *et al.*, "A 3.5 mW, 2.5 GHz Diversity Receiver and a 1.2 mW, 3.6 GHz VCO in Silicon on Anything," *IEEE Journal of Solid-State Circuits*, vol. 33, no. 12, pp. 2074–2075, Dec. 1998.

[209] J. Lin, "An Integrated Low-Phase-Noise Voltage Controlled Oscillator for Base Station Applications," in *IEEE International Solid-State Circuits Conf. (ISSCC)*, 2000, pp. 432–433.

[210] B. De Muer *et al.*, "A 1.8 GHz highly-tunable low-phase-noise CMOS VCO," in *IEEE Custom Integrated Circuits Conf. (CICC)*, 2000, pp. 585–588.

[211] T. Lui, "A 6.5 GHz Monolithic CMOS Voltage-Controlled Oscillator," in *IEEE International Solid-State Circuits Conf. (ISSCC)*, 1999, pp. 404–405.

[212] C. R. C. De Ranter and M. Steyaert, "A 0.25um CMOS 17GHz VCO," in *IEEE International Solid-State Circuits Conf. (ISSCC)*, 2001, pp. 370–371.

[213] H. Wang, "A 9.8GHz Back-Gate Tuned VCO in 0.35um CMOS," in *IEEE International Solid-State Circuits Conf. (ISSCC)*, 1999, pp. 406–407.

[214] A. Hajimiri and T. Lee, "Design Issues in CMOS Differential LC Oscillators," *IEEE Journal of Solid-State Circuits*, vol. 34, no. 5, pp. 717–724, May 1999.

[215] A. Kral *et al.*, "RF-CMOS Oscillators with Switched Tuning," in *IEEE Custom Integrated Circuits Conf. (CICC)*, 1998, pp. 555–558.

[216] B. Jansen *et al.*, "Silicon Bipolar VCO Family for 1.1 to 2.2GHz with Fully-Integrated Tank and Tuning Circuits," in *IEEE International Solid-State Circuits Conf. (ISSCC)*, 1997, pp. 392–393.

[217] B. de Muer *et al.*, "A 2-GHz Low-Phase-Noise Integrated LC-VCO set with Flicker-Noise Upconversion Minimization," *IEEE Journal of Solid-State Circuits*, vol. 35, no. 7, pp. 1034–1038, July 2000.

[218] M. Zannoth *et al.*, "A Fully Integrated VCO at 2 GHz," in *IEEE International Solid-State Circuits Conf. (ISSCC)*, 1998, pp. 224–225.

[219] J. Craninckx and M. S. J. Steyaert, "A 1.8-GHz Low-Phase-Noise CMOS VCO Using Optimized Hollow Spiral Inductors," *IEEE Journal of Solid-State Circuits*, vol. 32, no. 5, pp. 736–744, May 1997.

[220] P. Van de Ven *et al.*, "An optimally coupled 5 GHz quadrature LC oscillator," in *Symposium on VLSI Circuits*, 2001, pp. 115–118.

[221] T. Lui, "1.5 V 10-12.5 GHz Integrated CMOS Oscillators," in *Symposium on VLSI Circuits*, 1999, pp. 55–56.

[222] T. Wakimoto and S. Konaka, "A 1.9 GHz Si Bipolar Quadrature VCO with Fully-Integrated LC Tank," in *Symposium on VLSI Circuits*, 1998, pp. 30–31.

[223] P. Andreani, "A Comparison between Two 1.8GHz CMOS VCOs Tuned by Different Varactors," in *European Solid-State Circuits Conf. (ESSCIRC)*, 1998, pp. 380–383.

[224] B. Razavi, "A 1.8 GHz CMOS Voltage-Controlled Oscillator," in *IEEE International Solid-State Circuits Conf. (ISSCC)*, 1997, pp. 388–389.

[225] C. Lo and H. Luong, "2-V 900 MHz Quadrature Coupled LC Oscillators with Improved Amplitude and Phase Matchings," in *IEEE International Symposium on Circuits and Systems (ISCAS)*, 1999, vol. II, pp. 585–588.

[226] F. Herzel *et al.*, "Phase Noise in a Differential CMOS Voltage-Controlled Oscillator for RF Applications," *IEEE Transactions on Circuits and Systems II*, vol. 47, no. 1, pp. 11–15, Jan. 2000.

[227] C. Lam and B. Razavi, "A 2.6GHz/5.2GHz CMOS Voltage-Controlled Oscillator," in *IEEE International Solid-State Circuits Conf. (ISSCC)*, 1999, pp. 402–403.

[228] A. L. Coban *et al.*, "A Highly-Tunable 12 GHz Quadrature LC-VCO in SiGe BiCMOS process," in *Symposium on VLSI Circuits*, 2001, pp. 119–120.

Literature on ring oscillator designs

[229] A. Buchwald and K. Martin, "High-speed voltage-controlled oscillator with Quadrature outputs," *Electronic Letters*, vol. 27, no. 4, pp. 309–310, 1990.

[230] A. Hajimiri *et al.*, "Jitter and Phase Noise in Ring Oscillators," *IEEE Journal of Solid-State Circuits*, vol. 34, no. 6, pp. 790–804, June 1999.

[231] A. Pottbacker and U. Langmann, "An 8 GHz Silicon Bipolar Clock-Recovery and Data-Regenerator IC," in *IEEE International Solid-State Circuits Conf. (ISSCC)*, 1994, pp. 116–117.

[232] M. Meghelli *et al.*, "SiGe BiCMOS 3.3V Clock and Data Recovery Circuits for 10Gb/s Serial Transmission Systems," in *IEEE International Solid-State Circuits Conf. (ISSCC)*, 2000, pp. 56–57.

[233] R. K. Montgomery *et al.*, "10 and 26 GHz differential VCOs using Inp HBTs," in *Microwave Symposium Digest*, 1996, vol. 3, pp. 1507–1510.

[234] B. Razavi, "A Study of Phase Noise in CMOS Oscillators," *IEEE Journal of Solid-State Circuits*, vol. 31, no. 3, pp. 331–343, Mar. 1996.

[235] J. van der Tang *et al.*, "A 9.8-11.5 GHz I/Q Ring Oscillator for Optical Receivers," in *IEEE Custom Integrated Circuits Conf. (CICC)*, 2001, pp. 323–326.

[236] W. S. T. Yan and H. C. Luong, "A 900-MHz CMOS Low-Phase-Noise Voltage-Controlled Ring Oscillator," *IEEE Transactions on Circuits and Systems II*, vol. 48, no. 2, pp. 216–221, Feb. 2001.

[237] C. Park and B. Kim, "A Low-Noise, 900-MHz VCO in 0.6-um CMOS," *IEEE Journal of Solid-State Circuits*, vol. 34, no. 5, pp. 586–591, May 1999.

[238] J. van der Tang *et al.*, "A 2.7 V, 8 GHz Monolithic I/Q RC Oscillator with Active Inductive Loads," in *European Solid-State Circuits Conf. (ESSCIRC)*, 2000, pp. 304–307.

[239] S. Butala and B. Razavi, "A CMOS Clock Recovery Circuit for 2.5-Gb/s NRZ Data," *IEEE Journal of Solid-State Circuits*, vol. 36, no. 3, pp. 432–439, Mar. 2001.

[240] B. Razavi, "A 2-GHz 1.6-mW Phase-Locked Loop," *IEEE Journal of Solid-State Circuits*, vol. 32, no. 5, pp. 730–735, May 1997.

[241] J. van der Tang and D. Kasperkovitz, "A 0.9 -2.2 GHz Monolithic Quadrature Mixer Oscillator for Direct-Conversion Satellite Receivers," in *IEEE International Solid-State Circuits Conf. (ISSCC)*, 1997, pp. 88–89.

[242] D. Jeong *et al.*, "CMOS Current-Controlled Oscillators Using Multiple-Feedback-Loop Ring Architectures," in *IEEE International Solid-State Circuits Conf. (ISSCC)*, 1997, pp. 386–387.

[243] A. Buchwald *et al.*, "A 6 GHz Integrated Phase-Locked Loop Using AlGaAs/GaAs Heterojunction Bipolar Transistors," in *IEEE International Solid-State Circuits Conf. (ISSCC)*, 1992, pp. 98–99.

[244] L. Sun and T. A. Kwasniewski, "A 1.25-GHz 0.35-um Monolithic CMOS PLL Based on a Multiphase Ring Oscillator," *IEEE Journal of Solid-State Circuits*, vol. 36, no. 6, pp. 910–916, June 2001.

About the Authors

Johan Van der Tang was born on April 21, 1970 in Wolvega, the Netherlands. From 1988 until 1992 he studied Electrical Engineering at the Technical College in Leeuwarden (NHL). He subsequently studied Electrical Engineering at the University of Twente in Enschede, the Netherlands. He successfully concluded his studies at Twente University in 1995 with a M.Sc. project at the Institute for Micro-electronic Systems in Darmstadt, Germany. In September 1995 Johan van der Tang started as research scientist in the field of integrated transceivers at Philips Research Laboratories, Eindhoven. In this position he worked on analog integrated High-Frequency key building-blocks for satellite, television, radio and optical front-ends. In 1996 he joined the Centre of Technical Training (CTT) of Philips as a part-time lecturer. Since February 2000, Johan van der Tang works as an assistant professor in the Mixed-signal Microelectronics group at Eindhoven, University of Technology. In December 2002, he gained his Ph.D. degree after defending his thesis titled "High-Frequency Oscillator Design for Integrated Transceivers". For his thesis work he received the "Vederprijs 2002". His current research interests are design of innovative integrated transceivers varying from ultra low-power/low-bit rate transceivers to 30 GHz front-ends, reusable and programmable RF building blocks, and design methodology for analog RF circuits.

Dieter Kasperkovitz got his B.Sc. in Mechanical Engineering, his M.Sc. in Experimental Physics and his Ph.D. in Theoretical (Nuclear) Physics in Vienna, Austria. Before starting his industrial career he was engaged in a Postdoctoral Fellowship in Solid State Physics at the University of Cologne, Germany. He started his work at Philips Research in Eindhoven, the Netherlands, with the design of new functional devices like ultra-fast switches and IMPATT oscillators, bi-stable elements, image sensors, etc. Supporting innovative projects for the consumer market he designed the first digital tuning system for TV tuners. Other areas of interest were serial and random access mass memories, fully integrated FM receivers and Paging Receivers. Before his retirement from Philips Research he was also involved in the design of new architectures for integrated transceivers, in the development and evaluation of new technologies for RF-IC's, and in adaptive signal processing for transmission channels with high levels of noise and spurious signals. After retirement he co founded the innovative IC design company Semiconductor Ideas to the Market (ITOM).

Arthur H.M. van Roermund was born in Delft, The Netherlands in 1951. He received the M.Sc. degree in electrical engineering in 1975 from the Delft University of Technology and the Ph.D. degree in Applied Sciences from the K.U.Leuven, Belgium, in 1987. From 1975 to 1992 he was with Philips Research Laboratories in Eindhoven. From 1992 to 1999 he has been a full professor at the Electrical Engineering Department of Delft University of Technology, where he was chairman of the Electronics Research Group and member of the management team of DIMES. From 1992 to 1999 he has been chairman of a two-years post-graduate school for chartered designer. From 1992 to 1997 he has been consultant for Philips. October 1999 he joined Eindhoven University of Technology as a full professor, chairing the Mixed-signal Microelectronics Group. Since September 2002 he is member of the faculty board, with research portfolio. He is chairman of the board of ProRISC, a nation-wide microelectronics platform, and senior member of the IEEE. Specific areas of interest currently are AD and DA conversion and high-frequency transceivers.

Index